Professor Fokas emerges as an Uomo Universale, an authentic polymath, a true philosopher of Science, rather than the Homo Technologicus he criticises for their myopic, mechanistic view of life. The importance of the Arts and Letters in shaping the culture of scientists and technologists is a constant theme running through Fokas' extraordinary new book. In particular, he lauds the highly interdisciplinary Vienna Circle. The great risk to our civilisation is the misuse of scientific advancements. That interdisciplinarity is crucial to enabling key breakthroughs recurs through the book. Fokas sets a vast panoramic context in centuries of spectacular advances in Physics, Medicine and Computer Science. He suggests that these developments can be deeply explained only within the context of Neuroscience, and in particular using the fact that unconscious and conscious processes form a continuum. "Every conscious experience is preceded by an unconscious process," a surprising observation to the layperson. Regarding AI, which Fokas reminds us is developed at the Human Computer Interface, he suggests that the exponential increase in computational power creates trepidation in the layperson and societal panic, as the emerging patronising and opaque technology is advertised by evangelical proponents as omnipotent. Perhaps a trifle optimistically he suggests that its ringmasters should be obligated to abide by the universal principle of 'do no harm'. Importantly, he argues that 'machines cannot surpass human intelligence'. In particular, he emphasises that AI ignores the impact of the 'body proper' on the brain, the functional differences between the two cerebral hemispheres, and the vital role of glial cells which outnumber neurons six-to-one. The brain, Fokas explains, has an astonishing capacity for plasticity, abstraction, reduction and generalisation. The redundancy of neural circuits enables the phenomenal ability of the brain to make *associations*. Rather wonderfully, the ultimate expression of this sophistication is the *metaphor*, which again allows Fokas to go beyond Science and Technology and to emphasise the crucial importance of the Arts and Letters.

— C. Alan Short, LittD
President of Clare Hall Cambridge,
Professor of Architecture,
Emeritus, University of Cambridge

In the age of ultra-specialisation and accumulating advances in science and computing, TS Eliot's question of "Where is the wisdom we have lost in knowledge? and where is the knowledge we have lost in information?" never rung more true or urgent. Narrow and unquestioned thinking and mistaken belief in the triumph of new technology over the complexity and essence of human thought and wisdom, is a risk to us all, with far-reaching consequences to self and our environment. In a book of epic reach, Fokas – a physician, engineer, mathematician and, in my opinion, philosopher – straddles physics, computing, the arts, medicine and neuroscience, with remarkable acuity, to reveal the importance and value of inter-dependency. His gift is to be able to step back, take a detached "historical to the present" view that connects the dots between humanities and the sciences in prose which is eloquent, perspicacious and penetrating. The ability to place in context the AI revolution in a wider treatise on the human mind that integrates biology with philosophy is remarkable and important. This book is the work of a polymath and deserves to be read and interpreted by students and the public alike.

— Siddharthan Chandran
MacDonald Professor of Neurology,
Director of UK Dementia Research Institute

Thanasis Fokas has achieved what was seemingly unachievable. A hardcore mathematician, he has transformed himself into a masterful storyteller and has woven a profound multi-threaded story. Putting natural and artificial intelligence at the epicenter, he has created a colorful puzzle that brings together and connects a plethora of scientific and historical facts from several disciplines in the natural sciences, life sciences, and the humanities. His book combines the accuracy of science (grounded in numerous references and research citations), the excitement of discovery (of so many unexpected inter-disciplinary relations), and the emotional engagement of storytelling. Chapter 1 is the one closest to my own scientific background, yet in reading, I learned new things about Artificial Intelligence I was ignoring. I am certain that both expert and non-expert audiences will find the book eye-opening and enjoy reading it immensely.

— Yannis Ioannidis
President of the Association for Computing Machinery

Everyone is conscious of the unprecedented rapidity with which scientific discovery alters the human experience of reality; yet the means of developing an enhanced awareness of the scientific and technological advancements that change the nature of the world in which we live, remain elusive for most people. *The Embodied Mind*, in the depth with which it uncovers the development and significance of related fields of inquiry, and the lucidity with which it exposes the influence that advances in AI, Medicine, and Physics exert on the ways in which life is lived, is a work of truly exceptional power and relevance. The inter-related nature of the analysis at the heart of the book, and the accessible, lively, and trenchant style in which Fokas writes, will ensure that the work is of the utmost value to researchers of all disciplines, to students, and to the wider public.

— Charles Burdett
Professor of Italian and Director of the Institute of Languages,
Cultures and Societies, School of Advanced Study, University of London

In this new book, Fokas builds on his comprehensive analysis of the nature of consciousness and the vital importance of unconscious processes elaborated in *Ways of Comprehending*, to present a breathtaking canvas of human achievements in computing, medicine and physics. The ability of humans to compute has created marvels ranging from technology to civilization, achievements that are closely interlinked ever since Archimedes cried Eureka. By analyzing with deep insight both the promise and the dangers of AI, which together with quantum computing are the most powerful incarnation of computing invented by humans so far, Fokas gives us hope for exciting new advances, as well as reasons to pause and think critically about how humanity should deal with this unprecedented development. Are machines going to surpass human cognition? Will AI solve our most challenging problems like the stupendous advances in medicine and physics described in detail in the book? Or will it create new unthinkably difficult problems that will be impossible to address? There are no simple answers, and the book offers no illusions of such. But through a remarkable feat of a holistic and polymathic analysis of human thought as it evolved through millennia, from Einstein and other giants in Physics, to the recent tremendous advances in Immunology and the fight against cancer, Fokas offers us a clear framework for properly thinking about these issues – undoubtedly some of the most vexing we will have to face in coming decades. And in a turn of highly aesthetic sensitivity, the book wraps up by evoking the plight of Sophocles' *Antigone* as the timeless symbol of the human spirit's triumph in tragedy. I do not recall being so thoroughly awed and inspired by a book in a very long time.

— Efthimios Kaxiras
The John Hasbrouck Van Vleck Professor of Pure and Applied Physics,
Harvard University

The Embodied Mind
Unravelling AI,
Medicine, and Physics

The Embodied Mind

Unravelling AI, Medicine, and Physics

Athanassios Fokas

University of Cambridge, UK

World Scientific

NEW JERSEY · LONDON · SINGAPORE · BEIJING · SHANGHAI · HONG KONG · TAIPEI · CHENNAI · TOKYO

Published by

World Scientific Publishing Europe Ltd.

57 Shelton Street, Covent Garden, London WC2H 9HE

Head office: 5 Toh Tuck Link, Singapore 596224

USA office: 27 Warren Street, Suite 401-402, Hackensack, NJ 07601

Library of Congress Control Number: 2025036265

British Library Cataloguing-in-Publication Data
A catalogue record for this book is available from the British Library.

THE EMBODIED MIND
Unravelling AI, Medicine, and Physics

ISBN 978-1-80061-835-0 (hardcover)
ISBN 978-1-80061-848-0 (paperback)
ISBN 978-1-80061-836-7 (ebook for institutions)
ISBN 978-1-80061-837-4 (ebook for individuals)

For any available supplementary material, please visit
https://www.worldscientific.com/worldscibooks/10.1142/Q0540#t=suppl

Desk Editors: Aanand Jayaraman/Cian Sacker Ooi

Typeset by Stallion Press
Email: enquiries@stallionpress.com

*This book is dedicated to the memory of
Captain Vassilis C. Constantakopoulos, whose wisdom
and friendship are greatly missed.*

About the Author

Thanasis Fokas has a BS in aeronautics from Imperial College, a PhD in applied mathematics from Caltech, and an MD from the University of Miami. He also holds eight honorary degrees. In 1995 he was appointed to a Chair in Applied Mathematics at Imperial College, and in 2002, he became the first holder of the inaugural Chair of Nonlinear Mathematical Science at the Department of Applied Mathematics and Theoretical Physics of the University of Cambridge. He is a member of the Academy of Athens and of all three major European academies, including Academia Europaea. He is a fellow of Clare Hall College, Cambridge, of the Guggenheim Foundation, and of the American Institute for Medical and Biological Engineering. Among his many awards are the Naylor Prize of the London Mathematical Society, the Blaise Pascal Medal of the European Academy of Sciences, and Caltech's Distinguished Alumnus Award. Israel Gelfand wrote that 'Fokas is a rare scientist in the style of Renaissance'. This is consistent with his book, *Ways of Comprehending* (World Scientific, 2024), which is unprecedented in terms of breadth and depth.

Acknowledgement

I am deeply grateful to the publisher at World Scientific, Laurent Chaminade, for his crucial support, as well as to Konstantinos Tsoukalidis and Nicholas Ktistakis for their many important suggestions. I am also thankful to Evangelos Andreakos (Chapters 7 and 8), Anastasios Bountis (Introduction), Meletios Dimopoulos (Chapters 9 and 10), Petros Gelepithis (Part I), Vassilis Gorgoulis (Chapters 9 and 10), Yannis Ioannidis (Part I), Tim Kaxiras (Part III), Christos Papadimitriou (Part I), Yiannis Papanicolaou (Introduction), Marianna Politou (Chapter 12), and Vassilis Zoumpourlis (Chapters 9 and 10) for their valuable input in the respective parts of the book.

Contents

Introduction

The Industrial Revolution, which began in England at the end of the 18th century, was the main driving force of the transformation from feudalism to capitalism. It was marked by a leap in production, the accumulation of capital, and the exploitation of labour power. The Second Industrial Revolution, based mainly on electricity, brought Germany to the fore. A further surge in productivity and capital accumulation culminated in the Third Industrial Revolution, which was characterised by astounding scientific and technological breakthroughs. These brought about "the information age, the nuclear age, and the electronic age" (Toffer, 1980). We are now living in the Fourth Industrial Revolution, based on the dominance of the digital revolution, involving the internet, big data, artificial intelligence (AI), cyber-physical systems, etc.

As a mathematician, engineer, and physician, I have a great admiration for a variety of transformative scientific, technological, biological, and medical breakthroughs and their impact on society. However, I believe that these huge developments must be placed within the proper historical perspective. In this regard, it should be noted that in the 1500s, in some regions of the Global South, people lived in better conditions than in Europe. At that time, overall, there was no appreciable difference in income and living standards between Europe and the rest of the world. For example, at that time, the living standards in India were on par with those in Britain. What caused the current great disparity in gross domestic product (GDP) among different countries? The common answer is 'the differential effects in different countries of technological and industrial developments.' I believe that the complete answer to this important xxii *The Embodied Mind*

question is far more complicated. In particular, Europe's Industrial Revolution was significantly affected by the huge resources extracted from colonies. For example, gold, silver, and diamonds provided extra capital which was of crucial importance for industrial investments. It also allowed European countries to buy agricultural products from the East (for example, China accepted payment only in silver), which freed up labour power in their own countries to be moved from agriculture to industry. This was further aided by the availability of sugar and cotton produced in enslaved Africa and grain obtained from colonial India. Additional labour was made available via the exploitation of slaves from Africa and, importantly, from the displaced peasantry following the enforced system of 'enclosure'. A comprehensive analysis of the factors that led to the current huge disparity in GDP among different countries is beyond the scope of this book. Instead, I will concentrate on important aspects of the current situation, ignoring the factors that led to the existing reality.

The inability to predict the final impact of major developments taking place in synthetic biology, including gene editing, and especially in AI, including robotics, as well as the apparent imminent arrival of quantum computing, creates anxiety and fear, especially among the young. These developments, together with earlier and current devastating events, exacerbate feelings of despondency. Among such shocks are: the catastrophic acceleration of global warming; the waging of several wars, including those in Ukraine and Gaza, as well as various incidents of terrorism; the COVID-19 pandemic, whose effects continue to be felt to date, especially due to the occurrence of 'long-COVID-19' (Ely *et al.*, 2024); the occurrence of major economic crises in several parts of the world, notably the 2008 crisis[1]; and the unprecedented increase in population, from 4 billion in 1974 to 8 billion in 2022.

The income gap between the rich and the poor within a given country and across different countries continues to widen. Almost half of the world's GDP is owned by Fortune's Global 500 companies. In the US, the richest 1% of the population earns almost half the national income, while 40 million people live in poverty. After the US, China is the country exhibiting the most economic inequality; the top 10% possesses approximately 70% of China's wealth, whereas the bottom 50% possess only

---

[1]The national debt in many countries has reached unprecedented levels, which many fear will trigger a new economic crisis. In the US, the national debt is US$35 trillion; serving this debt is more costly than the defence budget.

about 8%.[2] The current unprecedented digital revolution, unfortunately, is a strong driver of inequality.

Extreme concentration of economic power is now as dominant as ever before. For example, it seems impossible that any company can challenge the hegemony of companies such as Apple, Google, or Amazon. In 2022, Apple was valued more than the total of all the companies listed in the UK's FTSE 100 Index.[3] Seventy percent of cloud computing services are provided by four companies.[4] Most of the chips worldwide are manufactured by Taiwan Semiconductor Manufacturing Company, which uses extremely precise machinery based on the technique of ultraviolet lithography; these machines are provided by a single supplier, the Dutch firm Advanced Semiconductor Material Lithography. The majority of the most advanced graphic processing units are manufactured by the American company Nvidia. Nearly 80% of high-quality quartz, a material essential for important technological applications from silicon chips to photovoltaic panels, comes from a single mine in North Carolina, US. Naturally, this extreme economic inequality leads to social resentment.

Certain corporations, by collecting, analysing, and taking advantage of their users' data, wield more power and influence than many sovereign states. At the same time, social media, apparently, exacerbate various malfunctions of developed societies. A study published in *Nature* (Lorenz-Spreen *et al.*, 2023), based on an analysis of nearly 500 published articles, substantiated that the use of digital media did play a positive role in enhancing political participation in autocracies and emerging democracies. However, in developed democracies, there is a strong correlation between the growing use of digital media and increasing polarisation, as well as growing distrust in politics and the increasing influence of populist movements.

In developed countries, the impact of social media on children is highly disturbing, bringing isolation and increasing the prevalence of anxiety and mental disorders. In the early 1990s, approximately half of the children in the US met, almost every day, with friends outside school; by

---

[2] Among the developed economies, the European Community (EU) exhibits a more even income distribution; the top 10% generates about 35% of the EU's income, and the bottom 50% generates about 20%.

[3] The Financial Times Stock Exchange (FTSE) 100 index consists of highly capitalised blue-chip companies listed on the London Stock Exchange.

[4] Amazon (31%), Microsoft (24%), Google (11%), and Alibaba (4%).

xxiv   *The Embodied Mind*

2017, this was reduced to less than 30%. The advent of the powerful iPhone further exacerbated children's dependence on social media platforms. Nowadays, children in the US spend, on average, 50 hours a week on their iPhone, depriving them of the time needed for meaningful activities that would allow them to become 'confident, competent adults' (Haidt, 2024). According to Haidt, many girls in the US spent the majority of their time concerned about 'visual social comparison and perfectionism'. In the UK, nearly half a million prescriptions for antidepressant medications are issued annually for under-18s.

The impact of addiction in developed countries is truly devastating. For example, the top two leading causes of death among middle-aged white Americans without a college degree are drug overdose and alcohol-related liver diseases; this group of people are dying younger than their great-grandparents. In the Canadian province of British Columbia, the leading cause of death for people aged 10–59 is drug overdose. In 2022, according to statistics provided by the Centers for Disease Control and Prevention, almost 110,000 people died in the US from drug overdose; among these deaths, 70% were due to synthetic opioids, mainly fentanyl. A more subtle form of addiction, namely the addiction to overconsumption, has contributed to the modern serious problem of obesity; in the US, 40% of the population is classified as obese. Since the prevalent use of the internet, other forms of addiction, including pornography addiction, have reached new heights, affecting the lives of millions of people, especially young men.

What will be the role of the young generation in the emerging world? What are the necessary skills required to be successful? More importantly, will there be any significant role left for humans to play in a society which is increasingly driven towards mechanisation, where emotional and affectional aspects are ignored, with emphasis only on purely algorithmic, machine-driven gadgets? Taking into consideration the current world situation, as well as the extreme difficulty in answering fundamental questions about the nature of the emerging world, it is not surprising that anxiety and depression among young people are increasing at an alarming rate in developed societies.

Overall, the current world situation makes it clear that communism in its existing manifestations is a failure, but also that capitalism, in its current form – with the obsession of maximising profits for shareholders and a complete disregard for human and ecological consequences – are incapable of providing a framework for fostering a just society based on human values, where citizens can strive towards *eudaimonia*.

Some Positive Developments

Many scholars paint an utterly bleak picture of the current situation; some even worry about the survival of the human race. Although I am very concerned, especially about the environmental catastrophe and the existence of unacceptable levels of hunger, suffering, poverty, and inequality, I believe that there exist positive trends, which provide hope. For example, the population is no longer increasing at a high rate. Actually, it is estimated that by the end of the century, the world population will level out, at around 10 billion (Roser and Ritchie, 2023).[5] Thankfully, the number of deaths among children under five declined from almost 13 million in 1990 to 6 million in 2015. Children dying from preventable causes declined from 17 million in 1990 to 8 million in 2013. Also, the likelihood of mothers dying during childbirth declined by almost 50% during the same period. Primary school enrolment has increased dramatically, while HIV and malarial infection rates have declined markedly. This means that some of the key goals set in the 2000 Millennium Development Goals were achieved.

Incidentally, in 2015, which was the year by which the implementation of these goals was supposed to have reached completion, the United Nations published a report claiming that two additional important goals, namely cutting the world's poverty by half and reducing hunger substantially, had also been achieved. The 2000 Millennium Development Goals, set by the world's heads of state at the United Nations headquarters in New York, were the world's first major public commitment to reducing poverty and hunger. Unfortunately, these 'success' claims are controversial. Regarding poverty, first, the starting point of analysis was moved from 2000 to 1990. This allowed for counting the gains made by China during the 1990s. Actually, a detailed analysis shows that almost all of the gains against poverty have occurred in one place, namely, China. Second, the poverty goal was modified from the earlier one of halving the *absolute* number of people in poverty to that of halving the *proportion* of people living in poverty. This is a more easily achievable goal since it takes

---

[5]The world is not expected to face a problem due to overpopulation but from the fact that, since 1950, the fertility rates have been on a constant decline (Siddig and Charalampous, 2024). The impact of this serious problem, coupled with the increasing ageing population, will be discussed in Fokas and Ktistakis (in preparation) in connection with the promising progress in the efforts to combat ageing.

advantage of the population growth. Third, a decision was made to consider only people living in poverty in developed countries, as opposed to the whole world, which takes further advantage of the 'proportion' goal, given that the population increased faster in underdeveloped countries. Similar considerations are valid for attempts to reduce hunger. In my opinion, the celebratory declaration of the United Nations regarding the successful reduction of poverty and hunger should have been replaced by a statement of inadequacy, especially considering that since 1995 there has been an unprecedented economic growth. Remarkably, the GDP per capita grew from approximately US$800 in 1970 to US$5,000 in 1995 and to almost US$13,000 in 2022.[6] This is further discussed in Gelepithis (2024).

The Recent Economic Growth

The sustained economic growth, as well as the fact that more people are fed nowadays than at any time, provide clear indications of the ability of humans to achieve difficult goals. The latter achievement is due to widespread market transformations (Lam, 2011) and, more importantly, the effects of the Green Revolution (the Third Agricultural Revolution),[7] namely the technology transfer initiatives that resulted in greatly increased crop yields. With regard to overall economic growth, the impressive result is apparently due to increased productivity achieved as a result of technological breakthroughs, as well as several economic factors, including the following: first, the liberalisation of international trade, especially after the 1995 agreements; and second, the adoption of the 'global value chain' model. According to this model, different parts of a given commodity have been offshored from developed countries to many parts of the developing world, especially to China. The impact of the know-how, gained as a result of the implementation of the global value chain model, on technological developments in China was huge; for example, since 2008, Apple has trained approximately 28 million workers, mostly in China (McGee, 2025).

---

[6]The poverty level in a given country is defined according to its GDP. For example, for the US, a person is considered to be in poverty if they live on approximately less than US$25 a day, whereas in Ethiopia the poverty line is at approximately US$2. Incidentally, the period 1999–2022, as a result of COVID-19 and related events, is considered a lost period in the fight against poverty.

[7]Among the key figures of the Green Revolution was the Nobel Laureate Norman Borlaug.

Returning to the impact of the digital revolution, it is worth noting that despite the recent decrease in international trade of relevant commodities due to the imposition of new tariffs and the banning of the production of strategic commodities outside the US,[8] the economies of both the US and China continue to grow (except for brief setbacks during the 2008 financial crisis and, more recently, the COVID-19 pandemic). This is apparently due to the significant capital investment in these two countries in data-driven digital technologies.[9] Incidentally, the economy of the EU, after the 2008 crisis, exhibits stagnation.

Sustained economic growth is undoubtedly a remarkable achievement. However, it should be remembered that GDP does not take into account the social cost of certain economic activities, such as the devastating impacts of increased productivity and unnecessary consumption on the environment.[10] The concept of measuring progress through the definition of GDP can be traced back to the 1930s when the economists Simon Kuznets and John Maynard Keynes attempted to design an economic index that would help policymakers avoid the occurrence of another Great Depression. Kuznets emphasised the importance of tracking net human welfare and advocated excluding economic activities generating external diseconomies, such as advertising, policing, and commuting. However, when the Second World War broke out, Keynes decided that *all* economic activities should be included in the GDP. In this way, Keynes was hoping to identify areas of increased productivity that could be used in the war effort. Surprisingly, an economic index designed under conditions of war was adopted by all countries as the most important measure of human

---

[8] These developments started after the 2008 crisis and have been accelerating since. Currently, there is a trade war between the US and China. For example, President Xi has ordered IBM and Dell equipment, along with Microsoft and Oracle tools, to be replaced by Chinese-made products; moreover, executives are not allowed to bring their Teslas and iPhones to the office. President Biden retained the tariffs imposed on products from China by President Trump and, importantly, gave billions in subsidies to American companies producing chips and other advanced technological necessities. President Trump is threatening to extend and intensify the 'tariff wars'; this could have a devastating global effect.

[9] In addition, China benefitted enormously from the transfer of the 'know how' as a result of the offshoring policy of the US and other developed countries. Also, China, apparently, benefitted from the loopholes of some of the trade rules of the World Trade Organization, which it became a member of after the initiative of the President Clinton administration.

[10] For example, cutting down a forest and selling the resulting timber will result in a positive contribution to the GDP, despite its devastating ecological effect.

progress! This misconception led to elevating the goal of increasing the GDP as the key objective of every country, independent of ideology, ignoring the enormous stress this effort places on human life, the environment, and other depletable resources. David Attenborough expressed clearly his concern by stating that, 'Anyone who thinks that you can have infinite growth on a finite planet is either a madman or an economist'.

Several objective measures of progress, including health, longevity, happiness, and education, can be achieved without a high GDP. For example, Costa Rica has a higher life expectancy than the US with a GDP of about a fifth of that of the US; furthermore, according to the United Nations' World Happiness Report, Costa Rica matches the US.[11] Apparently, there exist alternative indicators which depict human progress better than the GDP. They include the genuine progress indicator (GPI), which subtracts negative factors, such as pollution, crime, and resources depletion, adds positive factors, such as volunteer work and household labour, and adjusts for inequality. Comparing the graphs of GDP and GPI, it becomes clear that these two quantities had a parallel increase until the mid-1970s, but afterwards, GDP continues to increase whereas GPI remains at the value of the mid-1970s (Jason, 2017).

The Need for a Global Response

The magnitude of the current and emerging problems makes it clear that a comprehensive, global approach is needed as a matter of urgency. Such efforts must actively involve the most powerful and influential countries as part of the coordinating actions of international organisations. Global cooperation has led to some planet-saving interventions, such as: the moratorium on biological and chemical weapons; the signing of several treaties limiting the proliferation of nuclear weapons; and the 1987 Montreal Protocol, which phased out chlorofluorocarbons, the most important chemicals responsible for the ozone hole. There is hope that several current worldwide initiatives will be successful. In particular, there is guarded optimism that the moratorium on human gene editing will be respected and that, as a result of the Paris Agreement, decarbonisation

---

[11] The US continues to be the world leader in most areas of science, medicine, and technology, despite the huge progress made by China. This, in my opinion, in addition to US' high GDP, is also aided by the prevalence in the US of the ideology of 'free thinking'. It will be interesting to see if Chinese researchers will be able to overcome the ideology of rigid constraints, which is antithetical to 'thinking outside the box'.

will be accelerated and the increase in global average temperature will be kept below 2 degrees Celsius. In this direction, new initiatives, such as 'bio-energy capture and storage' and 'regenerative farming' could be of critical importance. For example, an article in *Science* (Lai, 2004) suggests that degraded soils can be regenerated by switching from intensive industrial farming to more ecological approaches; in turn, the regenerated soil could regain its ability to capture CO_2, sequestering up to 15% of global CO_2 emissions.

Individual Responsibility

The current and emerging problems require, on the one hand a coordinated global response, and on the other hand a positive response at the individual level. The psychologist Ferbert Gerjuoy wrote: 'Tomorrow's illiterate will not be the man who can't read; he will be the man who has not learned how to learn' (Toffler, 1970, p. 367). But even before we are able to learn, we must possess the tools to be able to comprehend the enormous complexity of the new world we live in. Clearly, this complexity requires an *interdisciplinary* approach to understanding. The merits and usefulness of such an approach were underlined in my book, *Ways of Comprehending* (Fokas, 2024), which covers diverse topics, including aspects of mathematics, engineering, physics, biology, neuroscience, medicine, philosophy, and painting. This broad exposure serves two purposes: first, it presents useful knowledge in several highly important areas, which I believe is necessary for comprehending our modern world; second, these different areas provide concrete examples of the usefulness of several cognitive tools introduced in Fokas (2024), which are helpful for understanding a variety of phenomena. *This dual aim is at the heart of the current book.* I strongly believe that, under certain conditions, a harmonious relationship between the modern world and its inhabitants *can* be achieved. The basic goal of this book is to help the reader achieve such a relationship.

AI and Breakthroughs in Medicine, Biology, and Physics

A global appreciation of several remarkable, highly encouraging developments in *technology*, *medicine*, and the *sciences* can be used to counteract the depressing worldwide problems mentioned earlier. The novel concept xxx *The Embodied Mind*

of metarepresentations and the tremendous impact of key metarepresentations, including language, mathematics, technology, and the arts, on our cultural evolution, were discussed in Fokas (2024).

With regard to technology, it is noted that the triptych of mathematics–computers–technology combines the most powerful existing metarepresentations (Fokas, 2024). AI constitutes the crowning achievement of this almighty triptych. The balanced analysis of AI presented in Part I brings hope: first, computers, in their current form in which silicon is not connected with neurons, cannot, in my opinion, reach human cognition; second, I believe that the existential threat of AI has been exaggerated. Actually, AI, in addition to providing more efficient ways of performing a variety of human tasks, in many cases, 'invents' new solutions which humans cannot perceive. This is related to a remarkable capacity of AI, which I have called 'new intelligence' and which is discussed in Chapter 6. Overall, I expect that AI will have a positive, transformative impact in many areas of science, technology, and medicine. The beneficial impact of AI on medicine is discussed in Chapter 2.

With regard to medicine and the sciences, it must be emphasised that the modern era is characterised by astounding medical-biological developments. These include the following: new understandings and harvesting of fundamental immunological mechanisms; truly life-saving breakthroughs in the treatment of many types of cancer; the development of a new technique for gene editing, which has already had a transformative impact in some genetic diseases and has also produced crop varieties that are more resistant to diseases and more efficient; promising progress in deciphering the mystery of ageing; our deepening understanding of epigenetics; the elucidation of the crucial role of microbiota; the discovery of new powerful medications for treating obesity, including the use of medications which stimulate glucagon receptors; and progress in understanding basic psycho-neurological mechanisms characterising addiction. The first three of these developments—namely, immunology, cancer, and gene editing—are summarised in Part II of the book; the remaining will be discussed in Fokas and Ktistakis (in preparation). In addition, a unified approach to the aetiology and treatment of certain common medical syndromes is also presented in this part of the book; this includes common interventional treatments for heart attacks and strokes.

After discussing in Parts I and II some key current developments in technology and medicine-biology, Part III reviews some of the most remarkable breakthroughs that ever occurred in the sciences; namely, it provides a comprehensive summary of physics. This includes the following: A discussion of the brilliant contributions of Einstein, which exemplify, in the best possible way, the creative impact of *associations*. A description of the journey from Newton to Maxwell and the mysteries of quantum mechanics provides ample evidence for the ability of the human brain to *generalise, mathematise,* and *unify*. A summary of the astonishing discoveries that took place at the Cavendish Laboratory of the University of Cambridge, where the dismantling of the atom was performed. An up-to-date description of experimental breakthroughs, including work at CERN which revealed that the proton and neutron are not elementary particles like the electron, but possess an internal structure, namely that they consist of quarks. The far-reaching implications of nuclear physics, from the emergence of electronic devices to medicine, particularly cancer treatment, and to the success of the Manhattan Project, are also discussed in this part; this illustrates the huge importance of *interdisciplinarity*.

The Grand Illusion

A deeper understanding of how we think and behave requires appreciating the crucial importance of unconscious processes. Remarkably, as discussed in Fokas (2024), the interaction with our environment is characterised by an illusion. For example, the champion Carl Lewis, as a result of extensive training, could begin running after a tenth of a second following the sounding of the starting pistol. He erroneously believed that he had begun running as a result of hearing the sound. In reality, he did not become *aware* of the sound for about a third of the second *after* the pistol was fired. He began running because *his brain perceived* the sound and gave the unconscious command, 'run'. Similarly, we are under the illusion that our *conscious self* is in charge of our communications, actions, feelings, etc. As evident from this example, awareness lags a considerable amount of time behind unconscious processes. This implies that our unconscious, and not our conscious, is in charge! I refer to this remarkable but largely unappreciated fact, as the 'grand illusion'. In the case of tactile and visual perceptions, the time interval required for us to become aware of unconscious processes is at least of the order of a third of a second. This interval was measured experimentally by the pioneer neuroscientist Benjamin Libet (2005). His transformative work is summarised in the first chapter of Part IV. The second chapter of this part discusses the impact of unconscious processes on health; this includes the effects of placebo and nocebo, as well as the emergence of several mysterious medical syndromes, which often spread like epidemics. Noting the fundamental role of unconscious processes in dreaming, I take the opportunity to also discuss in this part the crucial importance of sleep. The elucidation of the interaction of unconscious processes and conscious experiences requires the use of imaging techniques which have a time resolution of the order of a tenth of a second. Such techniques, which can actually provide almost real-time information about the functioning of the brain, are electroencephalography and magnetoencephalography, which are reviewed in the last chapter of Part IV.[12] Other techniques for mapping the brain are also discussed in the last chapter of this part.

The Crucial Role of Humanities

Appreciating astounding developments in technology, medicine-biology, and the sciences is, in my opinion, a necessary but not a sufficient condition for achieving a harmonious relationship with our external world and living a fulfilled life. It is imperative that we also make a sufficient investment in the arts and letters. The statement made in the 1950s by the American critic Lionel Trilling in his best-selling book, *The Liberal Imagination* (Trilling, 1950), that 'Literature is the human activity that takes fullest and most precise account of variousness, possibility, complexity, difficulty' is more pertinent than ever before. In particular:

> *nowadays, when sciences and technology have reached an unpresented level of power, and hence there exists huge potential for misuse and unthinkable destruction, it is necessary that the decisive impact of humanities on the culture of scientists, engineers, and technocrats is fully recognized and further explored.*

The cultivation of the arts and letters is the *only* way to avoid a 'mechano-morphic' approach to life, which will not only 'downgrade humans', as Yuval Noah Harari worries, but also could lead to the complete destruction of the human race. I fully agree with the statement of

---

[12]The functional imaging techniques of PET, SPECT, and fMRI, which are reviewed in Fokas (2024), are too slow for this purpose; their time resolution is of the order of a few seconds.

Barbara Maria Stafford that 'they [the neuroscientists] need to acknowledge that they have learned and, indeed, borrowed much from the humanities' (Stafford, 2007). Actually, many of the achievements of arts and letters are so deeply engraved in the human brain that not only neuroscientists but all scholars owe critical debt to the humanities.

A renewed appreciation of the arts and letters requires a global, innovative, multidimensional effort. This should include the active participation of philanthropists and the cooperation of major universities worldwide. For example, it is imperative, in my opinion, that the leading private universities of the US eliminate, or at least reduce substantially, tuition fees for degrees in the arts and letters.

It is very alarming that the phenomenal achievements of science and technology tend to overshadow the importance of the humanities. Many scholars, emphasising our astounding capabilities for scientific and technological achievements, characterise modern humans as *homo technologicus*. They even propose that the ultimate goal of our new technological era is to distil in an artificial algorithm our capabilities for the efficient solution of difficult problems.

The emphasis on the algorithmic aspects of our cognition is one-sided and hence inadequate to capture the essence of being human. What about our marvellous achievements in the arts and letters? How can our logical abilities be separated from our emotional, affectionate, and ethical self? I strongly believe that a human being is as much *homo atrium et litterarum, homo motus,* and *homo res ethica* as *homo technologicus*.

Fortunately, several experts on AI, in addition to emphasising the possible danger of 'machines controlling the human race', also advocate the need for finding new ways for personal satisfaction. For example, the computer scientist Stuart Russell, coauthor of the most widely used book in the industry on AI, *Artificial Intelligence: A Modern Approach* (Norvig and Russell, 1994), has called for the creation of a new discipline, which he named 'happiness engineering'. Russell states, 'We have to learn to be better humans' (Fortson, 2019). This brings to mind the last public address, 'The Old Humanities and the New Science' (1919), of the most influential physician of modern times, Sir William Osler. In the aftermath of the Great War, where Osler lost his beloved son, he expressed the hope that:

> "Through the Hippocratic combination of the love of humanity (*philanthropeia*) and the love of one's craft (*philotechnia*), humankind would acquire the wisdom (*philosophia*) to use science as a force for the human betterment".

Philosophy is vital for elucidating the essential role of arts in society. For example, Wittgenstein, echoing the founder of the Neoplatonic philosophy, Plotinus, and going further than Immanuel Kant's assertion that aesthetic judgement has a deep relationship with moral judgement, stated that 'ethics and aesthetics are one' (Wittgenstein, 1922, p. 6.421). Assuming the validity of this profound assertion, it follows that the arts are of critical importance for moulding the essence of humanity. Indeed, aesthetics is paramount in *any* human endeavour. This includes physics, engineering, and mathematics.

Kant's monumental work, the *Critique of Judgment*, begins with an account of beauty. In about a fourth of this book, Kant analyses four basic features of 'beauty judgement'. Perhaps the most interesting is the fourth, which states that beautiful objects appear to be 'purposive without purpose'. According to Kant, this means that beautiful objects should affect us as if they had a purpose. It may be that no particular purpose exists, and importantly, even if a purpose does exist, this is not important; beauty itself is enough.[13] In my opinion, this remarkable insight of the great philosopher captures the essence of humanities. It is imperative that we avoid the prevailing tendency of corporatising every important activity of our society and realise that the arts and letters prepare us for 'the beautiful and the virtuous'.[14]

Other important global effects of the humanities have been elucidated by several scholars. For example, the German poet and philosopher Friedrich Hölderlin, echoing the pre-Socratic philosopher Empedocles, elaborated on how magic and rituals serve as the predecessors of ethics and politics.[15] It is, of course, well known that a key element of the 'political' is the potential conflict between an individual and society. This is particularly pertinent today, when it is necessary for scientists and technocrats to learn how to navigate such conflicts, seeking a balanced approach between personal fulfilment and contribution to society at large. Humanities

---

[13] The other three features are: first, we must be *disinterested*, meaning that we take pleasure in something simply because we judge it to be beautiful, rather than judging it beautiful because we find it pleasurable. Second and third, such judgements are both *universal* and *necessary*, meaning that we expect others to agree with us.

[14] The importance of 'the beautiful and the virtuous' was fully appreciated by the ancient Greeks, καλός και αγαθός (*kalos kagathos*).

[15] Hölderlin was influenced by Hegel, whom he met during their schooldays in Tubingen. Actually, the influence was mutual.

provide a myriad of profound lessons on how this often difficult but vital task can be achieved. This is further discussed in the epilogue.

Concluding this introduction, I must confess that my greatest worry is not the direct effects of the relentless progress of technological advances, but their indirect impact on the triptych of information–knowledge–wisdom. Unfortunately, the digital world emphasises information rather than knowledge. For example, as explained in Chapter 1, deep learning, in contrast to the earlier form of logic-based algorithms of AI, does not rely on the mathematical formulation of models of reality. Hence, it does not depend on the knowledge needed to construct such models. Remarkably, starting without any knowledge input, it solves difficult problems simply via the process of 'training'. Since we can obtain a plethora of results by using the immense power of generative AI, why do we need to understand the intricate laws that dictate physical and biological reality?

How does information give rise to knowledge? This occurs only when information is placed in the proper context. However, the prevailing digital world is becoming increasingly less contextual.

Knowledge cannot become wisdom without being embedded in the proper historic and cultural framework. However, social media and the internet deprive individuals of the solitude needed for reflection, which, in my opinion, is an irreplaceable element required for transforming knowledge into wisdom, as well as for reaching deep insight and for achieving transformative discoveries.

I believe that our cultivation of the necessary fortitude to continue dedicating more and more of our solitude to sustained reflection, coupled with our immersion in the beautiful world of arts and letters, provides the only guarantees that the essence of humanity will prevail.

References

Ely, W., *et al.* (2024). Long Covid defined. *New England Journal of Medicine*, *391*, 1746–1753.

Fokas, A. (2024). *Ways of Comprehending: The Grand Illusion and the Essence of Being Human.* Singapore: World Scientific.

Fokas, A. and Ktistakis, N. (in preparation). Ageing and the quest for rejuvenation; diet, epigenetics, microbiota.

Fortson, D. (27 October 2019). The end of humanity: Will artificial intelligence free us, enslave us—Or exterminate us? The Berkeley professor Stuart

Russell tells Danny Fortson why we are at a dangerous crossroads in our development of AI. *The Sunday Times, 25.*

Gelepithis, P. A. (2024). *Unification of Artificial Intelligence and Psychology: Volume Two—Consequences*, Chapter 5. Berlin, Germany: Springer Nature.

Haidt, J. (2024). *The Anxious Generation.* New York: Penguin Press.

Jason, H. (2017). *The Divide: A Brief Guide to Global Inequality and Its Solutions.* New Hampshire, US: Heinemann.

Lam, D. (2011). How the world survived the population bomb: Lessons from 50 years of extraordinary demographic history. *Demography, 48*(4), 1231–1262.

Lai, R. (2004). Soil carbon sequestration impacts on global climate change and food security. *Science, 304,* 5677.

Libet, B. (2005). *Mind Time: The Temporal Factor in Consciousness.* Cambridge: Harvard University Press.

Lorenz-Spreen, P., *et al.* (2023). A systematic review of worldwide causal and correlational evidence on digital media and democracy. *Nature Human Behavior, 7,* 74–101.

McGee, P. (2025). *Apple in China.* New York: Simon and Schuster.

Norvig, P. and Russell, S. J. (1994). *Artificial Intelligence: A Modern Approach.* New Jersey: Prentice Hall.

Roser, M. and Ritchie, H. (2023). How has world population growth changed over time? Retrieved from OurWorldinData.org. https://ourworldindata.org/population-growth-over-time.

Siddig, E. and Charalampous, P. (2024). Global fertility in 204 countries and territories, 1950–2021, with forecasts to 2100: A comprehensive demographic analysis for the Global Burden of Disease Study 2021. *Lancet, 403,* 2057–2099.

Stafford, B. (2007). *Echo Objects: The Cognitive Work of Objects.* Chicago: University of Chicago Press.

Toffler, A. (1970). *Future Shock.* New York: Random House.

Toffler, A. (1980). *The Third Wave.* New York: Bantam Books.

Trilling, L. (1950). *The Liberal Imagination: Essays on Literature and Society.* New York: Viking.

Wittgenstein, L. (1922). *Tractatus Logico-Philosophicus.* London: Kegan Paul.

Part I

Artificial Intelligence and Related Neurophysiological Considerations

Computer science studies the process of computing as a way of transforming information. A basic element of theoretical computer science, pioneered by Turing, is to decide what can and what cannot be computed, as well as to quantify how costly (in terms of time and memory) it is to perform a given computation. Computer science, although based on the language of mathematics, is a distinct, extremely important area of study.

Generally, a formal language is characterised by its syntax, namely a set of symbols and specific rules of how these symbols can be manipulated. A formal language can gain meaning when certain basic notions are represented in terms of symbols and the entire language develops semantic structure. In this case, a formal language can carry information, which is embedded in the abstract structure of the language. Information is of course abstract and immaterial. The computer, by manipulating mathematical symbols, *transforms* information. The question of whether the computer can *generate* information and, more generally, knowledge, is discussed in Chapter 1. In addition, in this chapter, the basic steps in the development of artificial intelligence (AI) are reviewed, from logic-based algorithms to generative AI.

AI is based on a human–machine partnership; humans define a specific problem or, more generally, a goal, such as to write a poem about beauty, and then the computer delivers. In earlier forms of AI, every step of the process of solving a problem was well defined and well understood. In the new forms of AI, this is not the case; the computer, after being 'trained' with huge amounts of data, is able to go beyond the realm of human understanding. This profound departure from previous experiences is expected to have a far-reaching impact on our civilisation. Chapter 2 presents some of the remarkable implications of AI in medicine, whereas Chapter 3 discusses the potential synergy of AI with neurosciences.

The new forms of AI, especially generative AI, create anxiety and fear. Is it possible that the human–machine partnership will lead to the so-called 'intelligence explosion'—namely, a situation where machines become more intelligent than humans? Are we in the process of creating an all-knowing, all-controlling machine? This fear has led to several calls, even from pioneers of AI such as Geoffrey Hinton (Nobel Prize in Physics, 2024), for the prohibition of any further developments towards the construction of 'intelligent' machines. In my opinion, such calls are unrealistic. History suggests that technological developments are unstoppable. Until now, every attempt to place barriers against revolutionary technological advances has failed. For example, for a variety of reasons, including ideological concerns and the high cost of operating Arabic-language printers, the Ottoman Empire resisted endorsing the use of printers for a long time; but finally, in 1727, printing was officially accepted. Another well-known example is the failed efforts of the Luddites to prohibit industrial machinery. This suggests that further developments in AI cannot be stopped. Indeed, in 2020, start-ups in AI in the US, Europe, and Asia raised US\$38, US\$8, and US\$5 billion in funding, respectively; between 2024 and the end of 2025, the top seven US tech companies, namely Meta, Alphabet, Amazon, Microsoft, Nvidia, Tesla, and Apple, allocated US\$560 billion to spend on AI infrastructure.

Some scholars claim that machines will *never* become omnipotent. I believe that it is unwise to speculate that a given breakthrough will only have limited applicability. For example, in 1933, Ernest Rutherford (Nobel Prize in Chemistry, 1908), considered the 'father' of nuclear physics, claimed that it would not be possible to utilise atomic energy. Literally the next day following Rutherford's statement, the distinguished physicist Leo Szilard thought of the 'nuclear chain reaction'. In 1945, using this theoretical breakthrough, a nuclear device called Trinity, constructed within the Manhattan Project, was detonated in the New Mexico desert. More powerful explosions followed, including the detonation, in 1961, of an H-bomb in the Barents Sea with a mushroom cloud almost 60 miles wide.

Ethical considerations suggest that dangerous technological innovations should be restricted. However, in this regard, it is worth quoting John von Neumann, who wrote the following in connection with the Manhattan Project:

"What we are creating now is a monster whose influence is going to change history, provided there is history left; yet it would be impossible not to see it through. Not only for military reasons, but it would also be unethical from the point of view of the scientists not to do what they know it is feasible, no matter what terrible consequences may have".

Personally, I am in favour of a different point of view, which was expressed by the physicist Joseph Rotblat (Nobel Peace Prize, 2005). He argued that scientists and engineers need to think of the central element of the Hippocratic oath of physicians, namely *primium non nocere* (do no harm).

In my opinion, efforts should be directed, not towards abolishing AI, but towards ensuring that AI will be used for the benefit of society at large and not for its demise. For this purpose, Rotblat's position should be the guiding principle of any further developments. Challenges related to AI are discussed in Chapter 3.

The fear of whether machines might become omnipotent is related to the question of whether machines could reach and then surpass human intelligence. I believe that the discussion of the differences between artificial and human intelligence is incomplete unless it includes an analysis of the essential anatomical and functional characteristics of the brain. In this regard, the following three important elements are usually overlooked by proponents of AI: first, the impact of the *body proper* on the brain; second, the *lateralisation* of the brain, which refers to the anatomical and functional differences between the left and right cerebral hemispheres; and third, the vital role of the *glial* cells. How can AI be expected to reach the level of human thought if it ignores these fundamental elements of the generation of our cognition? These basic features of the brain and their impact on our thought process are discussed in Chapters 4 and 5. These two chapters are related to AI only indirectly; therefore, they can be read independently of Chapters 1–3 and 6, which deal directly with AI. In Chapter 6, several arguments are presented which suggest that there is, at the moment, an unbridgeable qualitative difference between artificial and natural intelligence.

Of course, currently, the most important question is not whether AI can reach the level of human thought, but how we can ensure that its ultimate impact will be beneficial to our living conditions and, especially, to our culture and the essence of humanity. Huge current problems is the vast inequality existing between different countries and within the same country, as well at the environmental catastrophe. Unfortunately, AI is exacerbating both these most serious problems (Stiglitz, 2012).

Another negative implication of the recent developments in AI is the exacerbation of intense antagonism between different countries, particularly between the US and China. In this regard, it is noted that the Chinese AI start-up DeepSeek shocked the world in January 2025 with an AI model, called R_1, that rivalled the top LLMs of OpenAI and Anthropic. R_1 was built at a fraction of the cost of the above models, using far fewer Nvidia chips, and was released for free. Later, a couple of weeks after OpenAI debuted its latest model, GPT-5, DeepSeek released an update to its V_3 model. According to experts, V_3 matched GPT-5 on tested benchmarks and is strategically priced to undercut it. This remarkable success, as well as a plethora of other astonishing achievements by China, should be viewed within a continuum of fundamental discoveries throughout history, as documented in the monumental work of Joseph Needham (1985). In addition, the two factors mentioned below are of crucial importance.

The first is the development by Chinese scientists of 'hypography', namely, an ingenious approach to using the Chinese language, which consists of tens of thousands of characters and no alphabet, as input to a standard QWERTY keyboard, which only has a few dozen keys. In the 1980s, Chinese scientists returned to QWERTY-style keyboards (which they had abandoned in the 1960s and 1970s) and constructed Chinese-compatible microprocessors, computer memory, graphics, and printers. The role of several individuals from IBM, MIT, the CIA, the Pentagon, and the Taiwanese military in the above success story is documented in Mullaney (2024).

The second factor is the scientific and technological know-how gained by China as a result of the implementation of the global chain model, already mentioned in the Introduction. This is exemplified by the close relationship between China and Apple (McGee, 2025). Although the candy-coloured computer iMac, designed in 1998 and credited with reviving Apple's fortunes (after the return of Steve Jobs in 1997), was first built in South Korea by LG, later versions were built in China by the Taiwanese assembly giant Foxconn. Also, although the first iPods, iBooks, and PowerBook laptops were made in Taiwan, by 2003 Apple, as well as many other Western companies had concentrated, its operations in China. The availability of thousands of labourers handcrafting hardware on a conveyor-belt production line under the guidance of American engineers allowed Apple to implement increasingly intricate designs. In 2003, Apple's global profits were US$69 million; by 2012, they had jumped to US$41.7 billion. Following Xi Jinping's rise to power, with his emphasis on 'in China for China', Apple re-adjusted. It imposed the rule that its suppliers could rely on Apple for a maximum of 50% of their revenues. Thus, as iPhone volumes soared from under 10 million in 2007 to more than 230 million in 2015, Apple's suppliers had to grow their non -Apple business just as quickly. This brought enormous benefits to China. Apple engineers taught Chinese how to perfect multi-touch glass and make the thousands of components within the iPhone. This technology was transferred to Chinese companies, led by Huawei, Xiaomi, Vivo, and Oppo.

President Trump thought he could bring iPhone production to America or at least move it to India. It appears that at the moment, this is impossible. After President Trump imposed formidable tariffs on China, Apple, which is the world's biggest company valued at US$3 trillion, lost US$800 billion in a matter of days. President Trump capitulated: he exempted smartphones and computers from his Chinese tariffs!

References

Frieden, T. R. (2025). Dismantling public health infrastructure, endangering American lives. *New England Journal of Medicine, 393*, 625–627.

Mullaney, T. (2024). *The Chinese Computer*. Cambridge: MIT Press.

Needham, J. (1985). *Science and Civilisation in China*. Cambridge: Cambridge University Press.

Stiglitz, J. E. (2012). *The Price of Inequality*. New York: Penguin.

Chapter 1

Artificial Intelligence

Artificial intelligence (AI) would not exist without the invention of the computer. In 1936, Alan Turing introduced the abstract notion of the 'Turing Machine'. This *imaginary* machine is very simple: it consists of an infinitely long tape of squares, with a 1 or a 0 on each square. The machine reads a square at a time, and then according to a table of rules which is contained in the machine, it implements a specified action, depending on whether the square is 1 or 0. Turing, with the help of the mathematician and computer scientist Alonzo Church, proved that, despite the simplicity of this abstract computer, if a problem cannot be solved by a Turing machine, it cannot be solved by any machine. Moreover, in analogy with Gödel's 'incompleteness theorem', Turing proved that there exist infinitely many well-defined problems, with each possessing a unique solution that cannot be computed via a Turing machine and hence by any machine.

For the abstract formulation of his computer, Turing was influenced by a lecture at Cambridge delivered by John von Neumann, which emphasised that a computer should have stored in it 'a program with instructions'. The first ever actual computer, named the Colossus which was built by Turing, did not have the characteristics of the abstract Turing machine. Actually, the Colossus, like all computers built before the discovery by von Neumann of the modern computer architecture, was a one-task machine incapable of being reprogrammed for any other task. The astounding impact of the computer became immediately evident with the Colossus; using this computer, Turing broke the Nazi Enigma code, leading to the Allied victory in the Battle of Britain despite a three-to-one advantage in favour of the German Luftwaffe.

In 1948, John von Neumann, after studying the brilliant work of Turing, introduced the architecture of the modern computer. His design, in addition to input and output channels, contained a 'memory unit', where programs (instructions) and data are stored. Moreover, his computer made crucial use of a 'central processing unit', where arithmetical and logical operations are carried out. The program stored in the memory unit could be modified, allowing the computer to be reprogrammed for different tasks. After the war, von Neumann, collaborating with several mathematicians and engineers, implemented his architecture by building at the Institute of Advanced Study at Princeton (where he had been a professor since 1933) an experimental electronic calculator, called JOHNNIAC. This machine was of crucial importance for the development of the modern computer.

It is worth noting that many of the features of von Neumann's architecture were introduced in the 1837 work of the Cambridge mathematician Charles Babbage, the 'father' of modern computing. Unfortunately, Babbage's computer, the 'Analytical Engine', never ran. Although it is unclear whether von Neumann was aware of this work, Babbage's ideas, which were certainly well ahead of their time, provided the basis for the development of the important field of 'software programming'. This is an area at the interface of mathematics and engineering which investigates appropriate 'languages' that can be used by the computer. One of the earliest researchers to make contributions in this area was Ada Byron (more widely known as Ada Lovelace), the sole legitimate child of Lord Byron, who wrote specific programs for Babbage's computer.

It was emphasised in Fokas (2024) that the human brain possesses an astonishing capacity for abstraction, reduction, and generalisation. The brain, in its insatiable quest for understanding, constructs two crucial metarepresentations. First, it builds abstract 'models', which capture essential features of the real world. These models rely on deep physical insights, as well as on the reduction of complicated entities into their basic constituents and on grouping similar parts into distinct categories. Second, it designs appropriate 'algorithms' for manipulating complex entities used for the description of the models. Mathematics provides the proper language for formulating the relevant models and their corresponding algorithms, as well as the tools to analyse them. The statement in Fokas (2024) that 'understanding is the elucidation of relationships' is exemplified perfectly by the formulation of models. Specifically, models express in mathematical form, relations between different entities. Algorithms, in the case of solving a problem with the aid of a computer, characterise the set of steps that the computer must follow for the solution of the given problem.

The combination of accurate models, ingenious algorithms, and powerful computers allows scientists and engineers to focus with great precision on particular aspects of reality and to attain highly complex goals. The apogee of this process is reached with AI, which provides the next step in the hierarchy of great human endeavours that began with writing and printing and continued with computing and the internet.

Several milestones in the history of AI are characterised by attempts to design artificial circuits that mimic aspects of the brain. In particular, in 1943, the early cybernetics expert and neurophysiologist Warren McCulloch, together with the electrical engineer Walter Pitts, motivated by the function of neurons, introduced the idea of a 'neural network'. In a sense, their thinking was the opposite of Shannon's; the latter, by introducing the notion of 'transferring information' removed meaning (and hence the thinking mind) from the concept of information; McCulloch and Pitts brought neurons (and hence the mind) back into play in the manipulation of information. In 1958, Frank Rosenblatt, a researcher at Cornell Aeronautical Laboratory, proposed that information could be encoded using an approach similar to the one used by the human brain, which encodes information by employing billions of its neurons with trillions of their synapses. In the same year, he introduced a set of mathematical relations mimicking the functions of a single layer of neurons and called this set 'perceptron'. The term 'artificial intelligence' was introduced in 1956 by John McCarthy, who defined it as 'machines that can perform tasks that are characteristic of human intelligence'.

Since its genesis, AI has been accompanied by hype. For example, in 1958, reporting on the invention of the perceptron, *The New York Times* wrote: 'an embryo of an electronic computer that will be able to walk, talk, see, write, reproduce itself and be conscious of its existence'. Although these exaggerated predictions were not realised, the early developments did give rise to a truly novel approach to computing, which in 1959 was called by Arthur Samuel, 'machine learning'. This novel approach was based on the realisation that, if the input to a computer – 'machine' – consists of a well-organised and sufficiently large set of data, then the computer could be programmed to 'learn' how to extract information from such a set.[1] This approach is conceptually different from standard logic-based algorithms, which are designed to follow well-formulated, strict rules. As will be explained in detail later, machine learning is based on the following simple idea: the different 'neurons' of a given neural network are connected via 'synapses', and the contribution of each synapse is determined by a specific number, called 'weight'; during the process of 'learning', the weights change until an optimal result is achieved.

Despite the early promise of machine learning, standard algorithms continued to dominate AI until the mid-1980s. The logic-based algorithms are highly effective for solving problems that can be formulated in terms of simple rules, but they perform poorly with problems that are difficult to describe formally. For example, considering that chess consists of 32 pieces moving between 64 locations, with each piece moving in a well-defined way, it follows that chess can be described in terms of a small set of rules. Hence, standard algorithms can be successfully used for devising an effective chess strategy. However, much of the world is not organised in a way that can be reduced to simple rules or symbolic representations. In such situations, logic-based algorithms do not perform well. An example of such a case is the recognition of a face in a photograph since this problem cannot be defined by a small set of rules. The inability of machine learning to make serious progress with the latter type of tasks led to the 'AI winter', that is, the decline of interest and investments in AI.

1.1 Deep Learning

The situation began to change in the mid-1980s with the introduction of 'multi-layered neural networks', named 'multilayer perceptron'. This was the result of the works of several researches, including the Russian mathematician Alexey Ivakhnenko, the Japanese engineer and neuroscientist Shun'ichi Amari, the American machine learning pioneer Paul Werbos, and especially Geoffrey Hinton (Nobel Prize in Physics, 2024). These neural networks make use of 'feedforward deep learning', meaning the information extracted from each layer is used as an input to the next layer. Machine learning for such networks was called 'deep learning'.

In the spring of 2012, Geoffrey Hinton and two of his students at the University of Toronto built the system AlexNet. In this neural network, if

---

[1]Examples of machine learning algorithms are Bayesian networks and random forest.

an error is spotted, the network, using a technique called 'backpropagation', adjusts its weights appropriately; this procedure is repeated again and again. In this way, the network 'learns' to recognise lines, shapes, and, finally, entire objects. Although in 2012, in an annual competition involving the recognition of a specific object in a given image, AlexNet beat the previous winner by only 10%, it became clear that a new epoch in AI had begun.

The presence of many layers allows for the manipulation of particular 'features' of the given data in such a way that *useful information contained in the data can be identified and extracted in an efficient, hierarchical manner*. For example, a four-layer neural network can recognise an image as follows. The image is segmented into a collection of small pixels (squares), with each pixel having a specific colour and brightness. This set of data is the input presented to the 'visible layer'. The remaining three 'hidden layers' extract increasingly useful information from the image. In particular, the first layer, by comparing the brightness of neighbouring pixels, identifies edges. Using this information, the second layer searches for corners and extended contours, which can be recognised as collections of edges. Then, the third layer, by searching for specific collections of contours and corners, can detect entire parts of a given object. In this way, this particular neural network can recognise specific objects contained in the image. The output provides a description of the image in terms of its parts.

It is worth noting that the level of difficulty in solving a given problem is expressed mathematically using the notion of 'complexity'. In computer science, the worst-case complexity occurs when the cost of computation, in terms of time or memory, increases exponentially with the magnitude of the input. Deep learning is particularly effective in some problems when dealing with very high algorithmic complexity.

Interestingly, the spectacular success of deep learning is not due to a dramatic improvement in the underlying software, which is actually similar to that of systems which performed poorly until the mid-1980s. Rather, it is the result of the following technical advances. First, machine learning requires a very large set of data. For example, in order for the computer to 'learn' to recognise the presence of a tumour in a given lung X-ray image, it must first be 'trained' with a very large dataset of similar X-ray images. A rough estimate is that a typical deep learning algorithm will achieve reasonable performance after it has been trained with a dataset of approximately 5,000 examples; it will match or exceed human performance after it has been trained with 10 million examples (Goodfellow *et al.*, 2019). In recent years, the processes of 'digitising' many aspects of our lives have been greatly accelerating, and this has resulted in the accumulation of precisely the type of 'big data' needed in machine learning. For example, as of 2017, ImageNet had more than 15 million labelled images; at the same time, YouTube's video library was growing by 300 hours of video every day, and Facebook's library already had billions of images (Topol, 2019, page 77). Second, video games motivated the development of appropriate graphic processing units that employ parallel computing architectures; these graphic capabilities are indispensable in machine learning. Third, the massive memory requirements of AI are accommodated by the development of 'cloud computing', which, as will be discussed in Part III, was motivated by the ever-increasing computing needs at CERN. Fourth, freely available open-source modules make AI codes easily accessible. Examples of such modules are Google's TensorFlow, Microsoft's Cognitive Toolkit, Facebook's PyTorch, and the University of California, Berkeley's Caffe.

Modern software infrastructure has resulted in faster and more powerful computers. In particular, it has led to a dramatic increase in the size of the underlying neural networks; actually, the size of the advanced neural networks is doubled approximately every 2.4 years. Also, it turns out that computation becomes half as expensive approximately every two years. This is roughly consistent with the so-called Moore's law. Gordon Moore predicted in 1965 that the number of transistors in an integrated circuit would double every two years. This was indeed the case from 1975 to 2012, with a slightly higher rate of increase in the late 1970's and a slower rate since 2013. In 1980, the number of transistors per square millimetre on a chip was of the order of a thousand; today, it is of the order of 100 million.

1.2 The Essence of Computation

Computation can be defined as the *process of transforming an input into an output*. The first step of a mechanical computation involves encoding the input in such a way that it can be 'read' by the computer. For example, if the input is an image, then it is segmented into small pixels, and the computer is provided with a quantitative measure of the colour and brightness of each pixel. The second step involves describing the transformation digitally and applying this transformation to the input. In the third step, the outcome of the transformation, which was implemented in the second step, is presented in an understandable way. For example, in machine translation, such as that performed by Google Translate, text in one language is encoded into appropriate bits, which are then transformed, and finally the resulting bits are presented as text in another language.

The heart of computing involves specifying the transformation. Mathematically, a transformation can be represented in terms of an entity called a 'function'. Functions can be thought of as rules, where given an input, a rule yields a unique output. An example of such a function is $f(x)=2x + 3$; in this case, the input $x = 0$ yields the output $f = 3$, the input $x = 1$ gives the output $f = 5$, etc. It is useful to recall that the hierarchical structure of mathematics is based on the brain's ability to generalise. Indeed, mathematics constructs extremely complicated structures from very simple ones. Hence, it is not surprising that *any* function can be built from certain elementary functions. An example of such an elementary function is the 'NAND gate'. It has only two inputs and one output, with each taking a value of either 0 or 1; if both inputs are 1, then the output is 0, whereas for every other case the output is 1. Computer scientists have proven that any computer capable of connecting an arbitrarily large number of such gates could become a 'computronium', meaning that it will be able to perform *any* computation.[2] Thus, according to the definition of Turing's machine discussed earlier, NAND gates are 'universal', that is, they can be used to build a Turing machine. Remarkably, it was proven in 1989 that simple 'neural networks' are also universal. Incidentally, the so-called 'cellular automata' are simple functions introduced by Stephen Wolfram, which can also be used to build universal machines.

A specific 'neuron' of a neural network has an output of either 1 or 0, depending on whether the sum of the inputs to this neuron from the other neurons of the network is above a certain threshold or below it, respectively. The output 1 corresponds to the physiological case that the neuron is sufficiently excited to fire, whereas the output 0 corresponds to the case that the neuron is inhibited. The different 'synapses' of the neural network that connect the neuron with its neighbouring neurons may have different significance. Hence, each synapse is assigned its own 'weight', quantifying the significance of this synapse.

---

[2] The name 'computronium' was introduced by the MIT researchers Norman Margolus and Tommaso Toffoli.

In the above discussion, *the function specifying the transformation is fixed, independent of the inputs*; this is precisely the *defining feature of standard logic-based algorithms*. In particular, for the case of neural networks, the relevant weights are mostly fixed. However, this is not how physiological neurons behave. As discussed in detail in Chapters 9 and 10 of Fokas (2024), learning at the molecular level is expressed via the *change* in the strengths and the number of synapses. This suggests that the weights of neural networks should not be *a priori* fixed; instead, their values should be adjusted according to what these neurons 'learn' during training. *The dependence of the transformation on the data is the essence of machine learning.* For the neural networks used in machine learning, 'neurons', following specific instructions, adjust their weights according to what they 'learn' during the process of being 'trained' with a large set of data.[3]

1.3 Achievements of AI

Progress in AI, coupled with the availability of powerful computers, led to the surprising development of machines defeating humans in a variety of games. IBM's Deep Blue triumphed in 1997 by overpowering chess champion Garry Kasparov, though it was not based on deep learning (despite the word 'Deep' in its name). Deep Blue's success was due to its advantages in memory and speed, which allowed it to quickly analyse the outcomes of a very large number of potential moves.

The use of deep learning yielded even more spectacular results. For example, the company DeepMind introduced in 2014 the so-called 'deep reinforcement learning'. This is based on the use of a trial-and-error strategy and appropriate 'encouragement', where a particular action is 'reinforced' if it brings success. In one of the most successful implementations of this type of 'learning', DeepMind's Deep Q-Network trained a specific neural network to play the game 'Breakout'. In this game, the player controls a paddle which bounces a ball up and down trying to knock out rows of coloured bricks. As more bricks are knocked out, the score increases. After extensive training, the network, through trial and error, learned to control the paddle and was able to knock out bricks row by row.

---

[3] Among such instructions are those defining 'backpropagation' and 'stochastic gradient descent'.

Remarkably, after playing a very large number of games, it 'invented' a new strategy: it targeted a single column as opposed to rows.[4]

Another spectacular achievement of AI occurred in 2016, when the DeepMind system AlphaGo won a five-match game of Go against the world's top player. In the game Go, players take turns placing black and white pieces, called 'stones', on a 19 × 19 board, with the goal of surrounding a larger territory than their opponent.[5] There is a gargantuan number of possible Go positions, so it is hopeless to try analysing possible sequences of future moves. Experienced players develop an intuition of how to choose advantageous positions. Remarkably, AlphaGo also developed such an 'intuition' by using a massive database of Go positions, as well as being trained by playing with itself. Initially, it learned to play Go by watching 150,000 games. Then, a very large number of copies of AlphaGo were created, and an algorithm was introduced which instructed AlphaGo to play against itself. In this way, the network could simulate millions of games and try out moves that had never been attempted by humans. In 2016, while participating in an international tournament, AlphaGo made the famous 'move 37', which was so unexpected that experts initially thought it was a wrong move. But it turned out to be a brilliant move, one that had never been used before. Interestingly, later versions eliminated any prior human knowledge; the system 'trained' by playing itself millions of times. AlphaGo's success was based on the following strategy: a deep neural network was 'trained' to predict, given a specific position, the probability of winning. Also, a different deep neural network was 'trained' to predict likely next moves. By properly combining this information and searching through a list of likely future sequences, the next move that would give the maximum advantage was identified.[6]

In late 2017, DeepMind announced AlphaZero, the successor to AlphaGo, which by learning simply via self-play defeated AlphaGo. Furthermore, after two hours of playing chess with itself, it defeated the best human chess player, and after four hours of self-play practice, it defeated Stockfish, which was until then the most powerful chess program in the world. Prior programs had relied on moves conceived by humans,

---

[4] Soon thereafter, such an AI system outplayed a human in 29 out of 49 Atari games.
[5] Go was invented in China more than 2,500 years ago. It is the oldest board game played uninterrupted until now.
[6] Soon after these successes, DeepMind was bought by Google but continues to be run independently. Meanwhile, Google had also hired Hinton and some of his colleagues.

i.e. human experience, knowledge, and strategy. AlphaZero had no prepro-grammed moves derived from human play. In this sense, AlphaZero did not have a strategy in the human sense; its only goal was simply to maxi-mise its proportion of wins to losses. It executed moves that humans had not considered and adopted such surprising tactics as sacrificing its queen!

Throughout the book, I use quotes to emphasise that words such as 'training', 'learning', and 'understanding', are used metaphorically. This is in contrast to many computer scientists who tend to use these notions liter-ally. Something analogous occurred regarding the word 'computations' in neuroscience. In 1999, it was correctly noted by the cognitive linguist George Lakoff that this analogy had become so strong that neuroscientists 'commonly use the Neural Computation metaphor without noticing that it is a metaphor'. What does it actually mean to 'train' a neural network? For example, what happens when a neural network is 'trained' to recognise the photo of a cat by being fed with a huge number of random photos? Each of these photos is supplemented with positive or negative reinforcement: positive feedback for cats and negative feedback for non-cats. The net-work uses probabilistic techniques to make 'guesses' about whether a given photo depicts a cat or not. These guesses, with the help of the power-ful technique of feedback, become increasingly accurate.

The achievements of AI in chess, Go, and video games have been extensively publicised and naturally have generated widespread interest. However, the main reason for the excitement generated by AI is its suc-cess not in games but in a variety of real-life situations, from the automa-tisation of routine labour tasks to the understanding of speech and images and to the mechanisation of certain medical diagnostics. There are many applications, ranging from voice, speech, text recognition, and translation of a large number of languages, to protein folding, discovery of novel antibiotics, and driverless cars. The impact of AI is already transformative in a variety of critical areas, including culture, education, scientific research, agriculture, transportation, manufacturing, defence, security, law enforcement, politics, and advertising. For example, governments are analysing social networks to prevent terrorist attacks; airport security soft-ware use neural networks to recognise faces on scanned passport photos; Google, by harnessing searches performed by others, can identify the preferences of millions of people; in agriculture, AI is facilitating the iden-tification of the most effective choice and administration of pesticides, detection of diseases, and the prediction of crop yields; Facebook uses neural networks to identify people in photos; Alexa employs them to inter-pret voice commands; and Google uses deep learning for translation.

Due to space limitations, only the impact of AI in medicine will be discussed in this book, but it should be remembered that, essentially, every human endeavour has been influenced by AI. For example, large law firms use 'virtual (digital) assistants', which are systems for searching, collecting, classifying, and processing a large amount of data and information contained in various legal documents (legislative regulations, judicial decisions, etc.); the role of these programs in out-of-court advocacy is particularly useful (for example, in the processing and drafting of contracts). The impact of AI on law continues to expand. For example, the program DONOTPAY, created by Joshua Browder (an English student at Stanford University), initially offered assistance to individuals concerned with road traffic law violations; within a two-year period, this program intervened in 375,000 cases of fines for illegal parking. Later, it was extended so as to provide assistance regarding consumer protection laws.[7] As an example of the possible role of AI in assisting judges, it is noted that at the Koblenz Court of Appeal (Germany), there is an ongoing pilot project aimed at using AI to facilitate and accelerate the processing of files (which are already digitised) for the particular case of building law (Travlos, 2023).

Regrading research in systems affecting 'scientific discovery', an example is provided by a system attempting to automate the scientific process presented in King *et al.* (2004). Starting with given observations, this system postulates appropriate hypotheses, devises experiments to test these hypotheses, physically runs the experiments using a laboratory robot, analyses the results, and, accordingly, repeats the cycle. This system was applied to the determination of a gene's function using deletion mutations in yeasts.

1.4 ChatGPT and Generative AI

A new breakthrough occurred in AI a decade after Hinton's landmark achievement; in 2022, OpenAI released ChatGPT, which belongs to a class of models called 'large language models' (LLMs). This remarkable development exploits the fact that language data come in a well-organised, sequential form. The network, by being 'trained' via reading an

---

[7]In 2023, in what would have been the first 'legaltech' case, an attempt was made by Browder to use his program in the actual proceedings of a legal case. After a massive backlash from US bar associations and under state threats of filing a lawsuit, Browder aborted this attempt (Travlos, 2023).

extremely large number of sentences, 'learns where to pay attention'. In this way, given a sentence, it can generate a prediction regarding what follows. For example, by the time the network reads the sentence 'Tomorrow there is going to be a major storm in Brazil', it has 'learned' that the key words in the sentence are 'tomorrow', 'storm', and 'Brazil'. A huge advantage of LLMs is that they can be trained directly on real-world data. This is to be contrasted with earlier deep learning models which were based on 'supervised' learning, namely they were trained using carefully prepared input data. Instead of such handmade data, LLMs can be trained using, for example, Wikipedia, millions of emails, or an arbitrarily large number of books.

A measure of the complexity of a network is the number of parameters that define its mathematical structure. GPT-3 was, at the time of its release, the largest neural network in existence, containing the astonishingly large number of 175 billion parameters. The parameter count of GPT-4 remains undisclosed. Apparently, at the moment, the largest neural network belongs to the Chinese company Alibaba, containing ten trillion parameters!

It should be emphasised that the capabilities of LLMs extend far beyond just language. Exploring the 'knowledge' obtained via the process of being trained using an unprecedented volume of data allows them to, for example, compose music, play various games, 'understand' financial statements, perform mathematical calculations, and assist judges. For example, in Colombia, Judge Juan Manuel Padilla based his decision regarding a claim by a minor with autism against a private insurance company on analysis obtained via ChatGPT (Travlos, 2023).

LLMs have given rise to a new form of AI, called 'generative AI'. This new development gave rise to a new type of AI expressed in the form of 'agents' that can perform better and faster anything that can now be done in a browser. OpenAI released in late 2024 two such agents: one, called 'operator', can implement basic tasks, such as online shopping; the second, called 'deep research', can generate within a few minutes research reports, apparently to the level of a PhD student. Similar 'agents' have been developed by Google, called Mariner, and by Anthropic. What is next? In an exciting recent development, it has been suggested that AI possesses self-programming capabilities. For example, neural networks, by being trained with scrambled and sorted sequences, can learn how to sort a list of numbers. Although this approach is still in its infancy, clearly it is conceptually very important since it brings AI closer to the functioning of the brain.

The dream of several AI researchers is to achieve 'artificial general intelligence' (AGI), meaning for the computer to attain and subsequently surpass human thought. Currently, this goal is pursued by a handful of companies, in particular by OpenAI and Anthropic. Luciano Floridi, the founding director of the Digital Ethics Center at Yale, claims that the enormous funding in this area is disproportionate to the expected gains and predicts that the 'bubble will soon burst'. In 2025, the overall profits of Microsoft are expected to reach US$10 billion, while the losses incurred by OpenAI, which is primarily funded by Microsoft, are expected to be of the order of US$5 billion (in 2022 alone, OpenAI's losses amounted to half a billion dollars). Mustafa Suleyman, the co-founder of DeepMind, believes that a current realistic goal is not AGI, but what he calls 'artificial capable intelligence'. This means that AI will be able to weave together sequences of actions that enable it to make long-term plans, such as those required to, for example, run a company.

1.5 Different Forms of AI and 'Training'

Classical logic-based algorithms instruct the computer to follow steps which produce *precise* results. In contrast, machine learning algorithms do not directly produce a specific outcome; they instruct the computer to follow steps which improve, iteratively, *imprecise* results until an acceptable result is obtained.

The first type of 'learning' used in deep learning was 'supervised learning', where inputs in a dataset are individually labelled according to the desired result. For the particular case of image recognition, the neural network is 'trained' on a set of pre-labelled images, 'learning' to associate an image with its appropriate label; for example, the image of a cat with the label 'cat'. Having encoded the relationship between images and labels and after sufficient 'training', the neural network is able to identify the image of a cat.

The second type of learning, called 'unsupervised learning', is based on the internet and the prevalence of digitisation. For example, when bankers want to analyse possible cases of financial fraud, they can directly search for inconsistencies in the huge volume of transactions recorded on files. 'Unsupervised learning' allows neural networks to identify anomalies based solely on the existing data. Another example of 'unsupervised learning' is provided by streaming video services, such as Netflix, which use appropriate algorithms to identify clusters of customers with similar viewing habits; then, the video services recommend to those customers specific streaming options. Interestingly, in analogy with unexpected, brilliant discoveries by autodidact scientists and artists, such as Albert Einstein and Arnold Schoenberg, deep learning systems, which are 'trained' without the specification of concrete outcomes, occasionally produce surprisingly innovative insights.

A third major category of machine learning is 'reinforcement learning', which was used by AlphaGo. In this case, a neural network trains itself, and instead of following a human strategy, it simply uses 'reinforcement' according to a specified reward function.

Considering that the subtleties of language prevent the specification of semantics via the use of simple rules, the impact of deep learning on language is not surprising. However, the development of LLMs provides a new, unexpected breakthrough. Indeed, there is a huge difference between translating texts and generating novel texts. Similarly, there is a qualitative difference between identifying a specific percept in a given image and creating an image of a human face that seems real.

The mind, by using the continuum of unconscious and conscious processes and by interacting constantly with the environment, induces actions and makes plans. This behaviour is autonomous and can be continuously modified. Currently, AI cannot represent 'common sense' knowledge and, importantly, cannot handle unanticipated problems by devising novel solutions. As AI becomes increasingly capable of formulating its own strategies based on the accumulation of 'knowledge' and the continuous interpretation and interaction of a myriad of messages from the environment, the gap between AI and human behaviour will diminish. The immense difficulty of building 'conscious machines' was noted by Stuart Russell, who was mentioned in the introduction: 'if you gave me a trillion dollars to build a consciousness machine, I'd give it back because I'd have absolutely no idea where to start' (Fortson, 2019). The question of whether AI can reach the level of human thought is discussed in the last chapter of this part.

References

Fokas, A. S. (2024). *Ways of Comprehending: The Grand Illusion and the Essence of Being Human.* Singapore: World Scientific.

Fortson, D. (26 October 2019). The end of humanity: Will artificial intelligence free us, enslave us — Or exterminate us? *The Sunday Times.*

Goodfellow, I., Bengio, Y., and Courville, A. (2019). *Deep Learning*. Cambridge: MIT Press.

King, R. D., *et al.* (2004). Functional genomic hypothesis generation and experimentation by a robot scientist. *Nature, 427*(6971), 247–252.

Topol, E. J. (2019). *Deep Medicine*. New York: Basic Books.

Travlos-Tzanetatos, D. (4 December 2023). Law in the face of the challenge of algorithms (in Greek: Το δίκαιο ενώπιον της πρόκλησης των αλγορίθμων). *The Road of the Left (Δρόμος της Αριστεράς)*.

Chapter 2

The Impact of Artificial Intelligence on Medicine

In 1970, William Schwartz wrote, 'Computing science will probably exert its major effects by augmenting and, in some cases, largely replacing the intellectual functions of the physicians' (Schwartz, 1970). However, almost two decades later, it was noted that 'the threat to physicians of a mathematical approach to medical decision simply has not materialized' (Pauker and Kassirer, 1987). Astounding developments in medical imaging, some of which were reviewed in Fokas (2024) (see also Chapter 21), have undoubtedly 'augmented' the cognitive impact of physicians. But certainly, almost 55 years after Schwartz's statement, the intellectual functions of physicians have not been replaced by computers. Is this about to happen now?

2.1 Current Successes of AI

The potential medical applications of AI are unlimited, from monitoring the glucose levels of diabetic patients and analysing images obtained via radiography, CT, ultrasonography and MRI, to improving robotic surgery via the development of robots possessing tactile sensation (Bartolozzi, 2018). In what follows, first, some striking medical implications of AI are highlighted.

AI algorithms have already been validated for, among others, the following: diagnosing skin melanoma; diagnosing pneumonia via the analysis of chest X-rays; interpreting certain pathology slides; and evaluating diabetic retinopathy. These algorithms perform such tasks at least as well as dermatologists, radiologists, pathologists, and ophthalmologists, respectively.

A landmark paper in 2017 showed that a deep neural network developed at Stanford University and trained with nearly 130,000 images of skin lesions could match, via image analysis, the accuracy of certified dermatologists for the identification of melanoma (Esteva *et al.*, 2017). Dermatologists diagnose cancerous lesions using the heuristic algorithm ABCDE, formed from the capital letters of the words Asymmetry, irregular Boarder, uneven distribution of Colour, large (greater than 6 mm) Diameter, and Evolving lesion. The AI algorithm may be particularly useful for general physicians, who are actually involved in two-thirds of skin lesion diagnoses with error rates as high as 50% (Topol, 2019a).

A group of computer scientists at Stanford University, led by Andrew Ng, trained a 121-layer neural network using more than 112,000 films from over 30,000 patients. Following this training, it could diagnose pneumonia from an X-ray image better than radiologists (Rajpurkar *et al.*, 2017).

Regarding advances in pathology, it is noted that until the development of digital technology, pathologists had to look at glass slides through a microscope, whereas now via the technique of 'whole slide imaging', they simply look at the entire tissue sample on a computer screen. It has been shown that the use of an algorithm developed by a group at MIT and involving a 27-layer deep network reduces the pathologist's error in diagnosing metastasis of breast cancer to lymph nodes (Wang *et al.*, 2017). Interestingly, the best results were obtained by combining the expertise of a pathologist with the information obtained via AI.

A recent breakthrough in AI is related to 'super-resolved fluorene microscopy'. This technique, for which Eric Betzig, Stefan Hell, and William Moerner were awarded the 2014 Nobel Prize in Chemistry, involves complex sample preparation and attaching fluorescence molecules to cells. This process renders subcellular features visible; however, it is time-consuming and may also harm cellular parts. Google, by training a deep neural network which compared fluorescent-labelled images with unlabelled ones, developed an open-source algorithm that can accurately predict how samples would fluoresce without the need for any fluorescence procedures (Sullivan and Lundberg, 2018).

IBM, using 35,000 retinal images, has reported an 86% accuracy of detecting early diabetic retinopathy (IBM, 2017). This is useful for evaluating diabetic patients since one of the pathognomonic findings in such patients involves changes in the blood vessels supplying the retina.

The English statistician and founder of nursing Florence Nightingale (1820–1910) used statistical graphs to revolutionise disease surveillance at a *local* level. Her data-driven analysis established that during the Crimean War more people died from poor sanitary conditions than in battle. AI has had an analogous impact at a *global* level. For example, HealthMap, using natural language processing, searches across the web for signals of emerging infection events. In its search, it is aided by information about prevalent pathogens in a given region. In 2009, this internet-based infectious disease surveillance system provided early evidence of the emergence of influenza A(H1N1). On 30 December 2019, HealthMap provided a warning of a 'cluster of pneumonia cases of unknown aetiology', just days after the first case of COVID-19 was identified.

The Enhanced Detection System for Healthcare-Associated Transmission (EDS-HAT) combines whole-genome sequencing with AI to automatically search patients' electronic medical records (EMRs) for data related to a given outbreak. The relevant algorithm compares the data of infected patients in a given hospital with those of patients in the same hospital who are not infected. In this way, similarities can be identified in the cases involving infected patients. This approach was used, for example, to identify a contaminated gastroscope as the source of a *Pseudomonas aeruginosa* outbreak among six patients in multiple units of the same hospital over a period of seven months.

The AI algorithm Eva, used by the Greek government to screen travellers for COVID-19 at the border, identified 1.35 times as many asymptomatic travellers who were infected as those identified with testing based on standard epidemiological metrics (which test travellers according to the number of cases or deaths per capita and the rate of infections of the country of origin) (Brownstein *et al.*, 2023).

AI, in addition to its contributions towards surveillance and the prevention of infections, is also having an impact on the treatment of infections. For example, by quantifying pixel intensity around Petri dishes, it provided a more accurate approach to the standard Kirby–Bauer disc diffusion test. In this technique, which is indispensable for choosing the correct antibiotic for treating a given bacterial infection, bacterial susceptibility is determined by measuring the area in which growth of the bacteria is inhibited around an antibiotic-treated disc.

In 2020, researchers at the Massachusetts Institute of Technology announced the discovery of a novel antibiotic, Halicin, which was able to kill strains of bacteria that had, until then, been resistant to all known antibiotics. Chemists use entities such as atomic weights and chemical bonds to capture important characteristics of molecules. Using these entities, standard research and development efforts for the discovery of new drugs take years of expensive, painstaking work: researchers start with thousands of possible molecules and, through trial and error and educated guesswork, reduce them to a handful of viable candidates. The AI approach for the discovery of Halicin was different: first, researchers developed a 'training set' of 2,000 known molecules; this set encoded appropriate data about each molecule, ranging from its atomic weight and the types of bonds it contains to its ability to inhibit bacterial growth. From this training set, the relevant neural network 'learned' the attributes of molecules expected to be antibacterial. Apparently, in this process, important attributes were identified that had neither been encoded nor noted by humans. Then, by analysing 61,000 molecules, it picked one that fit the criteria identified in the process of training. This neural network does not 'understand' why the specific molecules worked; unfortunately, it is still not known why Halicin is effective. Nonetheless, AI identified relationships that had escaped human detection or possibly cannot be detected by humans.

2.2 Medical Devices

AI has already led to the production of a variety of medically useful devices. For example, it has been rigorously established that cognitive behavioural therapy (CBT) can help individuals identify and subsequently change negative thought patterns. It has been reported that several CBT mobile apps, such as Lantern, Joyable, MoodGYM, and Ginger.io, can be helpful in mild depression (Topol, 2019a, 178).

Another example of AI's usefulness is provided by the approval by the U.S. Food and Drug Administration (FDA) of the Face2Gene app. This app, by using deep learning to analyse images of individuals with certain rare genetic diseases who have a distinctive constellation of facial features, can aid the diagnosis of more than 4,000 genetic conditions.

The immediate impact of AI in clinical medicine becomes evident by the FDA approval, in 2017, of AliveCor's Kardia band, which detects atrial fibrillation. This is a common cardiac arrhythmia associated with the risk of stroke. Moreover, AliveCor, by using 2.8 million electrocardiographs together with values for the potassium levels of hospitalised patients in the Mayo Clinic, has obtained FDA approval for a smartwatch that detects high levels of potassium. This condition can cause sudden death due to heart arrhythmia. This device is especially useful for patients with renal diseases.

AI has resulted in the development of more complex apparatuses. For example, the Swoop portable MRI system plugs into a standard electrical outlet, is controlled by an Apple iPad, and is easily usable. Also, portable ultrasound probes and smartphones are available, which can be operated by users without formal training in ultrasound.

Chatbot technology is now extensively used in a variety of areas, including medicine. A very early such device, named ELIZA, was developed in 1966 at MIT. A modern medical chatbot, based on GPT-4, is now available for use in clinical settings.

2.3 Precision Medicine

The crucial importance of AI becomes clear by its expected effect on 'precision medicine'. The implementation of this transformative trend in medicine requires the collection and analysis of a huge amount of information; in my opinion, this is impossible without AI tools. Indeed, a proper assessment of a patient should involve collecting a variety of data, including the following: the patient's medical history, family history, and social network environment; immune status and vaccinations; data regarding their microbiome, genome, and epigenome; data from wearable devices and biomonitoring; standard laboratory values and medical imaging information; and findings from physical examinations. The formulation of an effective treatment plan should include the analysis of all these data and pertinent knowledge from the corpus of medical literature. It is evident that the implementation of precision medicine will require huge resources and the collaborative efforts of diverse disciplines. An example of such an effort is the Chinese consortium iCarbonX. There is no doubt that if AI algorithms can be reduced to clinically useful 'apps', if they can be clinically tested, and if they can be used in the emerging environment of humanly unmanageable genomic, metabolic, and environmental data, then they will have a transformative impact in medicine.

By employing the machinery of AI, many of the tasks needed in precision medicine can be solved in parallel. Having collected appropriate data, the final solution will require the creative synthesis of the information obtained from the analysis of these data. For example, in order to maintain an updated medical history for each individual, it is important that every physician produces precise, standardised notes. This goal is aided, in principle, by the electronic hospital records. However, at the moment, such records are often not accurate, and a common practice is to construct them by copying earlier records.[1] The situation may worsen by the insertion of LLM-generated text directly into the medical records. In a *perspective* published in a leading medical journal, the authors expressed concern that such practice will reduce the informational quality of charts, rendering them less useful for both physicians and future AI models (McCoy *et al.*, 2024). On the other hand, AI could assist in producing accurate records: a voice assistant, like Alexa, can facilitate the process of keeping notes in an efficient manner, allowing physicians more time to communicate with patients and their relatives. Also, depending on the particular case, AI can be used for collecting additional pertinent data. For example, for mood disorders, comprehensive phenotype information should include the following: the tone, pitch, and intonation of voice; prosody, length of phrases, and sentiment of speech; eye contact, number of sighs, posture, and gestures.[2] Also, in addition to vital signs such as blood pressure, heart rate, and temperature, further data, such as galvanic skin response and breathing pattern, could be collected.

Precision medicine is expected to have a metamorphic impact on the treatment of cancer, where the required data should include the following: sequencing the patient's DNA, the tumour's DNA, and the tumour's

---

[1]Keeping such records has so far had a negative effect on medical practice: in the words of a Harvard physician, 'the electronic medical record has turned physicians into data entry technicians' (Brody, 2017). Attending to the keyboard and reducing quality time with patients has been implicated as a major contributing factor to physicians' high rate of depression (half of US physicians exhibit symptoms of burnout, and hundreds of suicides are reported every year, Oaklander, 2015).

[2]Each individual possesses specific features resulting from the interaction of genes ('genotype') and the environment. The set of observable characteristics, which is obviously affected by pathological processes, is called 'phenotype'.

RNA; characterising the immune status of both the tumour and the patient; and growing cancer cells *in vitro* to check the response to various drugs. In this regard, IBM's Watson and Tempus laboratories are currently attempting to employ deep learning on data from cancer cell cultures, medical imaging, and biopsies. Recent progress in the diagnosis and treatment of cancer is discussed in Part II.

An effective choice of medications and their precise dosage must also be personalised. Severe allergic reactions to many medications illustrate the importance of individual variations in drug administration, distribution, metabolism, and excretion. Specific genes are associated with the severe side effects of certain medications. It is also known that for many drugs, the therapeutic dose varies greatly across individuals, depending on their particular genome. For example, the genes CYP2C9 and VKORC1 are responsible for 30–40% of variations in the dose of warfarin, which is the most widely prescribed anticoagulant (blood thinner) in the US and in most of Europe (Manolopoulos *et al.*, 2010). This prompted the FDA to modify the label of warfarin, indicating that the patient's genotype should be taken into consideration when prescribing this medication. A similar warning is inserted in clopidogrel, which is a medication that inhibits platelet (thrombocyte) aggregation and, together with aspirin, constitutes a dual antiplatelet therapy, often used for preventing the formation of a clot (thrombus).

Although in some cases cancer can metastasise early in its development (Ghajar and Bissell, 2016), *screening for early detection* remains a cornerstone of cancer prevention and treatment. For example, identifying and monitoring individuals with a family history of breast cancer who carry certain mutations of the BRCA1 and BRCA2 genes increases the likelihood of early intervention if cancer does develop in these high-risk individuals. The BRCA genes produce proteins which suppress the development of cancer, whereas their mutated versions lose this capability.

An early cancer detection via a blood test, either via the identification of specific blood proteins indicating the presence of cancer in very early stages or via the DNA analysis of blood cells known as 'cell-free DNA' (cfDNA), remains the holy grail of early cancer intervention. An important step in this direction is the development of the screening method CancerSeek, which combines information regarding the presence of certain known blood proteins associated with some common cancers with data from the analysis of cfDNA. This method was applied to patients diagnosed clinically with one of eight common cancers, which were at an early stage and had not metastasised. The test was positive in a median of 70% of cases, with a 'specificity' of more than 99%, i.e. there were almost no positive results for healthy controls. Furthermore, an AI algorithm localised the cancer to its tissue of origin in 83% of cases (Cohen *et al.*, 2018).

A very important aspect of precision medicine is diet. In this context, it is noted that although there are many well-known dietary recommendations, most of them are not based on rigorous studies. However, there do exist relevant rigorous results. In particular, the Lancet Commission report of 2019 (Willett *et al.*, 2019) states that the best among the studied diets is the Mediterranean diet, which was typically followed in Crete, Greece, in the mid-20th century. The Cretan diet was low in red meat (the average intake of red meat and poultry combined was 35 g/day) and was largely based on plants. Interestingly, it was high in total fat (about 40% of energy), but this fat consisted mainly of olive oil, which is a monounsaturated fatty acid. It is worth noting that during this period, Greece had one of the longest life expectancies (Trichopoulou *et al.*, 2003). Consistent with this report, by combining data from 45 different studies, it has been shown that plant-based foods and whole-grain consumption are associated with a reduction in heart disease, cancer, and overall mortality (Aune *et al.*, 2017). Additional important results regarding diet are discussed in the following.

A number of rigorous studies have shown that coffee is associated with longevity (Gunter *et al.*, 2017). This is the result of a variety of different effects (van Dam *et al.*, 2020). In particular, in addition to caffeine, coffee contains hundreds of other biologically active phytochemicals as well as modest amounts of magnesium, potassium, and vitamin B_3. These compounds may reduce oxidative stress, have a positive impact on the gut microbiome, and modulate fat and glucose metabolism. In addition, coffee consumption decreases the risk of cardiovascular diseases (the lowest risk was observed for three cups a day), and also caffeine improves energy balance by increasing basal metabolic rate . Importantly, coffee consumption is associated with a slightly reduced risk of melanoma, breast and prostate cancers, as well as a substantially reduced risk of endometrial and hepatic cancers. Finally, coffee protects against liver diseases and gallstones, and there is a strong inverse association between caffeine intake and Parkinson's disease.

Incidentally, coffee, like aspirin and statins, induces a variety of different beneficial actions. In particular, it causes dilation of bronchi;

theophylline, a medication used in respiratory diseases including asthma, is related to caffeine. Also, it causes dilation of the coronary arteries. Coffee increases the heart rate, which leads to higher oxygen demand, but this is compensated by the dilation of the coronary arteries; therefore, even the consumption of up to five cups of coffee daily does not cause cardiac ischaemia. Paradoxically, caffeine causes constriction of the cerebral arteries; Cafergot, which was for many years the main medication for migraine headaches, contains caffeine. Of course, there is no substance without some side effects; coffee can cause anxiety, and as noted earlier, it also increases the heart rate.

Nut consumption (walnuts, hazelnuts, and almonds) is inversely associated with total mortality, with hazard ratios for death ranging from 0.89 for consumption of nuts once a week to 0.80 for consumption of nuts seven or more times a week (Bao *et al.*, 2013).

High carbohydrate intake increases the risk of heart disease and death (Dehghan *et al.*, 2017). This fact highlights the importance of defining the 'glycaemic index' of various foods, namely, the increase in blood glucose two hours after the digestion of a specific food. This index is a number that ranges from 0 to 100, with pure glucose arbitrarily assigned the number 100. Pioneering studies at the Weizmann Institute, summarised in the important book *The Personalized Diet: The Pioneering Program to Lose Weight and Prevent Disease* (Segal and Elinav, 2017), have shown that this index is highly personal. For example, the response to a slice of pizza varies among different individuals from below 30 to above 70. Furthermore, a person's microbiota is of crucial importance for determining the glycaemic index. An algorithm developed in the Weizmann Institute, called *Day Two*, predicts optimal dietary choices by using as input information from a gut microbiome sample. *These unexpected results make it clear that attempts towards a generic approach to diet in particular and to medicine in general, ignoring individual characteristics, cannot yield optimal results.* This is yet another indication of the potential importance of AI.

The relationship between diet and obesity will be discussed in the forthcoming book of the author with Ktistakis. Here, it is only noted that the recent discovery of medications for the treatment of obesity is expected to have a transformative impact in medicine, from sleep apnoea to diabetes and cardiovascular diseases. Such medications include semaglutide, which is an agonist of glucagon receptors, and tirzepatide, which is also a glucagon-receptor agonist and a glucose-dependent insulinotropic peptide; glucagon induces the feeling of satiation. For example, patients with a particular type of heart failure (characterised by preserved ejection fraction) who were prescribed tirzepatide lost weight and, importantly, had a significantly lower risk of a worsening heart failure event and of death from cardiovascular causes (Packer *et al.*, 2025).

2.4 Further Applications

The relationship between biological insights and medical applications is well-established. It is expected that recent spectacular applications of AI in biology will have a profound impact on medicine. One of these applications involves the solution of a long-standing open problem in biology: protein folding. As explained in detail in Chapter 19 of Fokas (2024), this is defined as follows: given the amino acid sequence of a given protein, determine 'how it folds', i.e. identify its three-dimensional structure. In 2021, DeepMind presented the deep learning neural network AlphaFold, which is capable of making highly accurate predictions of the geometrical structure of a given protein (Jumper *et al.*, 2021). Prior to this breakthrough, using crystallography, the three-dimensional structure of approximately 190,000 proteins, representing about 0.1% of all known proteins, was determined. The potential use of AI in biology becomes evident by the fact that AlphaFold determined the structure of approximately 200 million proteins – almost all of the known existing proteins. In 2022, AlphaFold was made available for public use. For this transformative discovery, Demis Hassabis and John Jumper were awarded part of the Nobel Prize in Medicine in 2024.

Combining the power of neural networks and the current capacity of reading the DNA sequence of very long molecules has led to a major improvement in 'variant calling', which refers to the process of determining where the DNA sequence from a patient differs from a reference sequence. This has led to the successful identification of rare inherited diseases. Further progress in this direction has been made by reading the patient's 'transcriptome', i.e. the set of all RNA transcripts. For example, this approach was used for the identification of a specific RNA variation responsible for the development of tremors and seizures in a 12-year-old girl (Gomes and Ashley, 2023). Deep learning, combined with the availability of a dramatic increase in data obtained via mass spectroscopy, has led to substantial progress in the analysis of 'proteomics' (proteins), in particular, and 'metabolomics' (the measurement of carbohydrates, fatty acids, lipids, amino acids, steroids, etc.), in general. This has facilitated, among others, the diagnosis of inborn errors of metabolism.

The impact of the recently developed LLMs on medicine is reviewed in Zhou *et al.* (2023) and Wang and Zhang (2024).

2.5 Challenges

Analysing existing datasets is important, but it is not the same as collecting and comparing data in a clinical setting. The procedure for establishing the usefulness of a new medication is well-established, but the standards for testing AI interventions are not clear. In particular, the question of evaluating the clinical importance of several remarkable achievements of AI in medicine remains largely open. A necessary step is designing appropriate large clinical trials. Among the limited number of conditions that have been clinically evaluated via such rigorous studies are the following: diabetic retinopathy, detection of wrist fractures in an emergency room setting, histological breast cancer metastasis, the presence of very small colonic polyps, and cataracts in a small group of children (Topol, 2019b).[3]

The training data used by AI may amplify biases present in society (Ferryman *et al.*, 2023). Recent efforts have ameliorated this problem, and actually, in certain cases, the use of AI has reduced bias (Pierson *et al.*, 2021). A related problem is to ensure that AI reflects human values in general, as well as patient values and goals in particular. Efforts to address this difficult challenge, especially in the current epoch of LLMs, are discussed in Yu *et al.* (2024).

The performance of many AI models deteriorates when applied to patients who differ from those used to train the associated neural network. This problem, known as 'data set shift', has been known at least since the early 1970s. The British surgeon F. T. de Dombal and his colleagues at the University of Leeds, motivated by the publication of an earlier paper in *Science* advocating the use of Bayes' theorem in medical diagnosis, developed the program AAPHelp. The goal was to help with the diagnosis of patients with acute abdominal pain. Using Bayes' theorem, a specific

---

[3] In this regard, it is stated in Topol (2019b) that 'the field is certainly high on promise and relatively low on data and proof', especially 'regarding readiness for implementation in patient care'.

mathematical formula was derived. This formula was based on clinical symptoms, including the location, quality, and severity of the pain, as well as on clinical signs, such as pulse and the response of patients to physical examination. AAPHelp generated the correct diagnosis in 91.8% of the cases. However, this number reduced to 65% when used, a few years later, in a hospital in Copenhagen. This was the consequence of significant epidemiological differences between the two sets of patients. For example, the presentation of pancreatitis was different, apparently, due to differences in alcohol consumption. Also, there were subtle discrepancies in the definition of what was meant by acute abdominal pain (November, 2012).

Some of the features of AI which endow it with its enormous power at the same time make it problematic from a mathematically rigorous statistical perspective. This is particularly important for medical applications. For example, it is not possible to trace the evidence from the data to the concrete outcome, and this makes verification and generalisation highly challenging. This problem is exacerbated by the fact that different AI algorithms often yield different conclusions. In addition, skilled applied statisticians and epidemiologists use tools which do not depend on the analysed data and hence cannot be captured by data-driven AI algorithms. Such tools include careful experimental design, a deep understanding of the study questions, and the tailoring of models to the relevant research question; in addition, ascertaining selection bias and careful model checking are crucial to the study (Hunter and Holmes, 2023).

AI algorithms are not free of errors. A highly worrisome example is provided by the 'IBM Watson for Oncology' algorithm. This program which is used by hundreds of hospitals around the world for choosing treatment for patients with cancer; unfortunately, it gave many erroneous recommendations.

A possible source of errors in machine learning algorithms is the emergence of 'instabilities.' In particular, in image reconstruction, such instabilities usually occur in three different forms: first, almost undetectable changes either in the image or sampling domains may result in severe artefacts in the reconstruction. Second, a small structural change, for example the presence of a tumour, may not be captured in the reconstructed image. Third, a counterintuitive type of instability, where training with more samples, instead of yielding better results, may yield poorer performance. Appropriate software capable of detecting some of these instabilities is presented in Antun *et al.* (2019). Uncorrected malfunctions of AI algorithms would give rise to an exponential increase in the currently occurring iatrogenic errors that AI is supposed to correct.

Physicians having used AI in making decisions may be subject to malpractice litigations. A similar situation can arise in cases which involve apparent malfunctions of software imbedded in a variety of medical devices. This means that physicians must now consider several unprecedented issues, including the following: first, what is the likelihood that errors can be identified regarding a specific AI application? For example, what is known about the differences between the training data and the patients for whom the AI model will be deployed? Second, would it be possible to detect such errors before they cause irreversible damage? Third, how serious are such errors? It is reasonable to expect that, in the near future, patients will be asked to consent when AI models are used in diagnosis or treatment (Mello and Guha, 2024).

An important limitation of AI is that, up to now, AI algorithms have a narrow focus. For example, given the image of an X-ray, AI can attain a specific goal, such as ruling out evidence of cancer. On the other hand, a radiologist follows a multi-task approach. For example, during the search for identifying features indicating the presence of a tumour, the radiologist also looks for evidence of other pathological conditions, such as the presence of rib fractures, calcium deposits, enlarged lymph nodes, pulmonary oedema (fluid collection), and cardiomegaly (heart enlargement). There is hope that soon the so-called 'generalist medical AI models' will be developed. The goal is that these models will be able to interpret a wide range of radiographic findings and place these interpretations within the given clinical context. In other words, the goal is to mimic the cognitive approach used by a radiologist. In addition to the great difficulty of developing such multi-task models, a Herculean effort will be required to 'train' them, validate them, and then establish their usefulness in rigorous clinical trials (Rajpurkar and Lungren, 2023).

A serious difficulty with the employment of AI algorithms is the lack of 'gold standards' in many clinical situations. For example, the question of deciding whether a given histopathology slide indicates an early-stage cancer is complicated by the lack of gold standards for a cancerous slide (Adamson and Welch, 2019). The process of using AI for this task begins with a set of digital images, which are marked by pathologists as 'cancer' or 'no cancer'. Then, using half of these images, the AI network is 'trained' to recognise the cancerous ones. The procedure is then validated by using the other half of the set. However, in many cases, there does not exist a histopathological 'gold standard' of what constitutes an early-stage cancer. This is related to a more fundamental question: 'what is early-stage cancer'? This question can be answered only dynamically in a proper clinical setting. Early-stage cancer is a pathological process which, if left untreated via local invasion or metastasis, will give rise to symptoms and may lead to death. The static picture obtained from histology is undoubtedly useful, giving important information about individual cells and the surrounding architecture, but it cannot replace the full clinical picture. There are well-documented cases of disagreements among pathologists regarding early-stage prostate and thyroid cancer, as well as breast and skin lesions.

Regarding the potential use of AI in assisting clinicians in complicated diagnoses, it should be remembered that the most difficult clinical cases arise in very rare medical presentations. In these situations, AI cannot provide assistance since AI requires prior training with thousand, if not millions, of similar cases.

Within the general field of AI, efforts are being made for the explicit consideration and modelling of human values. This is particularly important in medical applications, where it is imperative that the AI models used reflect accurately and explicitly patient values and goals, which of course depend on cultural, religious, social, and political factors (Yu *et al.*, 2024).

Medical educators are facing major challenges, from coping with the problem of maintaining high academic standards via persuading students not to use AI tools to complete assignments, to developing training programmes to prepare young physicians to cope with the coming wave of changes that will be induced by AI. Medical curricula must be radically revised to enable physicians understand the capabilities and limitations of AI, as well as to acquire the expertise to use a variety of AI modalities effectively. It is imperative that such curricula be rigorously assessed before being accepted at a global level .

References

Adamson, A. S. and Welch, H. G. (2019). Machine learning and the cancer-diagnosis problem-no gold standard. *New England Journal of Medicine*, *381*(24), 2285.

Antun, V., *et al.* (2019). On instabilities on deep learning in image reconstruction. Does AI come at a cost? *arXiv* 1902.05300.

Aune, D., *et al.* (2017). Whole grain consumption and risk of cardiovascular disease, cancer, and all cause and cause specific mortality: Systematic review and dose response meta-analysis of perspective studies. *BMJ, 353,* i2716.

Bao, Y., *et al.* (2013). Association of nut consumption with total and cause-specific mortality. *New England Journal of Medicine, 369,* 2001–2011.

Bartolozzi, C. (2018). Neuromorphic circuits impart a sense of touch. *Science, 360,* 966–967.

Brody, B. (October 2017). Why I almost fired my doctor. *New York Times.*

Brownstein, J. S., Rader, B., Astley, C. M., and Tian, H. (2023). Advances in artificial intelligence for infectious-disease surveillance. *New England Journal of Medicine, 388*(17), 1597–1607.

Cohen, J. D., *et al.* (2018). Detection and localization of surgically resectable cancers with a multi-analyte blood test. *Science, 359,* 926–930.

Dehghan, M., *et al.* (2017). Associations of fats and carbohydrate intake with cardiovascular disease and mortality in 18 countries from five continents (PURE): A prospective cohort study. *Lancet, 390,* 2050–2062.

Esteva, A., *et al.* (2017). Dermatologist-level classification of skin cancer and deep neural networks. *Nature, 542,* 115–118.

Ferryman, K., Mackintosh, M., and Ghassemi, M. (2023). Considering biased data as informative artifacts in AI-assisted health care. *New England Journal of Medicine,* 389(9), 833–838.

Fokas, A. S. (2024). *Ways of Comprehending: The Grand Illusion and the Essence of Being Human.* Singapore: World Scientific.

Ghajar, C. M. and Bissell, M. J. (2016). Metastasis: Pathways to parallel progression. *Nature, 540,* 528–529.

Gomes, B. and Ashley, E. A. (2023). Artificial intelligence in molecular medicine. *New England Journal of Medicine, 388*(26), 2456–2465.

Gunter, M. J., *et al.* (2017). Coffee drinking and mortality in 10 European countries: A multinational cohort study. *Annals of Internal Medicine, 167,* 236–247.

Hunter, D. J. and Holmes, C. (2023). Where medical statistics meets artificial intelligence. *New England Journal of Medicine, 389*(13), 1211–1219.

IBM. (2017). IBM machine vision technology advances early detection of diabetic eye disease using deep learning.

Jumper, J., *et al.* (2021). Highly accurate protein structure prediction with AlphaFold. *Nature, 596,* 583–589.

Manolopoulos, V. G., Ragia, G., and Tavridou, A. (2010). Pharmacogenetics of coumarin oral anticoagulants. *Pharmacogenetics, 11,* 493–496.

McCoy, L. G., Manrai, A. K., and Rodman, A. (2024). Large language models and the degradation of the medical record. *New England Journal of Medicine, 391*(17), 1561–1564.

Mello, M. M. and Guha, N. (2024). Understanding liability risk from using health care artificial intelligence tools. *New England Journal of Medicine, 390*(3), 271–278.

November, J. (2012). *Biomedical Computing.* Baltimore, MD: John Hopkins University Press.

Oaklander, M. (August 2015). Doctors on life support. *Time.*

Pauker, S. G. and Kassirer, J. P. (1987). Decision analysis. *New England Journal of Medicine, 316*(5), 250–258.

Packer, M., *et al.* (2025). Tirzepatide for heart failure with preserved ejection fraction and obesity. *New England Journal of Medicine, 392*(5), 427–437.

Pierson, E., Cutler, D. M., Leskovec, J., Mullainathan, S., and Obermeyer, Z. (2021). An algorithmic approach to reducing unexplained pain disparities in underserved populations. *Nature Medicine, 27*(1), 136–140.

Rajpurkar, P. and Lungren, M. P. (2023). The current and future state of AI interpretation of medical images. *New England Journal of Medicine, 388*(21), 1981–1990.

Rajpurkar, P., *et al.* (2017). CheXNet: Radiologist-level pneumonia detection on chest-X rays with deep learning. *arXiv.*

Schwartz, W. B. (1970). Medicine and the computer: The promise and problems of change. In *Use and Impact of Computers in Clinical Medicine*, pp. 321–335. New York: Springer.

Segal, E. and Elinav, E. (2017). *The Personalized Diet: The Pioneering Program to Lose Weight and Prevent Disease.* New York: Grand Central Life and Style.

Sullivan, D. P. and Lundberg, E. (2018). Seeing more: A future of augmented microscopy. *Cell, 173*, 546–548.

Topol, E.J. (2019a). *Deep Medicine.* New York: Basic Books.

Topol, E.J. (2019b). High-performance medicine: The convergence of human and artificial intelligence. *Nature Medicine, 25*(1), 44–56.

Trichopoulou, A., Costacou, T., Bamia, C., and Trichopoulos, D. (2003). Adherence to a mediterranean diet and survival in a Greek population. *New England Journal of Medicine, 348*(26), 2599–2608.

van Dam, R. M., Hu, F. B., and Willett, W. C. (2020). Coffee, caffeine, and health. *New England Journal of Medicine, 383*(4), 369–378.

Wang, D. and Zhang, S. (2024). Large language models in medical and healthcare fields: Applications, advances, and challenges. *Artificial Intelligence Review, 57*(11), 299.

Wang, D., *et al.* (2017). Deep learning for identifying metastatic breast cancer. *arXiv.*

Willett, W., *et al.* (2019). Food in the anthropocene: The EAT–Lancet Commission on healthy diets from sustainable food systems. *The Lancet, 393*(10170), 447–49.

Yu, K. H., Healey, E., Leong, T. Y., Kohane, I. S., and Manrai, A. K. (2024). Medical artificial intelligence and human values. *New England Journal of Medicine, 390*(20), 1895–1904.

Zhou, H., *et al.* (2023). A survey of large language models in medicine: Progress, application, and challenge. *arXiv preprint* arXiv:2311.05112.

Chapter 3

Opportunities and Challenges

3.1 The Interaction of AI with Neuroscience

There already exists a variety of brain–computer interface devices. This transformative new technology is based on the direct communication between the brain and external computing devices. It involves recording cortical activity using an interface consisting of an array of electrodes that detect electrical activity from the brain during a specific cognitive function, such as the excitation corresponding to movement or speech. Then, the use of powerful AI algorithms allows the translation of these signals to an actual movement or speech production.

A recent breakthrough with respect to this technique has raised realistic hopes that soon many paralysis victims will be liberated from their wheelchairs. Specifically, an array of electrodes implanted in the cortex of Gert-Jan Oskam, who lost the ability to use his legs 12 years ago following an accident, allowed researchers to monitor the brain activity of this patient. By decoding this activity using AI and then sending a stream of signals to a second set of electrodes stimulating Gert-Jan's spinal cord, the flexion of appropriate muscles was achieved. In this way, in 2023, Get-Jan was able to walk again.

Regarding speech, two patients are described in Card *et al.* (2024), who were suffering from amyotrophic lateral sclerosis; one had severe dysarthria (difficulty in speaking) and the other anarthria (inability to articulate speech). Four micro-electrode arrays, each consisting of 64 electrodes, were implanted in the area of the brain dealing with vocal tract representations, which is located in the precentral gyrus. This allowed the patients to produce speech of approximately 32 words per minute with an accuracy of 95%. A variety of additional applications of brain–computer interfaces are summarised in Essmann and Mueller (2022).

There is hope that in the future, brain surgery needed for the implantation of electrodes will be avoided. For example, a helmet constructed by researchers at Imperial College, called Cogital, contains 21 electrodes, which can be used to obtain appropriate EEG data corresponding to a specific behaviour. For example, there is a particular type of brain excitation, and hence corresponding EEG data, not only when making a fist but also when *thinking* of making a fist. Sending this information to the spinal cord could allow a paralysis patient to be able to make a fist.

Humans can learn to make complex associations from a small amount of data, whereas currently, machine learning requires a very large number of examples. On the other hand, using a very large set of data, AI is able to make associations which are *beyond* our cognitive abilities. *The deeper scientists and engineers penetrate the secrets of fundamental neuronal processes, the higher the probability that aspects of brain functions will be mimicked by AI. This will give rise to more efficient deep learning algorithms, which in turn will be useful for further investigating brain functions.* In this way, I expect that AI will be of critical importance for understanding the greatest existing information processor: the human brain.

As mentioned in Chapter 1, various aspects of AI have had a neuronal motivation. Actually, there exist interesting relationships between neural networks and *actual* neuronal systems. Researchers at the Allen Institute for Brain Science, using 300 live human brain cells derived from surgical samples, studied how neurons respond to incoming signals and how they generate appropriate outputs. This revealed several similarities with the behaviour of specific artificial neural networks (Koch, 2017).

A concrete example of observed analogies between deep learning and biological neural networks is provided by an AI algorithm based on 'nearest-neighbour interactions.' The computational strategy of this algorithm is analogous to the one used by the fruit fly in olfaction (smell), where learning an odour facilitates recognising a similar one (Dasgupta *et al.*, 2017). Another example of the usefulness of mimicking neuronal algorithms is provided by a type of artificial neural network called 'memory-aware synapses', which are designed to mimic the fundamental property of synapses that the occurrence of repetitive, synchronised firing strengthens a given synapse (Aljundi *et al.*, 2017).

An example of the use of AI for elucidating functional properties of the brain is provided by a particular neural network introduced by DeepMind. In this regard it is noted that the so-called grid cells play an important role in space navigation. Indeed, these cells are implicated in 'vector-based navigation', meaning that they are crucial for the neural 'computation' of the distance and direction between two points; this enables an individual to identify the shortest route between these two points. Interestingly, grid cells exhibit perfectly *hexagonal firing patterns*. Apparently, these patterns emerge as an equilibrium state in competitive neural networks, where certain key cells *inhibit* all other cells in their vicinity. DeepMind has used deep reinforcement learning in a particular virtual environment involving the localisation of 'virtual rodents.' Interestingly, the best results were obtained using neural networks employing vector-based navigation, and remarkably, this occurred when the relevant neurons exhibited hexagonal grid-like representations. Furthermore, these patterns appeared when these cells were connected via *inhibitory* interneurons. This provides direct support for the theory that the hexagonal firing pattern of grid cells is indeed involved in vector-based navigation (Savelli and Knierim, 2018; Banino *et al.*, 2018).

3.2 Neuromorphic Circuits

I have argued in Fokas (2023) that AI cannot reach the level of human thought; the relevant arguments are discussed in detail in Chapter 6. However, my analysis is based on the current situation, where AI is mostly concerned with 'software', i.e. the design of appropriate algorithms. The situation could change dramatically if the underlying 'hardware' is also modified, namely, *if silicon (artificial) neurons are combined with actual neurons.* This process, which has already begun, will certainly be aided by the fact that many brilliant scientists and engineers, as well as many exceptional undergraduate and graduate students, are now involved in AI research.

In the 1980s, the Caltech scientist and engineer Carver Mead introduced the concept of 'neuromorphic electronic systems', which refers to the idea of constructing chips to *structurally* mimic the brain. The natural generalisation of this concept is the creation of biohybrid computers via the combination of silicon neurons and biological neural networks. A concrete implementation of this concept, which takes the form of a neuromorphic interface, is discussed in Broccard *et al.* (2017). A promising approach, where both the hardware and software are novel, has been pioneered by IBM and used for image recognition. It involves a neural network consisting of 200,000 artificial synapses of two different types; these two types mimic short-term and long-term processes. This system has an efficiency of more than two orders of magnitude better than typical networks, executing 28 billion operations per second per watt (Ambrogio *et al.*, 2018). This is consistent with the fact that the brain is highly energy-efficient, using only 10 watts of power (less than a small household light bulb).

Another idea with far-reaching consequences involves combining AI with nanotechnology. The latter allows the manipulation of single atoms for the purpose of constructing materials with desirable properties.

3.3 Challenges

Every great development in science and technology has been associated with concerns about its impact on the human race. The enormous benefits of technology raise hopes for further technological advancements, but at the same time technological breakthroughs create anxiety. Humans trust nature much more than they trust artificial creations. After all, as Aristotle understood, 'Nature does not create anything useless', whereas some of humankind's creations are not only useless but also potentially catastrophic.

For many scientific and technological advances, their positive impact far outweighs their negative potential. A prime example is the synthesis of ammonia, achieved by Fritz Haber and Carl Bosch (Nobel Prizes in Chemistry, 1918 and 1931. Since nearly 50% of the nitrogen found in human tissues originated from the Haber–Bosch process, the discoveries of these two Nobel laureates allowed the increase of the population from 1.6 billion in 1900 to 8.2 billion in 2025. This is the reason why the scientist Vaclav Smil considers ammonia, together with cement, steel, and plastics, the four material pillars of our civilisation. Ammonia is also used for chemical weapons, but fortunately, this potential effect has been largely suppressed. The judgement regarding breakthroughs in nuclear physics in more difficult. It certainly has had life-saving implications in nuclear medicine and could have a transformative effect on energy, despite the fact that this important potential benefit has not been explored sufficiently. On the other hand, it led to the inhuman devastation of Hiroshima and Nagasaki, to the enormous waste of resources for the sake of nuclear armament, and to several nuclear accidents; importantly, it has also caused great anxiety to millions of people over a possible nuclear catastrophe. To a large extent, nowadays this anxiety is replaced by the fear of the potential negative impact of AI and of the other key scientific and technological developments mentioned in the introduction. As noted earlier, this fear, together with the Damoclean sword of environmental catastrophe causing 'eco anxiety', contributes decisively to the growing prevalence of depression in the new generation.

Several famous individuals have expressed their concerns about the possible catastrophic impact of AI. Sam Altman, the CEO of the company OpenAI, has stated that the risks of an uncontrollably evolving progress in AI are comparable to those of a pandemic or even a nuclear war. According to Henry Kissinger, AI could cause a rupture in history and unravel the way civilisation works. Bill Gates has stated that it is potentially more dangerous than a nuclear catastrophe. My late colleague at the University of Cambridge, Stephen Hawking, was fearful that the development of AI could spell an end to the human race. Personally, I am in agreement with the Stanford University computer scientist Andrew Ng, who, responding to the technology entrepreneur Elon Musk's suggestion that the potential rise of killer robots provides a strong motivation for humans to colonise Mars, noted that 'Fearing the rise of killer robots is like worrying about overpopulation in Mars before populating it' (Dowd, 2017). I strongly believe that the dangers of a nuclear or an environmental catastrophe are far more realistic than the dangers arising from AI. In this regard the serious negative impact on the environment of an 'AI revolution' should not be underestimated.

Arguments presented in Chapter 6 suggest that we should not worry about the superior intelligence of AI. Perhaps, instead of fearing AI, we should embrace its great potential for contributing to a better society, from the decrease of mortality and morbidity caused by diseases and car accidents to the reduction of injustice and poverty. However, the final impact of AI on society will indeed be positive if and only if leading researchers in this area, as well as key national and international organisations, concentrate on finding ways to overcome several serious challenges.

3.4 Regulation and More

In my opinion, the basic questions regarding the societal impact of AI can be formulated as follows: *do there exist appropriate technical expertise, as well as legal and political mechanisms and especially ethical norms, to ensure that the new technology is used for the benefit of society?*

Regarding technical expertise, it is encouraging that technology itself can solve some of the serious problems appearing in advanced AI models, including, for example, bias in data; when researchers in GPT-2 prompted the network with the incomplete sentence, 'the white man worked as…', GPT-2 generated 'police officer', 'judge', 'prosecutor', or 'the President of the United States'. By changing white to black, GPT-2 generated the answer 'pimp'. In order to address such racist aspects of GPT, researchers arrange multi-run conversations with the associated model, prompting it to express offensive positions. After bias was identified, appropriate human insights were integrated into the system. Incidentally, it is advocated in Ferryman *et al.* (2023) that biased clinical data should be used as 'artefacts.' Namely, in the same way that artefacts are used in an archaeological and historical context in order to provide information about social structure, the biased clinical data could be used to reveal current uncomfortable truths.

With respect to legal interventions, it is worrisome that governments are often mismanaged and slow to react to crucial developments; this raises concerns about their ability to manage and regulate new technologies in the best interests of their populations. In addition, there exists the following objective conflict: on the one hand, governments appreciate the strategic importance of AI, as well as of robotics, gene editing, quantum computing, etc. Thus, having designated cyberspace as a domain in which they must innovate in order to prevail over their rivals, they tend to concentrate on development, ignoring safety and regulatory issues. On the other hand, existential fears force them to take steps towards safety and regulation. This conflict expresses itself most clearly in China's AI policy, where civilian companies are strictly regulated, while military-industrial companies are subject to considerably less regulatory control.

Regarding concerns about governmental interventions, there currently exist mixed developments. On the one hand, the European Union proposed in 2021 and voted as a binding European law in December 2023 an ambitious piece of legislation: the AI Act.[1] This European law treats AI research and deployment according to a risk-based scale. Specifically, technologies causing direct harm or having 'unacceptable risk' will be prohibited. However, military applications are not explicitly excluded. Apparently, the final version of the AI Act was significantly weakened by

---

[1]Although the law was voted after the emergence of ChatGPT, it was formulated before ChatGPT.

the intervention of high-tech companies. If AI affects human rights or critical aspects of the proper functioning of society, including basic infrastructure, public transport, healthcare, and welfare, then it will be classified as 'high risk'. This type of AI will be subject to a high level of accountability and must be 'transparent, secure, subject to human control and properly documented'. On the other hand, one of the 78 executive orders rescinded by President Trump in his first few days back in the Oval Office in January 2025, was an executive order regarding AI signed by President Biden in October 2023. This order provided guidelines for responsible security and safety standards on AI development and use, addressed the impact of AI on civil rights and consumer and worker protection, and advocated international collaboration. In addition, Vice-President JD Vance warned against 'excessive' restrictions that would 'kill a transformative industry'.

It seems that in addition to the strict implementation of appropriate laws, additional legal steps are needed. For example, Mustafa Suleyman has proposed that it is imperative for companies involved with sophisticated AI to be required to obtain special permission in order to be allowed to operate. The more general-purpose a given AI system and the higher the probability it can become autonomous, the easier it is for the AI to escape human control. Companies involved with this type of AI, before they become operational, should be required to obtain a specific licence. For its approval, these companies should demonstrate that they possess the proper technical expertise, as well as the organisational structure to be subject to the legal norms specified, for example, by the European Act.

In my opinion, it is necessary that a substantial portion of an AI company's revenues should be directed towards safety, with the specific aim of improving the chances that the final impact of AI will be beneficial to society. Concrete steps should include the following: first, research efforts should be intensified towards understanding how the AI models achieve their goals. At the moment, the fundamental question of how deep learning algorithms reach their 'decisions' remains open. Without making progress towards answering this question, we cannot gain a deep understanding of many of the processes investigated by AI; moreover, we cannot learn from AI its new form of 'intelligence', which could then be used for the solution of a variety of other difficult problems. Second, revenues should be invested towards ensuring that AI systems do not escape human control. Each company should have an appropriate committee with the obligation to report to an international organisation which should be created for this purpose. Such an organisation should have enforcement capabilities; otherwise, it would be ineffective. Incidentally, an analogous international organisation in synthetic biology, called 'Secure DNA', aims to identify and prohibit the creation of pathogenic DNA sequences. In this regard, it is recalled that photocopiers shut down when attempting to copy or print money. Although it is unrealistic to expect such complete forms of safety in AI, research efforts should be made to identify 'dangerous' forms of AI. Third, an international organisation should be created where AI accidents would be reported. An analogous structure in synthetic biology, called 'AI Incidental Database', following the practices of the cybersecurity industry, documents, analyses, and reports safety events. Fourth, sufficient resources should be dedicated towards investigating ways that the standard financial goal of every company, namely to maximise shareholder returns, is reconciled with the need to contribute to social equality. It is worth noting that in the Middle Ages, 'profit for the sake of profit' was not acceptable. Overall it is crucial that the computer is not allowed to make decisions simply for the sake of efficiency. It is imperative that the process of computation is not reduced to a closed, opaque implementation, where we can only control the parameters of the system.

In addition to the above steps, efforts should be intensified towards 'security', i.e. guarding not only against the development of malicious software but also against hacking. In this regard, it must be remembered that the more powerful a system, the more devastating the effect of a possible cybernetic attack. Earlier attacks provide vivid reminders of potential dangers. For example, on 12 May 2017, several countries were hit by a ransomware attack called WannaCry. Exploiting a specific vulnerability of an older Microsoft system, hackers were able to incapacitate the computers, among others, of the National Health Service of the UK, Hitachi, FedEx, Telefonica, and the Chinese Ministry of Public Security. WannaCry tricked some users into opening an email, which released a virus that replicated and infected, in just one day, a quarter of a million computers across 150 countries. This system was based on the cyberattack unit called EternalBlue, built by the National Security Agency of the US. A group of hackers, called the Shadow Brokers, somehow stole this unit and put it up for sale. It was obtained by North Korean hackers (perhaps state-sponsored) who launched the attack. A 22-year-old British hacker noted a strange-looking domain on the viral code, which was unregistered. He bought it for less than £11, which allowed him to take control of the virus, while Microsoft corrected the vulnerability of its software. A few months later, a new version of this unit, called NotPetya, attacked Ukraine's national infrastructure, infecting 10% of the country's computers, including those monitoring the radiation at the Chernobyl power station.

Preventing such attacks is of paramount importance for our survival. What will happen to our global communication system, which depends on space-based systems, if vital satellites are attacked? What will be the effect of a cyberattack on the digital networks responsible for the running of our energy, transport, and banking systems?

Incidentally, cybersecurity is one of the central issues in the current intense antagonism between the US and China regarding 5G. China is currently experimenting with 6G. The new networks have a huge advantage over 4G in terms of speed and capacity, and they also allow the simultaneous connection of hundreds of electronic devices. This implies that, in addition to concerns regarding the harmful effect of 5G on health and the environment, there is also the danger stemming from the fact that the interconnection of many devices makes it easier to collect personal data of the user.[2]

Several 'ethical questions' must be addressed, which should include the following: is it possible to define the objectives of AI in an ethically acceptable manner? Can we ensure that we avoid the so-called problem of 'goal misalignment', which refers to the situation where intelligent machines pursue goals that are counter to our well-being, and we cannot stop them? Evidently, the latter question is of existential importance.

Father Paolo Benanti, the Pope's advisor on AI, has expressed the fear that we are becoming an 'algocracy', namely a society ruled by algorithms. Increasingly, algorithms control every aspect of our lives, from our shopping choices to the specific programmes we watch on TV. In 2020, the University of Oxford established the Institution for Ethics in AI, directed by the philosopher John Tasioulas, with the aim of engaging philosophers, historians, and lawyers to investigate how AI should be used and controlled.

The 'smarter' AI becomes, the more pressing it is to invest in the latter question. The goal-oriented behaviour exhibited by humans is the result of the interaction of a multitude of factors, including survival instincts, as well as *tradition*, whose great contribution to our culture is the propagation of *collective wisdom* across generations. There are attempts to model

---

[2] It appears that Europe is planning to use the 5G technology of Denmark's Nokia and Sweden's Ericsson, instead of the 5G technology of China's Huawei.

the goal-oriented behaviour of humans by designing AI systems that maximise an appropriate 'reward function'. Designing systems whose tasks will be consistent with the goals defining humanity remains a challenge of fundamental importance in the modern era of AI.

I am very concerned that advances in technology in general and AI in particular will further increase existing social inequalities, which may have a negative, if not devastating, effect on democracy. It is argued in the book *Power and Progress* (Acemoglu and Johnson, 2023) that, as a result of the church's control over the economy in the Middle Ages, the agricultural revolution did little to improve people's living standards. In the first half of the 19th century, the main beneficiaries of the Industrial Revolution were factory owners. After the Second World War, the economic situation of labour improved dramatically. Since the 1980s, the split of the proceeds of economic growth between labour and capital has shifted towards the latter; this process has been accelerating with the digital revolution. Enormous economic power in the hands of a very few individuals undermines democracy. Another great threat to democracy is the control of social media by a handful of plutocrats.

Stuart Russell, who has been campaigning for a ban on the use of AI for creating autonomous weapons, has been studying the fundamental problem of how to align AI with human goals and interests. He admits that early on, there was so much excitement about how to understand and create intelligence that no one asked the important question of 'what is this intelligence going to be used for?' Can we now make sure that this human-made intelligence can be controlled? (Russell, 2019).

Some of the challenges currently emerging in AI have been considered in the past by physicists, biologists, and chemists. For example, the Russell–Einstein Manifesto, issued in London in 1955 by Bertrand Russell and signed, among others, by Albert Einstein a few days before his death, highlighted the dangers of nuclear weapons and promoted the peaceful resolution of international conflicts. A few days after the publication of this manifesto, the philanthropist Cyrus Eaton offered to sponsor a conference promoting these goals, which was held two years later in Pugwash, Nova Scotia (Eaton's birthplace). This was the first of the series of *Pugwash Conferences on Science and World Affairs.* In 1995, the Pugwash conferences and the physicist Joseph Rotblat were awarded the Nobel Peace Prize. Similarly, distinguished biologists and chemists campaigned strongly that their discoveries be used for medications and useful materials and not for biological and chemical weapons. In this direction, conferences promoting a 'beneficial AI', such as the one organised in Puerto Rico in 2015, are important.[3]

It is imperative that deep learning does not lead to a new arms race. What will be the impact on the global balance of power when leading world powers use the ability of AI to 'perceive' patterns that human soldiers and strategists cannot? Soon, AI will be able to pilot or copilot a variety of vehicles in the air. In this regard, it is understandable but worrisome that, in 2017, the Pentagon budget requested US$12–15 billion for AI-related projects. In the DARPA program AlphaDogfight, AI fighter pilots outperformed humans in simulated combat by executing manoeuvres beyond the capabilities of human pilots. There is no doubt that through the piloting of jets to fight wars or drones to deliver commodities, AI is about to have a transformative impact on the future of both military and civilian aviation. Unfortunately, AI is already being used on the battlefield. For example, the robot gun that in 2020 killed the scientist heading Iran's efforts to achieve nuclear capabilities used AI to adjust automatically its aim.

The impact of AI on employment must be investigated. For example, McKinsey & Company has estimated that 35% of the work of law office assistants could be automated. Appropriate steps must be taken to address this potential danger. For this purpose, a new centre was founded in the UK in 2018, led by my colleague in the Academy of Athens, Christopher Pissarides (Nobel Prize in Economics, 2010). Dimitris Travlos-Tzanetatos, an expert on labour law, began analysing in the early 1970s the impact of AI on the particular case of employment related to the practice of law. He identified specific benefits of AI which would increase the effectiveness of lawyers and judges but also highlighted several serious limitations and the danger of 'automation of justice'. He emphasised that law is based on *values*; therefore, the axiomatic basis of mathematics which underlies AI cannot by itself render justice (Travlos-Tzanetatos, 1971). Recently, Travlos-Tzanetatos presented a thorough analysis of the risk of professional degradation and technological unemployment arising from the use of AI. By appealing to Marx's theory of surplus, he argues that an increase in technological unemployment will be avoided since human

---

[3] This conference was organised by the MIT physicists Max Tegmark and Frank Wilczek (Nobel Prize in Physics, 2004), the University of Berkeley computer scientist Stuart Russell, and Stephen Hawking; other participants included Elon Musk, Vernor Vinge, and the co-founder of DeepMind, Demis Hassabis.

labour is necessary for the extraction of surplus value (Travlos-Tzanetatos, 2019). He supports his argument by pointing out that there is a global increase in wage labour, albeit one that is accompanied by 'deregulation'. He states that, at the moment, the main emphasis of the labour laws should be on protecting fundamental rights of working people, including the right to work, leisure leaves, personal data protection, and dealing with crowd-working as well as mobile work, whose effect is the dramatic decrease in leisure time. Interestingly, in the last chapter of his book, the use of technological breakthroughs, including AI, is discussed within the context of accelerationism, namely the call by certain scholars for the drastic intensification of capital growth. In this context the influential book *Empire* by the post-Marxist philosophers Michael Hardt and Antonio Negri (2000), as well as the positions of the political scientists Jeremy Rifkin and Robert Reich and that of the social philosopher Andre Gorz, are discussed.

Every effort must be made to ensure that AI is used for the elucidation and propagation of truth and not for the facilitation of 'fake news' and the manipulation of public opinion. A relevant, highly worrisome example is provided by a video created via deep learning of a 'speech' by President Obama, which is almost impossible to recognise as fake. In 2025 an Italian businessman was tricked by the AI-generated voice of Italy's defence minister into transferring one million euros.

There are also serious concerns regarding privacy. For example, using an upgraded type of Apple Watch, it would be possible to upload the position, health status, and conversations of every individual wearing such a device. Facial reading algorithms, such as Google's FaceNet, Apple's Face ID, and Facebook's DeepFace, can recognise one face among a million others. Half of the US' adult population have their facial images stored in a database available to the police. Ideally, a balance must be achieved between, on the one hand, privacy and draconian laws prohibiting the use of personal data for any malicious purpose, including opinion manipulation, and, on the other hand, protection against crime and terrorist attacks.

In my opinion, taking into consideration the resources and emotional cost of care providers associated with saving just one life, the insanity of losing many innocent lives in a terrorist attack makes surveillance a necessity, even if this threatens privacy. The fact that a very large number of such attacks has been prevented is an illustration that society often has the ability to solve, or at least to reduce, problems that the society itself has created. Indeed, the potency of such attacks is the result of technological advancements. But at the same time, a great number of such attacks have been prevented due to highly advanced forms of surveillance that have also been created as a result of technological innovations.[4] Of course, counterterrorism provides a 'treatment' as opposed to the 'cure' to this serious problem. The only way for such acts of violence to be reduced radically is for societies at large to address in a systematic way, poverty, ignorance, injustice, intolerance, racism, and religious fanatism.

References

Acemoglu, D. and Johnson, S. (2023). *Power and Progress.* New York: PublicAffairs.

Aljundi, R., *et al.* (2017). Memory aware synapses: Learning what (not) to forget. *bioRxiv.*

Ambrogio, S., *et al.* (2018). Equivalent-accuracy accelerated neural-network training using analogue memory. *Nature, 558,* 60–67.

Banino, A., *et al.* (2018). Vector-based navigation using grid-like representations in artificial intelligence. *Nature, 557,* 429–433.

Broccard, F. D., *et al.* (2017). Neuromorphic neural interfaces: From neurophysiological inspiration to biohybrid coupling with nervous system. *Journal of Neural Engineering, 14,* 041002.

Card, N. S., *et al.* (2024). An accurate and rapidly calibrating speech neuroprosthesis. *New England Journal of Medicine, 391*(7), 609–618.

Dasgupta, S., Stevens, C. F., and Navlakha, S. (2017). A neural algorithm for a fundamental computing problem. *Science, 358,* 793–796.

Dowd, M. (2017). Elon Musk's billion-dollar crusade to stop the AI apocalypse. *Vanity Fair.*

Essmann, B. and Mueller, O. (2022). *25-AI-Supported Brain-Computer Interfaces and the Emergence of Cyberbilities.* Cambridge: Cambridge University Press.

Ferryman, K., Mackintosh, M., and Ghassemi, M. (2023). Considering biased data as informative artifacts in AI-assisted health care. *New England Journal of Medicine, 389*(9), 833–838.

Fokas, A. S. (2023). Can artificial intelligence reach human thought? *PNAS Nexus, 2*(12), pgad409.

Hardt, M. and Negri, A. (2000). *Empire.* Cambridge: Harvard University Press.

---

[4]This observation suggests that, hopefully, society will also be able to cope with the problems created by AI.

Koch, C. (2017). To keep up with AI we'll need high-tech brains. *Wall Street Journal.*

Russell, S. (2019). *Human Compatible: AI and the Problem of Control.* New York: Viking.

Savelli, F. and Knierim, J. J. (2018). AI mimics brain codes for navigation. *Nature, 557,* 313–314.

Travlos-Tzanetatos, D. (1971). The impact on law of electronic information (in Greek). *Newspaper of Greek Lawyers, 38,* 381.

Travlos-Tzanetatos, D. (2019). Labour law in the fourth industrial revolution. Digitization, robotics and artificial intelligence (in Greek).

Chapter 4

The Embodied Brain and Laterisation

Many of the unconscious processes begin in the body proper as opposed to the brain. Indeed, the brain's basic functions are hugely influenced by its obvious topological attribute, namely that it is embodied.

The remarkable effects of the body proper on the brain are discussed in the first section of this chapter.

An important feature of the brain, often ignored by researchers of artificial intelligence, is that the brain exhibits *laterisation*, meaning that the functions of the two hemispheres are not identical. There is an anatomical asymmetry between the two hemispheres, which suggests a functional asymmetry. In particular, the right hemisphere is longer and wider than the left, except for the posterior parieto-occipital region. The functional asymmetry is exemplified by the presence of the Broca and Wernicke areas only in the left hemisphere. A broad framework underlying neuronal mechanisms that give rise to metarepresentations was presented in Chapters 5 and 6 of Fokas (2024). It will be shown in the second part of this chapter that the two brain hemispheres have different effects on the genesis of metarepresentations.

4.1 The Embodied Brain

The brain, in addition to the global connectivity exemplified by the thalamocortical system and the parallel-unidirectional connectivity found in the cerebellum, basal ganglia, and hippocampus (both of these important types of connectivity were discussed in detail in Fokas, 2024), possesses a third type of topological connectivity. This consists of a diffused, highly complex set of connections resembling a large multicomponent fan. These connections begin in a variety of *nuclei*, namely in specific collections of specialised neurons, located in the brainstem and the hypothalamus. The names of these nuclei are related to the substances they release. They include the noradrenergic *locus coeruleus*, the serotonergic *raphe nucleus*, and the dopaminergic *nucleus accumbens*, as well as the cholinergic and histaminergic nuclei. Neurons from these nuclei diffuse into large parts of the brain, thereby influencing billions of synapses. These neurons are particularly sensitive to unexpected changes, such as a flash of light or a loud noise.

There exist a large number of molecules, named *neuromodulators* and *hormones*, which play a crucial role in ensuring proper interactions between the body proper and the brain. Neuromodulators, which include dopamine and noradrenaline, are certain types of neurotransmitters that modify the effects of other neurotransmitters. Hormones are produced in a variety of glands and are transported by the circulatory system. Hormones appear in various forms, including eicosanoids such as *prosta-glandin*, *thromboxane*, and *leukotrienes*, steroids such as testosterone, progesterone, and corticoid, and proteins. Some of these molecules are produced in the brain, and their main targets are organs outside the brain; others are produced outside the brain, and they directly affect the function of the brain; and some are produced both in the brain and outside the brain. Several examples are discussed in the following.

Oxytocin is a nine-amino-acid peptide that is structurally related to endorphin. It is released in the posterior pituitary, acts both as a hormone and a neuromodulator, and has various effects. In particular, first, it affects the amygdala and the nucleus *accumbens*. Second, it is important for cir-cadian homeostasis, i.e. the control of body temperature, metabolic rate, and wakefulness. Third, it surges during sexual orgasm as well as during the final stages of childbirth. Fourth, among its actions outside the brain are its vital effects on labour and lactation. Oxytocin, which is often referred to as the 'hormone of love', increases affection, generosity, and tranquillity, produces loving feelings, encourages bonding, suppresses fear, and enhances the ability to decipher facial clues. It has been shown that the intranasal administration of oxytocin caused a substantial increase in trust among humans (Kosfeld *et al.*, 2005).

The hormone *prolactin*, which is released both in the anterior pituitary and the uterus, is crucial for milk production; it induces gratification after sexual acts and promotes nesting behaviour. For example, male rats after being given prolactin began building nests. Before the child is born, prolactin levels in men increase and their testosterone levels decrease, reducing aggression.

Somatostatin, which is released in the pancreas and the gastrointestinal system, as well as in the hypothalamus, suppresses cortical activity. *Cortistatin,* despite its structural similarity to somatostatin, is not released outside the brain like somatostatin, but only in the cortex, where it induces slow-wave sleep and suppresses overall cortical neuronal activity.

The impact of the body proper on the brain is exemplified by the fact that 95% of the important neurotransmitter serotonin—which, as discussed in Chapter 16 of Fokas (2024), is enhanced by antidepressant medications—is produced in enteric neurons, where it facilitates gut movement (Damasio, 2018, 136).

The human brain is the result of billions of years of evolutionary improvements, which gave rise to a superb organ capable of coordinating appropriate actions for survival and reproduction, as well as for achieving competitive advantages with respect to other species. In particular, the brain ensures that the neurological and immune systems work together with microbiota and mycobiota to achieve optimal results (Trowsdale, 2024).

Neuromodulators and hormones affect every aspect of an organism's function, from immunity to behaviour. The impact of these molecules is immense. For example, the vital role of steroids is exemplified by corticoid. This hormone increases with stress, anxiety, and fear. There are patients, such as those suffering from Addison's disease, who, as a result of insufficiency of their adrenal glands, do not produce enough corticoid. In such patients, any significant stressor, such as an infection, if not accompanied by an increase in externally administered steroids, may exhibit corticoid insufficiency, leading to a life-threatening situation known as 'adrenal crisis'. In this case, the organism decompensates as a result of several different effects, including the following: the loss of the normal suppressive action of glucocorticoids on the pro-inflammatory molecules cytokines results in a rapid rise in the level of the latter, causing fever, malaise, anorexia, and bodily pains; the loss of the synergetic action of cortisol with catecholamines on vascular tone leads to vasodilation and, hence, to severe hypotension; reduced glucogenesis in the liver causes hypoglycaemia; and mineralocorticoid deficiency causes sodium and water loss, as well as potassium retention, giving rise to dangerously high levels of potassium that cause cardiac arrythmias. In the 1950s, glucocorticoid replacement therapy became available, which provided a life-saving intervention for patients with Addison's disease. President John Kennedy was the most famous patient suffering from Addison's disease.

The importance of eicosanoids becomes clear through their vital role as regulators of immunity. In this regard, it is noted that testosterone, which is an anabolic steroid, suppresses the response of the immune system. Testosterone is low in monogamous individuals with children and high in promiscuous individuals without children. This perhaps explains why happily married people have better immunity. Corticosteroids also suppress immunity, which perhaps explains the fact that religiosity and those medications that reduce stress are associated with better immunity since they lower corticosteroids by reducing anxiety (Trives, 2013). Incidentally, the ring finger being longer than the index finger indicates exposure to high levels of testosterone early in life. This has been associated with decisiveness, directness, dominance, and self-confidence (Swaab, 2014).

The effect of the neuromodulators on behaviour is exemplified by the action of noradrenaline, which regulates arousal, as well as by the effects of dopamine on the anticipation of pleasure and rewards. High levels of dopamine are associated with the tendency to seek novelty and adventure (Trives, 2013). In addition, noradrenaline enhances vigilance to environmental threats, while dopamine is involved in motivating individuals towards achieving their goals regardless of obstacles. There is evidence that the balance between noradrenaline and acetylcholine dictates whether an individual faced with a difficult decision abides by a preexisting belief or abandons it (Cohen and Aston-Jones, 2005).

Neuromodulators and hormones not only modulate the brain's activity but also affect neuroplasticity, which refers to long-term changes in the structure and function of the brain. This implies that it is impossible to begin understanding human behaviour without taking into consideration these molecules. The American anthropologist Helen Fisher, the author of the book *Anatomy of Love*, after reviewing a large number of studies, claimed that oestrogen, oxytocin, testosterone, and dopamine have different impacts on the formation of personalities from that of serotonin. The latter is associated with the characteristics of concreteness, cautiousness, and orderliness, as well as a prominent sense of duty.

The understanding of the impact of the body proper on cognition continues to expand. For example, it has now been established, at least in mice, that the processes of memory loss due to ageing or due to Alzheimer's disease are based on completely different regions of the brain and involve different mechanisms. In particular, the former involves a structure in the hippocampus called the *dentate nucleus*, whereas memory loss due to Alzheimer's disease, as stated in Chapter 14 of Fokas (2024), begins in the *entorhinal* cortex. Age-related memory loss involves the CREB group of proteins. These CREB genes, which were mentioned in Chapter 9 of Fokas (2024) in the discussion of learning, are crucial for turning short-term memory into long-term memory (Pavlopoulos *et al.*, 2013). In this regard, the following recent finding further underlines the relationship between the body proper and the brain. The hormone *osteocalcin*, which is excreted by bone, increases the production of CREB proteins (at least in mice), and hence it enhances memory (Kandel, 2018, 116). Since exercising increases bone mass and, therefore, the production of osteocalcin, these studies suggest a beneficial effect of exercising on memory. Incidentally, it had been shown earlier that osteocalcin influences the production of serotonin, dopamine, and other neurotransmitters (Oury *et al.*, 2013). These remarkable studies, together with the fact that exercise promotes the release of endorphins and enkephalins, justify the Roman aphorism 'a healthy mind in a healthy body'.[1] It is worth noting that ancient Greek philosophers are characterised as *peripatetic*, which means 'walking up and down', emphasising the fact that they preferred to discuss and argue while walking. Although the walks of the ancient philosophers were of a leisurely type, they would have nonetheless been delighted with the recent findings that not only exercising but even a brisk walk promotes neurogenesis (Yau *et al.*, 2014).

The action of neurotransmitters and hormones on cells is characterised by a high level of complexity. In this regard, the so-called G-proteins (guanine nucleotide-binding proteins) are of key importance. These are a family of proteins residing inside cells, which act as molecular switches, facilitating the transmission of a variety of signals from outside a cell to its interior. For example, when adrenaline binds to a cell receptor, it does not directly stimulate enzymes inside the cell but stimulates a G protein, which then stimulates an enzyme. For this discovery, Alfred G. Gilman

---

[1] The original Latin statement is '*Mens sana in corpore sano*'. Usually, Roman aphorisms are translations of statements made in classical Greece. However, in this case, the opposite occurred: the above Roman aphorism was later translated into Greek: '*νούς υγιής εν σώματι υγιεί*'.

and Martin Rodbell were awarded the Nobel Prize in Medicine in 1994; Brian Kobilka and Robert Lefkowitz were also awarded the Nobel Prize in Chemistry in 2012 for work related to G proteins.

As another example of the enormously complex web of interactions between the body proper and the brain, it is noted that the cells lining the interior of cerebral microvessels, called endothelial cells, have several distinguishing features which allow these cells to act as a 'physical barrier'. This so-called 'blood–brain barrier' facilitates the entry of essential nutrients into the brain but excludes potentially harmful compounds (Abbott *et al.*, 2006). However, there are several locations where this barrier is not present. For example, the pineal gland (epiphysis) secretes melatonin directly into the systemic circulation. Also, a particular set of sensory neurons, called dorsal root ganglia, which transmit information from the limbs and body to the central nervous system, are devoid of the above barrier. The blood–brain barrier is part of a multicomponent structure that is affected by inflammation and a variety of transient or chronic pathological conditions, including Alzheimer's and Parkinson's diseases.

It is surprising that the importance of the brain's embodiment was mostly ignored by philosophers until the 19th century. Indeed, until the writings of Schopenhauer, Hegel, and Nietzsche, Western philosophy, with a few exceptions such as Heraclitus, treated the brain as if it were disembodied! Incidentally, most of these philosophers dogmatically based their analysis on the *rational* as opposed to the *transcendental*. Edmund Husserl (1859–1938), who studied mathematics and physics before he concentrated on psychology and philosophy, correctly criticised earlier philosophical approaches for their failure to analyse the nature of experience (for which the body is crucial), for their emphasis on 'mad rationalism' ('*verirrenden Rationalismus*'), and for their 'blindness to the transcendental' ('Blindheit fur das Transzendentale'), as quoted in Levin (1999, 61). Also, he highlighted the uniqueness of the human body as a material object which allows humans to 'live' their embodied existence, sharing some of these experiences with other individuals. Similarly, Maurice Merleau-Ponty (1908–1961), under the influence of Henri Bergson's thesis that the human body is a mediator that allows consciousness to engage with the world, emphasised the central role of the body in the brain's construction of reality.[2] The most eloquent recent corrobora-

---

[2]Henri Bergson also noted the importance of empathy. In addition to Bergson, Merleau-Ponty was influenced by his close friendship with Paul Sartre, with whom he shared the

tors of these positions are George Lakoff and Mark Johnson, who have stated:

> "The mind is not merely embodied but embodied in such a way that our conceptual systems draw largely upon the commonalities of our bodies and of the environment we live in [...]. Our conceptual systems are not totally relative and not *merely* a matter of historical contingency [...] truth is mediated by embodied understanding and imagination" (Lakoff and Johnson, 1980).

They also proposed that 'the very structure of reason itself comes from the details of our embodiment' (Lakoff and Johnson, 1999, 4). Moreover, they wrote, 'Since reason is shaped by the body, it is not radically free, because the possible human conceptual systems and the possible forms of reason are limited' (Lakoff and Johnson, 1999, 4).

Many studies provide support for these positions, emphasising the vital importance of embodiment. For example, there is evidence that there is a relation between the sounds of words generated by the larynx and the type of objects these words represent. In particular, this so-called 'kiki/bouba' effect is named after the words '*kiki*' and '*bouba*', which suggest a spiky-shaped and a softly rounded object, respectively. The leading neuroscientist Rodolfo Llinás (2002) has suggested that the universal structure of syntax may be related to a specific sequence of limb movements during running (Vowles, 1972).

4.2 Left versus Right Hemispheres and Their Impact on Forming Metarepresentations

It has been argued by Andrey Vyshedskiy that hominids, after being pushed out of trees by deforestation, had to rely on early identification of predators in order to survive since they were neither fast nor strong enough to protect themselves from the savanna's predators. This made it

---

common ideology of Marxist determinism. However, whereas both Merleau-Ponty and Sartre condemned the repression of the Stalinist period, Merleau-Ponty went further. He questioned whether it was possible to refute Stalinism without condemning Marxism itself. On the other hand, Sartre refused to withdraw his statement that 'An anti-communist is a dog'. In this way, the break in their friendship became inevitable.

imperative to improve their *visual analysis*, namely their ability to perceive *local details*. Vyshedskiy claims that this happened at the expense of holistic recognition. Indeed, various experiments suggest that animals have superior holistic recognition than humans (Vyshedskiy, 2014, 76, 165).

The left hemisphere has less white matter and hence is less interconnected overall than the right. This suggests that the left focuses on *local* parts of a given percept or situation, whereas the right hemisphere takes a more *global* view. This implies that the left should be better than the right for *reduction*, i.e. the deconstruction of a percept or concept into its ingredients. Also, the left is best suited for understanding elements *abstracted* from the given context. On the other hand, the right hemisphere excels in Gestalt approaches, placing a given situation into perspective. Indeed, the left hemisphere is more involved in the identification of simple, generic shapes, whereas the right is more important for identifying complex figures with individual characteristics (Blakeslee, 1990).

The right hemisphere is also better at processing spatial information and capturing broad signals from the environment. Then, depending on the situation, it proposes bonding strategies or defence mechanisms. These features are consistent with the fact that the right is responsible for prolonged, global attention, whereas the left, particularly the orbitofrontal cortex and the basal ganglia, excels in focused attention. Imaging studies in hypnosis, which is characterised by focused attention on a specific situation, show left hemisphere activation. Also, the association of the right hemisphere with broad attention may explain why it is difficult to recall a forgotten word when *concentrating* on it; after placing less attention on this word, the process moves to the right hemisphere, where the availability of more connections and hence more *associations* allows for the word to be recalled.

Awareness of an input begins in the right hemisphere and is then further processed in the left. In particular, learning new information or skills begins in the right, including information of verbal nature, despite the fact that language is mostly associated with the left. Then, once it has become familiar through practice, it is shifted to the left, including the skill of playing a musical instrument even though music is associated with the right hemisphere.

The expertise of the right hemisphere for an integrated, global view is supported by experiments in split-brain patients, which have shown that the right hemisphere attends to the entire visual field, whereas the left hemisphere attends only to the right. This is consistent with the fact that the 'hemi-neglect phenomenon', in which patients cannot perceive their left-hand side, occurs mostly after a right-sided stroke.[3] Since self-consciousness requires a global integration, these arguments suggest that the right hemisphere is more important than the left for aspects related to self-consciousness. Indeed, the right prefrontal cortex is crucial for self-recognition. In addition, *anosognosia*, which refers to the inability of stroke patients to recognise deficits in parts of their bodies, occurs after a right-sided stroke. Similarly, *prosopagnosia*, which is discussed in Chapter 12 of Fokas (2024), and *asomatognosia*, i.e. the inability to recognise parts of the body, occur after a right-sided deficit. The right hemisphere is also more important for deciphering and expressing emotions and feelings, especially social emotions. In particular, psychopathy, which is characterised by a lack of guilt, shame, and responsibility, is associated with right frontal lobe deficits. Moreover, the right hemisphere is more important for the so-called 'theory of mind', namely the ability to place oneself in another person's position.

Overall, the left hemisphere treats objects and the body as an assemblage of parts, dissociated from context. It specialises in abstraction, in vocabulary and syntax, and in achieving goals via the use of algorithmic procedures and suitable tools. In particular, *apraxia*, which is characterised by the inability to employ tools, is usually associated with left-sided deficits. In addition, the left hemisphere plays a major role in our communication with the external world. For example, although certain basic emotions activate the right amygdala, the corresponding feelings activate the left (Morris *et al.*, 1998).

Michael Corballis, in his tribute for his collaborator Roger Sperry, writes that 'the left side represents the fruits of human invention, including language, mathematics and a partwise way of representing objects' (Corballis, 1998, 1085).

From the above brief remarks and especially from Corballis' quote, one may reach the simplistic conclusion that the basic metarepresentations of language, mathematics, science, computers, and technology are mostly creations of the left hemisphere. However, in my opinion, this is

---

[3] It is usually assumed in clinical neurology that semi-neglect occurs *only* after a right-sided stroke. However, the article by Vallar and Calzolari (2018) contains two references which discuss the appearance of a semi-neglect after s left-sided stroke giving rise to right-half neglect. In this article, clinicians are alerted not to ignore this possibility for patients with a left-sided stroke.

not the case. As one would expect from evolutionary considerations, the functions of the two hemispheres are *complementary*. Hence, both hemispheres contribute to the genesis of every metarepresentation. For example, regarding language, understanding the meaning of a whole sentence and its emotional significance, as well as metaphors, irony, and humour, is associated more with the right hemisphere. This becomes clear from the fact that patients with a right-sided deficit cannot appreciate emotional prosody and have difficulty distinguishing humour from insults and lies (Winner *et al.*, 1998). Also, the right hemisphere is more important than the left for non-verbal communications, and there is right hemisphere involvement when associating distantly related words (Seger *et al.*, 2000).

Mathematics, science, medicine, and technology are of course crucially dependent on the employment of algorithms and the use of tools, all of which are dominated by the left hemisphere. However, the development of these hugely important disciplines requires deep insight and the formation of quite remote associations, which involve the right hemisphere more than the left. Interestingly, calculating prodigies use more of their right hemisphere than their left.[4] Also, as it will be discussed in my forthcoming book, *From the Ancient Greece to Viennese Modernism: Lessons in Creativity*, several studies have shown that the right anterior temporal and right prefrontal areas are crucial for insight (Kounios *et al.*, 2008). Importantly, it should be noted that novel artistic metarepresentations involve largely unconscious processes, and apparently such processes involve the right hemisphere more than the left.

The complementarity of the two hemispheres expresses itself in various forms, which include the presence of a harmonious balance of inhibition and excitation between left and right brain structures. As already mentioned in Chapter 3 of Fokas (2024), an interesting manifestation of the disturbance of this balance is the presentation of the so-called 'paradoxical functional facilitation'. Namely, the lack of inhibition after brain injury exerted by the damaged part of the brain on its counterpart can enhance a specific function of the latter part (Kapur, 1996). Paradoxical functional facilitation may also occur due to *compensatory plasticity*. Namely, the normal part of the brain overcompensates for its injured counterpart.

Another manifestation of the disturbance of the balance between right and left hemispheres is the tendency of patients with right-sided stroke to

---

[4]Multiplication activates the left hemisphere, but simple addition and subtraction activates the right.

minimise their own symptoms and even to joke about their predicament. In these cases, the pragmatic and somewhat melancholic right hemisphere, due to its injury, cannot exert its inhibition on the optimistic and joyous left hemisphere.

Despite the complementarity of the two hemispheres, Iain McGilchrist, in his highly erudite book, *The Master and the Emissary* (McGilchrist, 2010), attempts, erroneously in my opinion, to explain a variety of important occurrences in the history of the Western world by appealing exclusively to one or the other hemispheres. For example, he writes:

"Zeno's paradoxes rest on the adoption of the left hemisphere's view that every flowing motion in space or time can be resolved into a series of static moments or points that can then be summed to give back the living whole" (McGilchrist, 2010, 139).

He adds:

"Reductionism is an inescapable consequence of a purely left-hemisphere vision of the world, since the left hemisphere sees everything as made up from fundamental building blocks" (McGilchrist, 2010, 424).

In my opinion, the origin of Zeno's paradoxes is not the 'fragmented' formulation of the relevant motions, but rather the inability of the ancient Greeks to sum up series with infinitely many terms. For example, consider Zenon's particular paradox that one will never cover the distance of 'one unit' by halving in every step the remaining distance. It turns out that the summation of the corresponding series, namely $1/2 + 1/4 + \cdots$, is indeed 1, and hence this paradox is immediately resolved. Regarding reduction, it is worth noting that the remarkable creations of Mark Rothko, which constitute the epitome of the process of reduction, are so emotionally evocative that they drive many viewers to tears, and as stated earlier, emotions are primarily dealt with in the right hemisphere rather than in the left.

McGilchrist, recalling that spatial perspective and melancholy are attributes of the right hemisphere, associates the Renaissance with right-hemispheric features. In particular, he emphasises the dominance of perspective in paintings, as well as the occurrence of melancholy that is expressed in the ecclesiastical music of that era and in the preoccupation with death. In particular, he notes that both Erasmus and Sir Thomas More had on their desks a skull as a reminder of death (*memento morti*). He also points out that Hans Holbein, in his famous painting *The Ambassadors* (1533), presents two men with symbols of knowledge and prosperity, but he also includes in the foreground a distorted depiction of a skull. It appears that, in his analysis, McGilchrist neglects the fact that an important attribute of the Renaissance was the synergy of optimism with human's unlimited creativity, and optimism is a feature of the left hemisphere. In this regard, it is noted that the left has more dopaminergic neurons and that the closer a lesion of a stroke patient is to the left prefrontal cortex, the higher the probability that the patient will exhibit depressive symptoms. Similar arguments can be made for a variety of other positions presented by McGilchrist, including his biased analysis of 'the hubristic movement which came to be known as the Enlightenment' (McGilchrist, 2010, 329).

References

Abbott, N. J., Ronnback, L., and Hansson, E. (2006). Astrocyte-endothelial interactions at the blood-brain barrier. *Nature Review Neuroscince*, *7*, 41–53.

Blakeslee, T. R. (1990). *The Right Brain.* London: Macmillan.

Cohen, J. D. and Aston-Jones, G. (2005). Decision amid uncertainty. *Nature*, *436*(7050), 471–472.

Corballis, M. C. (1998). Sperry and the age of Aquarius: Science, values and the split brain. *Neuropsychologia*, *36*, 1083–1087.

Damasio, A. (2018). *The Strange Order of Things.* New York: Pantheon Books.

Fokas, A. S. (2024). *Ways of Comprehending: The Grand Illusion and the Essence of being Human.* Singapore: World Scientific.

Kandel, E. (2018). *The Disordered Mind.* New York: Farrar, Straus and Giroux.

Kapur, N. (1996). Paradoxical functional facilitation in brain-behaviour research: A critical review. *Brain*, *119*(5), 1775–1790.

Kosfeld, M., Heinrichs, M., Zak, P. J., Fischbacher, U., and Fehr, E. (2005). Oxytocin increases trust in humans. *Nature*, *435*(7042), 673–676.

Kounios, J., *et al.* (2008). The origins of insight in resting-state brain activity. *Neuropsychologia*, *46*(1), 281–291.

Lakoff, G. and Johnson, M. (1980). *Metaphors We Live By.* Chicago: University of Chicago Press.

Lakoff, G. and Johnson, M. (1999). *Philosophy in the Flesh: The Embodied Mind and Its Challenge to Western Thought.* New York: Basic Books.

Levin, D. M. (1999). *The Philosopher's Gaze: Modernity in the Shadow of Enlightenment.* Oakland, CA: University of California Press.

Llinás, R. (2002). *I of the Vortex, from Neurons to Self.* Cambridge: MIT Press.

McGilchrist, I. (2010). *The Master and the Emissary.* Connecticut: Yale University Press.

Morris, J. S., Öhman, A., and Dolan, R. J. (1998). Conscious and unconscious emotional learning in the human amygdala. *Nature, 393*(6684), 467.

Oury, F., *et al.* (2013). Maternal and offspring pools of osteocalcin influence brain development and functions. *Cell, 155*, 228–241.

Pavlopoulos, E., *et al.* (2013). Molecular mechanism for age-related memory loss: The histone-binding protein RbAp48. *Science Translational Medicine, 5*, 200ra115.

Seger, C. A., Desmond, J. E., Glover, G. H., and Gabrieli, J. D. (2000). Functional magnetic resonance imaging evidence for right-hemisphere involvement in processing unusual semantic relationships. *Neuropsychology, 14*(3), 361.

Swaab, D. (2014). *We Are Our Brains.* London: Penguin Books.

Trives, R. (2013). *Deceit and Self-Deception.* London: Penguin Press.

Trowsdale, J. (2024). *What the Body Knows.* Connecticut: Yale University Press.

Vallar, G. and Calzolari, E. (2018). Unilateral spatial neglect after posterior parietal damage. In *Handbook of Clinical Neurology*, Vol. 151, pp. 287–312. Elsevier.

Vowles, D. V. (1972). *The Psychobiology of Aggression.* Scotland: Edinburgh University.

Vyshedskiy, A. (2014). *On the Origin of the Human Mind.* Greatspace Independent Publications.

Winner, E., Brownell, H., Happé, F., Blum, A., and Pincus, D. (1998). Distinguishing lies from jokes: Theory of mind deficits and discourse interpretation in right hemisphere brain-damaged patients. *Brain and Language, 62*(1), 89–106.

Yau, S. Y., Gil-Mohapel, J., Christie, B. R., and So, K. F. (2014). Physical exercise-induced adult neurogenesis: A good strategy to prevent cognitive decline in neurodegenerative diseases? *BioMed Research International.* 403120.

Chapter 5

The Vital Importance of Glial Cells

Many researchers of artificial intelligence attempt to mimic the functions of the neurons. However, 85% of the brain's cells are glial cells. The plethora of these cells, together with Aristotle's aphorism that 'Nature does not create anything useless', suggests that these cells must be very important.[1] This is certainly the case, as will be discussed in this chapter.

A clear indication of the incompleteness of even a rudimentary understanding of the brain's basic functions is provided by the observation that the crucial impact of glial cells, until very recently, was largely ignored. New progress in this area is the result of the development of new tools, particularly *calcium imaging*. The success of this imaging technique is based on the fact that calcium signals provide the primary means by which all cells transmit information across the cell membrane from outside the cell into the cytoplasm, i.e. the cell's interior. Moreover, calcium is also used for propagating information within the cell. Since glial cells do not possess synapses and do not propagate electrical impulses, they cannot be studied using the usual tools employed for the study of neurons; however, glia's activation *can* be visualised via calcium imaging.

As shown in detail in the excellent book by R. Douglas Fields, *The Other Brain* (Fields, 2009), glia's role is far more encompassing than the one designated as 'supportive'. The importance of glia was foreseen by Fridtjof Nansen (1861–1930). In 1886, this Norwegian scientist, explorer, and humanitarian, after discovering that the ratio of glia to neurons increases as one ascends the ladder towards higher vertebrates, speculated

---

[1]According to Aristotle, 'οὐθὲν γάρ, ὡς φαμέν, μάτην ἡ φύσις ποιεῖ'.

that glia might be 'the seat of intelligence' (Galambos, 1961). Incidentally, Nansen was truly multitalented. After learning the silver-staining technique during his visit to Golgi's laboratory, he studied for his PhD certain marine life, where he was able to establish the discrete nature of nerve cells. This work provides the basis of the 'neuron doctrine', later advocated by Cajal, and therefore it is surprising that the 'father' of this doctrine failed to cite Nansen's publication (Edwards and Huntford, 1998). In 1888, Nansen led a team that made the first crossing of the Greenland interior using cross-country skis, and his techniques greatly influenced Arctic and Antarctic expeditions. He was awarded the Nobel Peace Prize in 1922 for his work on behalf of the refugees who emerged after the First World War. In particular, he introduced the 'Nansen passport' for stateless people, which continues to benefit countless individuals.

Interestingly, Marian Diamond, a neuroanatomist at the University of California, Berkeley, who analysed samples from the prefrontal and inferior parietal regions of Einstein's brain, found that, although on average there were as many neurons in these parts of Einstein's brain as in the cortex of other men, there were twice as many glia. Remarkably, the ratio of glia to neurons in these regions of Einstein's brain was 2:1, instead of the usual ratio of 1:1.

5.1   Origin of the Glia

Microglia derive from the cell line which gives rise to cells involved in immunity. The brain's isolation deprives it of access to the vital immune system, and for this reason the brain co-develops together with its own 'exclusive guard', the microglia. These cells, depending on the situation, can release a variety of chemicals, including cytokines, glutamate, reactive oxygen, and nitrogen species.

The other three main types of glia, namely astrocytes, oligodendrocytes, and Schwann cells (with the last one found outside the central nervous system, or CNS), derive from the same parentage that gives birth to neurons. Actually, during development, the unfathomable web of neural connections with their enormous number of synapses (there exist approximately 87 billion neurons, and each may have thousands of synapses) is guided by glia. Indeed, the scaffolding for this astounding creation is provided by a particular type of short-lived glia, called radial cells. Soon after an infant is born, radial cells begin to fade away, but some of them are transformed into astrocytes, whereas some remain; these cells have the potential to transform into either glia or neurons later in life. Master glial cells not only direct the path of neurons, but they also *generate* them. The progenitors of glia produce glia moving along radial cells, and during their journey some of these cells are *transformed* into neurons.

The principle of continuity suggests that, since glia can be transformed into neurons, there must be cells which are neuron–glia hybrids, sharing some of the characteristics of both glia and neurons. This is indeed the case. In addition to microglia, astrocytes, and oligodendrocytes, a fourth type of CNS glia has been discovered, called NG2. These cells are developmentally arrested oligodendrocytes and, remarkably, can revert to a progenitor state later in life. Although like other glia they do not have an axon, in contrast to typical glia they do have synapses and can communicate electrically (like neurons). Incidentally, in multiple sclerosis, NG2 cells do not differentiate into oligodendrocytes, but they become promyelinating oligodendrocytes, and for this reason they fail to wrap myelin, which is the protective layer around the neurons (Longo *et al.*, 2018).

As discussed in the earlier chapter, the brain is protected by the blood–brain barrier. Further evidence of the importance of glia is provided by the observation that neurons, despite their extensive requirements for glucose and oxygen, are not in contact with arterial supplies. Neurons are supplied with energy by the astrocytes, which themselves are in direct contact with tiny arteries called capillaries.

Mature neurons cannot divide except for hippocampal neurons, which are produced by neural stem cells throughout life. Studies in genetically identical adult mice have established that the ability to actively exploit a larger territory correlates positively with individual differences in hippocampal neurogenesis (Freund *et al.*, 2013). Incidentally, in all mammals except in humans, neurogenesis also occurs in the olfactory bulb, which controls the sense of smell.

Interestingly, one of the early indications for the existence of neurogenesis arose as a result of nuclear weapons proliferation! Indeed, following aboveground atomic bomb testing, the amount in the atmosphere of the radioactive ^{14}C isotope almost doubled, reaching peak levels in 1963. Plants use carbon in photosynthesis, and since humans acquire carbon from plants, when a human cell is 'born' through the division of a precursor cell, it can incorporate ^{14}C into its DNA. Investigators in Sweden analysed postmortem brain tissue and found that the great majority of neurons from people born before 1955 had low levels of ^{14}C in their DNA. However, a subpopulation of neurons in the hippocampus had higher levels of ^{14}C than the levels that existed prior to 1955, suggesting that these neurons had been born via the process of division (Bhardwaj *et al.*, 2006).

In contrast to typical neurons, glia can divide. In particular, microglia and astrocytes react to invading microbes and brain injury by dividing and mobilising an appropriate cellular army. Taking into consideration that cancer is uncontrolled proliferation, it is not surprising that the great majority of brain cancers originate from glial cells. Indeed, 51% of all brain cancers are the dreaded *glioblastoma multiforme*, which originates from the progenitors of astrocytes and oligodendrocytes; 25% of brain cancers originate from astrocytes, while 5% originate from oligodendrocytes. Incidentally, the skin-like cells lining the surface of the brain and the hollow fluid-filled cavities inside the brain called *ventricles* can, on rare occasions, proliferate abnormally. The resulting tumours—called, respectively, *meningiomas* and *ependymomas*—are usually benign.

It is worth noting that brain cancer spreads differently from cancers in other organs. Indeed, cancer usually spreads by using various enzymes to dissolve the extracellular space, thereby opening up a pathway between cells. Apparently, cancer in the brain is spread via an alternative route: extracellular space is formed via the shrinkage of cells. This is achieved by causing the malfunction of chloride channels which control salt-concentrations. It is not surprising that these channels have been a target in the search for an effective treatment of glioblastoma.

5.2 Astrocytes

As mentioned earlier, calcium imaging has played a major role in deciphering the plethora of effects of the glial cells. Although the typical glial cells do not have synapses and do not generate action potentials, studies using calcium imaging have established that *glial cells 'monitor' and, more importantly, affect neural communication.* How is this achieved? Astrocytes have receptors capable of detecting a large number of signalling molecules, including all the neurotransmitters used by neurons for synaptic communication. For example, they have receptors for glutamate and GABA, which are, respectively, the most important excitatory and inhibitory neurotransmitters of the nervous system. Astrocytes also possess receptors for a variety of other molecules. When such a molecule is attached to a glial cell's receptor, calcium channels across the glial membrane open up, allowing the influx of calcium. This influx signals glia to release a variety of molecules, including neurotransmitters. In other words, astrocytes, in addition to their crucial importance for delivering energy to neurons and providing a physical matrix for structural support, are vital for synaptic communication. Indeed, the current view of a synapse is that it forms *a tripartite consisting of the pre- and post-synaptic components together with surrounding astrocytes.* Actually, it is most likely *a tetrapartite, also involving astroglia.* For example, the activation of the glutamate receptor NMDA requires the presence of the amino acid D-serine, which is synthetised by astrocytes from the ubiquitous L-serine found in diet and is then released in the synaptic cleft.[2] In addition to glutamate and GABA, astrocytes can also release adenosine triphosphate (ATP), which, after it is released, can quickly lose its three phosphates to become the inhibitory neurotransmitter adenosine. This neurotransmitter, as will be discussed in Chapter 20, is critical for sleep. Importantly, the neuroactive molecules secreted by astrocytes substantially influence neighbouring neural networks (Eroglu and Barres, 2010). Interestingly, the chemicals secreted by astrocytes are contained in vesicles located throughout their cell bodies; these 'sacks' are similar to the synaptic vesicles found in neurons.

Astrocytes use an additional mechanism for controlling neural communication; namely, *they can change the number of functional synapses.* Indeed, glial 'fingers' can withdraw from their contact areas with dendrites, exposing more synapses. In particular, this has been observed in the enormous 'bushy' astrocytes that populate, in a tile-like manner, the hippocampus. Similarly, during pregnancy, as a result of the withdrawal of astrocytic membranes, the number of synapses of oxytocin-containing neurons in the hypothalamus is doubled (these large neurons are called *magnocellular cells*). Another example of this important action of glia is provided by studies in the cerebellum involving defective glutamate. In this case, cerebellar astrocytes (called Bergmann glia) withdrew from surrounding certain neurons 'like receding tiles' (Iino *et al.*, 2001). In this way, the number of synapses increased; in addition, the available glutamate increased further due to its reduced absorption by glia.

---

[2] Recent genetic studies have found that some schizophrenics have a defect in the D-serine synthetising gene.

Remarkably, *glial cells also have a direct effect on synaptogenesis.* Astrocytes can release the protein *thrombospondin*, which is of vital importance for the formation or removal of synapses (Ullian *et al.*, 2004). For example, Wesley Thomson of the University of Texas at Austin and Jeff Lichtman of Harvard University have shown that Schwann cells control the development, repair, and removal of synapses at the neuromuscular junction (Kang *et al.*, 2014). Another example is provided by the results of studies in individuals who are exposed to high-intensity sound, such as rock music. In this case, first, astrocytes attempt to protect hearing neurons by covering some of their synapses and by absorbing the excess glutamate, which, at high concentrations, can be neurotoxic. At the same time, they release neurotrophic factors to protect neurons from this neurotoxicity. If the effect on these neurons is prolonged, then astrocytes enlarge, become 'gliotic', and begin eliminating damaged synapses.

It is worth noting that a clinically important channel expressed in astrocytes is the water channel aquaporin 4 (AQP4). This channel plays a critical role in water balance, neuroexcitation, neuroinflammation, and astrocyte migration. In the autoimmune condition known as 'neuromyelitis optica spectrum disorder', AQP4 channels located in the optic nerve and in extensive spinal cord segments are targets of autoantibodies, causing extensive inflammation. This results in a marked loss of astrocytes, which affects the myelination of the associated neuronal axons.

Another important function of astrocytes is to collect and subsequently discard from the space around neurons potassium ions, which are released after the neurons fire electric pulses. For this purpose, astrocytes employ a vast multicellular network connected through certain protein channels called *gap junctions*. For this purpose, astrocytes, using specialised fine structures called 'end-feet', dump the accumulated potassium into the bloodstream (Fields, 2009). This process generates small voltage potentials across the astrocytes' membranes, which in turn creates slow brain waves that can be detected using EEG.[3] Astrocytes, in addition to potassium, also absorb other substances, including dopamine, glutamate, GABA, and various other neurotransmitters. As some of them are charged, their absorption also creates weak electric currents.

Some of the actions of astrocytes are implemented via an important component of the immune system called the *complement system*, which

---

[3] For example, a light flash makes retinal glial cells (called Muller cells) to be positively charged by eight millivolts.

will be discussed in Chapter 7. Here, it is only noted that astrocytes induce neurons to massively increase their production of a protein called C1q, which in turn activates the complement system.

In summary, glial cells communicate directly via gap junctions, as well as indirectly via the spread of calcium waves. The latter communication is slow (of the order of seconds) as opposed to the fast (of the order of milliseconds) electrical communication employed by neurons. The slow, steady, non-specific, global communication of glia complements and enhances the fast, specific, and local communication of neurons. This provides yet another mechanism for plasticity, which, as explained in Chapter 10 of Fokas (2024), is crucial for learning and memory. Furthermore, it provides another illustration of the importance of the inter-action between local and global processes. Overall, glial cells are necessary for achieving a perfect homeostatic balance with respect to neurons.

The multitude of substances released by glia provides hope for the discovery of new medications, as well as possible explanations for the effects of various—currently unknown—neuronal mechanisms. For example, the growth factor *neuregulin* controls the growth of the myelin-producing Schwann and oligodendrocyte cells. It turns out that many schizophrenics have a defect in their oligodendrocytes as a result of the fact that the gene governing neuregulin is defective in some of these patients. Researchers are currently attempting to develop medications stimulating the growth of oligodendrocytes. As will be discussed in the following, such medications may also be beneficial to patients with spinal cord injuries. Also, efforts are being made to manufacture the astrocyte-produced growth factor *glial-derived neurotrophic factor*, which (at least in culture) stimulates the production of dopamine and hence could be used to treat Parkinson's disease.[4] Similarly, there is hope that the growth factor *brain-derived neurotrophic factor*, which is produced in increasing quantities by astrocytes exposed to beta-amyloid, could be used in Alzheimer's disease to rescue neurons from beta-amyloid toxicity. As discussed in Chapter 14 of Fokas (2024), a basic characteristic of Alzheimer's disease is the overproduction or impaired clearance of beta-amyloid; this process certainly involves microglia, which normally actively absorb the excessive beta-proteins, but under conditions of severe inflammation this

---

[4] As will be discussed in Chapter 11, many neurodegenerative diseases are accompanied by the accumulation in neurons of misfolded proteins. Remarkably, similar blobs of abnormal proteins are observed in the neighbouring astrocytes.

capacity diminishes. Also, in the presence of beta-amyloid, microglia produce certain molecules known to be neurotoxic.[5]

Regarding the possible role of glia in a variety of unknown neural mechanisms, it is noted that during epileptic seizures, calcium rushes into glia, which results in the release of a large amount of glutamate, causing hyperexcitability. Perhaps, glutamate is also released from glia following electroconvulsive therapy, which could provide an explanation for the beneficial effect of this treatment in severe depression. In this case, the stimulated glial cells also release other substances which may have an additional beneficial effect.

Unfortunately, some of the substances excreted from glia are deleterious. For example, the glutamate released from astrocytes following a stroke causes the death of additional neurons long after the initial stroke, since, as stated earlier, glutamate in high concentration is neurotoxic. Glutamate is also implicated in the devastating disease 'amyotrophic lateral sclerosis' (ALS). In the classical form of this disease, which has afflicted, among others, Lou Gehrig and my late colleague Stephen Hawking, only motor neurons, i.e. the neurons from the spinal cord to corresponding muscles, are destroyed. However, in recent years, it has been documented that some patients with ALS also have cognitive deficits. In 1993, a genetic abnormality was identified in several ALS patients; namely, there is a defect in the gene producing an antioxidant enzyme, called *superoxide dismutase*. Astrocytes and microglia possessing this mutation produce an as yet unknown substance which is selectively neurotoxic to motor neurons. In addition, defective astrocytes are unable to absorb excessive glutamate, which may also contribute to neurotoxicity.

5.3 Myelin

Myelin for neurons of the CNS, i.e. those in the brain and the spinal cord, is produced by oligodendrocytes. The name 'white matter' correctly reflects the fact that this part of the brain takes its colour because of

---

[5]As discussed in Chapter 14 of Fokas (2024), the reduction in beta-amyloid has so far yielded only a minimal cognitive improvement in Alzheimer's patients. Even more surprising, with regard to the lack of substantial improvement, is the following result. In early Alzheimer's patients, beta-amyloid production was reduced by inhibiting an enzyme (BACE-1) cleaving the beta-amyloid precursor protein, which paradoxically led to cognitive deterioration (Knopman, 2019).

myelin, but the name oligodendrocyte, which in Greek means 'few branches', is a misnomer. Actually, the cellular processes of oligodendrocytes extend to large distances, and each of these processes forms a long 'tentacle' that grasps a single neuron, wrapping several times around it.

Myelin for peripheral neurons is formed by Schwann cells, which occur in three different types: the myelinating ones that form the myelin of large axons, the non-myelinating Schwann cells that embrace the small axons, and the terminal Schwann cells which engulf the tip of the axon, namely the part of the axon near its target, e.g. the part of a motor neuron that reaches the synapse connecting to a muscle fibre.

The myelin sheath is crucial for the uninterrupted, smooth electrical communication of neurons. Incidentally, invertebrates lack myelin, but they compensate by having enormous axons, which for the squid have a diameter of about a millimetre. The importance of myelin becomes evident by noting that defective myelination of neurons gives rise to multiple sclerosis, which is the most prevalent chronic central *neuroinflammatory* disease—namely a disease where inflammation occurs in the CNS. Part of the neuroinflammation is caused by the infiltration into the CNS of blood-derived immune cells. In this regard, the following question arises: how do typical immune cells bypass the blood–brain barrier and reach the CNS? Apparently, they enter by infiltrating the meninges, which are the membranes that line the skull and vertebral canal. From there, they reach the CNS parenchyma either through the meningeal blood vessels or via the system that produces the cerebrospinal fluid (which is called the choroidal plexus system) (Kipnis, 2016). The exit of immune cells from the CNS is, apparently, achieved via the recently discovered meningeal lymphatic vessels (Louveau *et al.*, 2015). An additional important contributing factor to neuroinflammation observed in multiple sclerosis is the activation of the CNS's own immune cells, namely microglia.

It is worth noting that neuroinflammation has been suggested as a 'unified feature' of neurodegenerative diseases, including Alzheimer's and Parkinson's diseases. This hypothesis gained support following a study showing that the use of non-steroidal anti-inflammatory medications during midlife had a protective effect against the development of Alzheimer's disease later in life. However, subsequent treatment trials with such drugs, or with steroids, were not effective. Also, and more importantly, in contrast to multiple sclerosis, in most neurodegenerative diseases, there are only low-to-moderate levels of inflammatory mediators in the CNS (Ransohoff, 2016).

Multiple sclerosis, in which three-fourths of the patients are female, is characterised by defective electrical communication between neurons, manifested clinically in a variety of symptoms which appear in the pathognomonic fashion of affecting different parts of the body at different times (a symptom is pathognomonic if its presence is characteristic for the given disease). Multiple sclerosis lesions can appear throughout the CNS, including the cortex, the spinal cord, and the optic nerve. Although approximately half of the cortical lesions are around blood vessels, the most easily recognisable cortical lesions are in the white matter, where they take the form of focal areas of demyelination, inflammation, and glial reaction. Similarly, the spinal cord involvement is also around blood vessels and in the form of circumferential demyelination. Overall, tissue damage results from a complex, dynamic, and poorly understood interplay between the immune system, glia, and neurons. With regard to the immune system, bloodborne macrophages infiltrate active multiple sclerosis lesions and remove myelin debris and inflammatory by-products. Microglia are abundant in these lesions, but it is not known whether their role is protective or pathogenic. The above lesions cause a variety of symptoms ranging from tingling, numbness, and mild visual problems to paralysis and blindness.

In patients suffering from multiple sclerosis, typical episodes of neurological disability are fully or partially resolved after days or weeks. These episodes characterise the 'relapsing course', which, unfortunately, after 10–20 years is usually followed by the 'progressive form' of the disease; in about 15% of patients, this development appears very early. Relapsing multiple sclerosis is primarily a disorder of the adaptive immune system (B and T lymphocytes), whereas progression is mediated by both the adaptive and innate immune systems, namely by B and T lymphocytes, microglia, and macrophages. The recent deeper understanding of immunological processes, which will be reviewed in detail in Chapter 7, has led to dramatic progress in the treatment of this, formerly devastating, disease. Indeed, for the treatment of the relapsing form of multiple sclerosis, more than a dozen therapies have been approved, including five preparations of interferon beta and several monoclonal antibodies. In particular, there exist specific monoclonal antibodies (called anti-CD-20) acting against B and T lymphocytes. The effect of these therapies can be monitored using MRI (enhanced by gadolinium), which can image the focal breakdown of the blood–brain barrier. For the progressive form, treatment choices are limited to the chemotherapy agent mitoxantrone and the monoclonal antibody ocrelizumab, which acts against B-lymphocytes.[6]

The serious and varied effects of multiple sclerosis provide clear proof of the importance of myelin. There are also claims that myelin affects intelligence. Indeed, studies via 'diffusion tensor imaging' comparing white matter development in children ages 5–18 have directly correlated higher development of white matter structures with higher IQ (Schmithorst *et al.*, 2005). Incidentally, recent studies suggest that from middle age onwards, white matter is severely affected by obesity. Actually, studies comparing the white matter of lean and overweight individuals of the same age established that the brains of lean individuals were similar to those of overweight individuals 10 years earlier. However, despite this substantial difference in the white matter, no difference in cognitive ability was observed (Ronan *et al.*, 2016).

It should also be remembered that the completion of myelination in the forebrain in the early 20s coincides with reaching maturity. Interestingly, Marian Diamond has correlated the enriched environment of mice with the development of a thicker cerebral cortex, which is the result of enhanced myelination (Diamond *et al.*, 1964).

A most important function of myelin is the control of the speed of propagation of neural pulses. Remarkably, both oligodendrocytes and Schwann cells can sense electrical activity in their axons, despite the fact that, in contrast to astrocytes, these cells have no receptors for most neurotransmitters. It turns out that they sense electricity by sensing ATP, which is produced by the activated axons (Fields, 2009, 286–290). The crucial importance of the action potentials is discussed extensively in Chapter 2 of Fokas (2024). The speed of propagation of this generic form of neuronal communication is determined by the thickness of myelin and by the number of the so-called *nodes of Ranvier*. The latter, acting as electrical repeaters, have a saltatory effect of greatly increasing the speed of action potentials, which can reach 100 metres per second. By controlling the number of myelin wraps and by determining the location of the nodes of Ranvier, glial cells control the speed of

---

[6]B-lymphocytes play a prominent role in multiple sclerosis. Evobrutinib is an oral inhibitor of the so-called Bruton's tyrosine kinase, which is a molecule transmitting signals between myeloid cells and B-lymphocytes. The action of evobrutinib inhibits the release of cytokines from B-lymphocytes, which in turn inhibits the activation of the proinflammatory M_1 macrophages, resulting in clinical improvement (Montalban *et al.*, 2019).

propagation of action potentials along the axon. This has a vital global effect on neural networks since it provides a way for various signals to be synchronised, which is crucial for both unconscious and conscious processes.

The nodes of Ranvier have been implicated in the pathogenesis of aggressive inflammation in both peripheral and central neurons. In this regard, it is noted that up to 30% of patients with acute or chronic inflammatory polyneuropathy harbour autoantibodies. Although specific targets (antigens) of such autoantibodies have been identified only in 13% of the cases, it has been hypothesised that most of these targets are located on the nodes of Ranvier or in paranodal domains, i.e. in nearby regions (Stathopoulos *et al.*, 2015). Indeed, these regions contain a large number of structural proteins, channels, and adhesion molecules, which provide ideal sites for autoantibody 'attacks'.[7] In some cases, the resulting inflammation may cause paralysing weaknesses, which are rapidly reversible when the associated autoantibodies are inhibited with specific immunotherapies. Until the beginning of the 2010s, such conditions, which are now called 'paranodopathies', were thought to be caused by demyelination. However, myelin lesions would be inconsistent with the rapid onset of recovery via immunotherapy since remyelination takes several months to have an effect. Interestingly, there is evidence that the mechanisms responsible for paranodopathies may play a part in the pathogenesis of multiple sclerosis.

Another important contribution of glia is that by shrinking or swelling they can control the extracellular space in the brain. With age, some of this space is lost, which perhaps impedes the diffusion of messenger molecules across neurons. This impediment may have important implications since studies suggest that 'volume transmission', i.e. the communication based on the diffusion of neuroactive substances across the extracellular brain space, affects learning abilities (Vargová and Syková, 2014).

Glial cells participate in many additional functions, including neuropathic pain, which typically occurs after a spinal cord injury, and addictions. The former effects are discussed in the following.

---

[7]Autoantigens identified at the nodes of Ranvier include neurofascin 186 and gliomedin in the nodal domain, as well as contactin-1, Caspr1, and neurofacsin 155 in the paranodal domain.

5.4 Neuropathic Pain

Following certain injuries, some patients develop chronic pain, called 'neuropathic'. This is partly caused by a variety of pain-inducing chemicals that are produced by the hyperactivity of glial cells. In general, pain sensory neurons from the skin penetrate the top surface of the spinal canal; there, each of these neurons (called dorsal root ganglia) forms a synapse with a pain neuron travelling to the brain. This simple anatomical arrangement implies that it is possible to interfere with pain sensations at three different levels: first, at the skin, using for example Novocain; second, in the spinal canal, using for example an epidural injection, such as the one employed to ensure painless labour, or *nonselective* sodium-channel inhibitors, such as lidocaine and carbamazepine; third, at the cerebral cortex, using for example opioid medications. Opium, marijuana, and nicotine activate the same receptors that are used by the endogenous molecule endorphin. Marijuana, in addition to these receptors, also activates cannabinoid receptors, which are of two types: CB1, which mediate the psychoactive effects of marijuana, and CB2, which mostly have an anti-inflammatory action.[8] Blocking CB2 receptors prevents microglia from becoming hyperactive and hence prevents the secretion of various pain-inducing chemicals, such as cytokines. A different strategy for the central control of pain is to inhibit the activation of cytokines using minocycline, which is an antibiotic similar to tetracycline but somehow has an anti-inflammatory effect.[9]

In a recent promising development, a *selective* sodium-channel blocker has been developed (called VX-548). Although the results of a rigorous clinical trial testing this medication were rather disappointing, with the difference between its effect and that of a placebo being clinically insignificant (Jones *et al.*, 2023), this trial provides an early foray into a very promising new class of medications.

---

[8] The medication Sativex, acting on cannabinoid receptors, has been approved in several countries for pain in multiple sclerosis.

[9] In addition to CB2 receptors, microglia also possess the ATP receptor P2X4. ATP is released when neurons are damaged. Following such an injury, the number of P2X4 receptors in the spinal cord increases, making them hyperalert.

5.5 Spinal Cord Injury

Peripheral neurons regenerate; however, as stated earlier, neurons in the CNS generally do not. The *nerve growth factor* produced by Schwann cells is critical for the regeneration of damaged peripheral neurons, but unfortunately, the use of this factor to stimulate CNS neurons is ineffective. There are a multitude of different mechanisms preventing the regeneration of CNS neurons. For example, following an injury to a peripheral neuron, the associated Schwann cells usually survive and provide a scaffold for regrowth. In contrast, oligodendrocytes, which are the CNS analogue of Schwann cells, continue to die for the next six hours following injury.[10] Another impediment for CNS repair following injury is that microglia and astrocytes have both positive and negative effects: on the positive side, they wall off the region around the injury, absorb debris, and release neurotrophic substances supporting neurons; but at the same time, they produce various pro-inflammatory chemicals which are deleterious to oligodendrocytes and neurons. Moreover, glial cells form scar tissue, which prevents axons from moving past the damaged site.

A crucial role in the regeneration of peripheral neurons is played by the tip of the damaged neuron, which is called the 'growth cone'. Unfortunately, as soon as this cone is touched by some of the substances produced by CNS glia, it collapses. In particular, this happens if the growth cone touches myelin produced by oligodendrocytes.[11]

Despite the above difficulties, the deepening understanding of the many different factors preventing CNS neuron regeneration raises hope that it may soon be possible to overcome some of the underlying obstacles, so that the devastating effects of spinal cord injury can be alleviated. For example, it may be possible to use a viral gene transfer strategy similar to that used for spinal muscular atrophy (Mendell *et al.*, 2017). Specifically, the central canal of the spinal cord is lined by ependymal stem cells. Following spinal cord injury, these cells are activated; however, although they produce astrocytes, unfortunately, they do not generate oligodendrocytes, which are crucial for remyelination. In a recent breakthrough, investigators using genetic engineering were able to

---

[10]Remarkably, CNS neural regeneration has been achieved by grafting peripheral neuron segments into CNS and using them to guide regeneration.

[11]Two particular myelin proteins, called Nogo and MAG, responsible for this collapse have been isolated; the search for additional proteins is ongoing (Schwab, 2010).

express in the stem cells of mice the gene *Olig 2*. This prompted the ependymal stem cells to generate a large number of oligodendrocytes; the resulting myelination gave rise to improved conduction velocity through the injured spinal cord (Llorens-Bobadilla *et al.*, 2020).

To conclude, it should be noted that the recent appreciation of some of the remarkable functions of glia vindicates some of the ideas of the 'reticulists' discussed in Chapter 2 of Fokas (2024) and reminds us of the importance of open-mindedness. A plethora of chemicals produced by glia exert a global neuro-modulatory effect on large collections of neurons. In this sense, the abstract idea of the existence of a (virtually) 'topological neural network' complements the 'neuron doctrine', which is based on the assumption that neurons communicate only via their synapses. The synapses and the chemicals provide concrete illustrations of the interplay between local and global processes, respectively.

References

Bhardwaj, R. D., *et al.* (2006). Neocortical neurogenesis in humans is restricted to development. *Proceedings of the National Academy of Sciences*, *103*(33), 12564–12568.

Diamond, M. C., Krech, D., and Rosenzweig, M. R. (1964). The effects of an enriched environment on the histology of the rat cerebral cortex. *Journal of Comparative Neurology*, *123*(1), 111–119.

Edwards, J. S. and Huntford, R. (1998). Fridtiof Nansen: From the neuron to the North Polar Sea. *Endeavour*, *22*(2), 76–80.

Eroglu, C. and Barres, B. A. (2010). Regulation of synaptic connectivity by glia. *Nature*, *468*(7321), 223–231.

Fields, R. D. (2009). *The Other Brain: From Dementia to Schizophrenia, How New Discoveries about the Brain Are Revolutionizing Medicine and Science.* New York: Simon and Schuster.

Fokas, A. S. (2024). *Ways of Comprehending: The Grand Illusion and the Essence of Being Human.* Singapore: World Scientific.

Fokas, A. S. and Ktistakis, N. (in preparation). Ageing and the quest for rejuvenation; diet, epigenetics, microbiota.

Freund, J., *et al.* (2013). Emergence of individuality in genetically identical mice. *Science*, *340*(6133), 756–759.

Galambos, R. (1961). A glia-neural theory of brain function. *Proceedings of the National Academy of Sciences of the United States of America*, *47*(1), 129.

Iino, M., *et al.* (2001). Glia-synapse interaction through Ca2+-permeable AMPA receptors in Bergmann glia. *Science*, *292*(5518), 926–929.

Jones, J., *et al.* (2023). Selective inhibition of NaV1. 8 with VX-548 for acute pain. *New England Journal of Medicine, 389*(5), 393–405.

Kang, H., Tian, L., Mikesh, M., Lichtman, J. W., and Thompson, W. J. (2014). Terminal Schwann cells participate in neuromuscular synapse remodeling during reinnervation following nerve injury. *Journal of Neuroscience, 34*(18), 6323–6333.

Kipnis, J. (2016). Multifaceted interactions between adaptive immunity and the central nervous system. *Science, 353*, 766–771.

Knopman, D. S. (2019). Lowering of amyloid-beta by β-secretase inhibitors - Some informative failures. *New England Journal of Medicine, 380*, 1476–1478.

Llorens-Bobadilla, E., *et al.* (2020). A latent lineage potential in resident neural stem cells enables spinal cord repair. *Science, 370*(6512), eabb8795.

Longo, D. L., Reich, D. S., Lucchinetti, C. F., and Calabresi, P. A. (2018). Multiple sclerosis. *New England Journal of Medicine, 378*, 169–180.

Louveau, A., *et al.* (2015). Structural and functional features of central nervous system lymphatic vessels. *Nature, 523*, 337–341.

Mendell, J. R., Al-Zaidy, S., Shell, R., Arnold, W. D., Rodino-Klapac, L. R., Prior, T. W., ... and Kaspar, B. K. (2017). Single-dose gene-replacement therapy for spinal muscular atrophy. *New England Journal of Medicine, 377*(18), 1713–1722.

Montalban, X., *et al.* (2019). Placebo-controlled trial of an oral BTK inhibitor in multiple sclerosis. *New England Journal of Medicine, 380*(25), 2406–2417.

Ransohoff, E. (2016). How neuroinflammation contributes to neurodegeneration. *Science, 353*, 777–782.

Ronan, L., *et al.* (2016). Obesity associated with increased brain age from midlife. *Neurobiology of Aging, 47*, 63–70.

Schmithorst, V. J., Wilke, M., Dardzinski, B. J., and Holland, S. K. (2005). Cognitive functions correlate with white matter architecture in a normal pediatric population: A diffusion tensor MRI study. *Human Brain Mapping, 26*(2), 139–147.

Schwab, M. E. (2010). Functions of Nogo proteins and their receptors in the nervous system. *Nature Reviews Neuroscience, 11*(12), 799–811.

Stathopoulos, P., Alexopoulos, H., and Dalakas, M. C. (2015). Autoimmune antigenic targets at the node of Ranvier in demyelinating disorders. *Nature Reviews Neurology, 11*(3), 143.

Ullian, E. M., Christopherson, K. S., and Barres, B. A. (2004). Role for glia in synaptogenesis. *Glia, 47*(3), 209–216.

Vargová, L. and Syková, E. (2014). Astrocytes and extracellular matrix in extra-synaptic volume transmission. *Philosophical Transactions of the Royal Society B: Biological Sciences, 369*(1654), 20130608.

Chapter 6

Artificial Intelligence versus Human Thought

Telephone and television figured prominently in my childhood's personal recollections, while later I enjoyed video movies and CDs. Then, nothing dramatic occurred during my life for many years until the appearance of the internet. The lives of my three children, born in the period 1990–2001, are greatly affected by the mobile phone and social media. My prediction is that the impact of these technological inventions, which at the moment appears dramatic, is minor in comparison with the forthcoming massive changes that will be caused by the expected progress in AI.

As noted in Chapter 3, the remarkable achievements of deep learning have led some scholars to raise the question of whether AI can reach the level of artificial general intelligence. The milestone, when achieved, was called by Vernor Vinge 'reaching the singularity'. The term 'singularity' within the context of human history was first introduced in 1958 by John von Neumann. A year after this brilliant man's death, the mathematician and von Neumann's close friend Stan Ulam wrote that von Neumann made the following statement:

> "the ever accelerated progress of technology [...] gives the appearance of approaching some essential singularity in the history of the race beyond which human affairs, as we know them, could not continue".

This concept has been popularised by the futurist scholar and director of Google Engineering, Ray Kurzweil (2005).

The question of whether a computer can be 'creative' has a long and illustrious history. For example, the pioneering programmer Ada Byron stated that a computer cannot 'think' creatively. Turing thought deeply about this problem and, in his famous 1950 paper, introduced the 'Turing Test', which supposedly provides a way of ascertaining whether a computer has reached human-level intelligence. According to him, what matters is not the process but performance; therefore, he proposed an 'imitation game': if a machine operated so proficiently that observers could not distinguish its behaviour from that of a human, then the machine should be recognised as intelligent.

Von Neumann, in his 1958 book, *The Computer and the Brain*, in an effort to answer this fundamental question, presented a detailed comparison between the functional properties of the brain and the basic operational principles of a computer. In particular, he noted that the speed of neural processing is extremely slow in comparison to the speed of digital computing. Also, he emphasised that the computational accuracy of modern computers is very high, whereas the accuracy of neuronal computations is quite low. This is related to the fact that modern computers are 'digital' machines, i.e. their operation is based on the abstract, binary system of 0 and 1. Incidentally, this is to be contrasted with older computers, which were 'analogue' machines, i.e. their operation was based on the use of a physical (as opposed to abstract), continuous (as opposed to discrete) quantity. Specifically, in older computers, each number was represented by a particular value of the voltage, which is a physical, continuous quantity. As in the case of these older computers, neuronal computations are based on analogue processes. For example, neurons use the summation of the continuous variation of their voltage to 'decide' whether to be activated, namely, whether to generate an action potential (Fokas, 2024). Assuming that the accuracy of an individual neuronal computation is related to the frequency of the action potential signal, one expects an accuracy of the order of 1% or 2% of the maximum frequency variation. This maximum is of the order of 100; hence, one expects an accuracy of at most two decimal places. On the other hand, the computer can easily achieve an accuracy of 13 decimal places.

Von Neumann pointed out that the overall function of the brain may be based on a hybrid of analogue and digital processes. For example, the strength of the contraction of a given muscle and the secretion of a specific hormone follow an analogue mechanism, whereas the 'all or nothing' nature of the generation of an action potential may follow a digital process.

He also emphasised that the brain is energy-efficient. Indeed, since the energy consumption of the brain is of the order of 10 watts, and the number of neurons is of the order of 10^{10}, it follows that each neuron dissipates 10^{-9} watts, which compares very favourably with the dissipation of 10^{-1} watts of a transistor.

How does the brain compensate for its slowness and lack of accuracy at the individual neuronal level? According to von Neumann, the answer is 'by using a massive parallel processing'. Specifically, he claimed that the brain compensates for his lack of 'logical depth' by using 'logical breadth'.

As a result of recent developments in neuroscience, this astonishing feature of neuronal 'computations' has been further elucidated. We now know that the input to a given neuron is achieved via a large number of synapses; in many neurons, this number is of the order of 10,000. After integrating this input, the given neuron 'decides' whether to 'fire', i.e. to be activated or not. The 'parallel' nature of the brain's function means that all synapses can be active simultaneously. Since each synapse is active 100 times per second, and since the brain possesses approximately 10^{14} synapses, it follows that the brain has the capacity of processing 100 multiplied by 10^{14}, i.e. 10^{16}, operations per second. This is a remarkably large number of operations. For example, the analogous number for a modern desktop machine is 10^9. It must be emphasised that the processing of information at synapses is only one of the plethora of mechanisms used by the human brain for generating and manipulating information. Some of these mechanisms are discussed in Chapter 10 of Fokas (2024); moreover, the newly understood role of glia as a 'slower but global' modulator of information was discussed in the previous chapter.

Von Neumann claimed that two important features of the brain, namely, first, that it has plasticity—which is the marvellous capacity to 'learn' from experiences—and second, that, regarding its computations, it is an analogue 'machine', can both be emulated by a computer. First, although a computer does not learn from experience, there exist methods (especially with the development of AI) for continuously restructuring its software. Second, a digital computation can emulate analogue values to any desired degree of accuracy. These arguments, together with Turing's concept of the universality of a Turing machine, led von Neumann to the claim that a computer can emulate the brain.

Are the operations of the brain based on mathematical principles? In the last two and a half sentences of his pioneering book, von Neumann writes enigmatically that whatever is the 'language' that the brain uses, 'it cannot fail to differ considerably from what we consciously and explicitly consider as mathematics'.

Returning to the question of artificial general intelligence, there is in my opinion a serious *methodological problem* concerning the possible proof that a machine has surpassed the human level of intelligence. This would require proof that for *every* conceivable human goal, the machine achieves better performance. So far, this has been accomplished for *particular* goals involving the direct competition between a human expert and a machine. For example, this has happened in the goals of winning a chess or Go game. However, it is apparent that such an approach cannot be used for an uncountably large number of possible situations. This suggests that unless a different methodology is introduced, the question of proving whether artificial general intelligence has been reached is not well defined.

In any case, in order to address this question, it is necessary first to provide a definition of intelligence. The cosmologist and leading AI exponent Max Tegmark attempted to provide such a definition. He defined intelligence as 'the ability to accomplish complex goals'. In his book, *Life 3.0: Being Human in the Age of Artificial Intelligence* (Tegmark, 2017), it is claimed that this definition encompasses the Oxford Dictionary's definition of 'the ability to acquire and apply knowledge and skills', as well as several of the definitions proposed in the *Nobel Week Dialogue 2015: The Future of Intelligence*. In this conference, among the definitions proposed was that intelligence is 'the capacity for problem solving, learning, logic, and planning'. According to Tegmark, acquiring knowledge, learning new information or a new skill, solving specific problems, employing logical algorithms, and designing concrete plans can all be considered as processes subsumed by the phrase 'accomplishing complex goals'.

In my opinion, Tegmark's definition suffers from 'mechano-morphism'. Namely, in the same way that humans' explanation of various phenomena is 'anthropomorphic', i.e. it is biased by human (*anthropos*) thinking, computer scientists tend to be influenced by 'machine thinking'. An illustration of the effect of mechano-morphism is provided by Tegmark's introduction of Lives 1.0, 2.0, and 3.0' . Tegmark, motivated by the fact that every computer stores suitable software, namely, programs containing a list of operating instructions, defines human life as Life 2.0; the defining feature of this life is that human 'hardware' has been created via evolution, but the corresponding 'software' is mainly designed after birth via learning. In Life 1.0, exemplified by bacteria, both hardware and software have been created via evolution. According to Tegmark, artificial general intelligence represents Life 3.0, in which both hardware and software are designed. I believe that this classification, which is clearly motivated by notions in computer science, provides a useful model for discussing qualitative developments in algorithms, but it is far too simplistic to represent essential features of life.

It will be argued in the following that although Tegmark's definition of intelligence is adequate for machines, it does not capture the essence of human thought. Actually, in my opinion, this definition is appropriate for technology, which can be defined as:

> *a collection of devices and engineering practices as a means to achieve a complicated goal.*

In what follows, well-known human features not imitated by machines will be discussed; these include deep understanding and emotional involvement. Then, certain arguments which appear novel will be presented.

6.1 Understanding

Even if, for the sake of argument, we assume that Tegmark's definition *is* adequate, it is important to note that human understanding differs from that of machines. This has serious implications for the *types of goals* at which machines can be superior to humans. For instance, in machine translation, the computer, after being trained with massive amounts of data, discovers complex relations among words, from which it can refer other relations. For example, the computer can 'understand' that the pair 'king' and 'queen' has analogies with the pair 'husband' and 'wife'. However, the type of relations established by the computer is of a completely different nature than to those established via human associations, whose crucial importance in our thought has been emphasised throughout my previous book (Fokas, 2024) and in this book.

Conscious and unconscious associations establish relations on the basis of a deep understanding of the underlying constituent parts, whereas the computer does not understand the meaning of these parts.

In contrast to the human brain, the computer, by following specific instructions, manipulates symbols without understanding the content of those instructions or the meaning represented by the symbols it manipulates. As the Turing Prize-winning computer scientist Joseph Sifakis points out, AI can be characterised by the statement 'performance without understanding'. AI can solve very complicated problems, but in contrast to the scientific knowledge generated by the brain, AI does not yield deep understanding of the underlying processes (Sifakis, 2020). For example, regarding language, the computer forms syntactic relations as opposed to semantic ones. For this reason, it is not surprising that AI still performs poorly in the so-called 'Winograd Schema Challenge' (Davis *et al.*, n.d.). This is exemplified by the challenge to determine what the word 'they' refers to in the following two sentences: 'The city councilmen refused the demonstrators a permit because *they* feared violence, and The city councilmen refused the demonstrators a permit because *they* advocate violence'. This raises questions of whether a machine can pass a carefully designed 'Turing test', namely, whether a machine can converse in writing well enough to trick a person into thinking that they are conversing with another human.

Further arguments regarding the differences in 'understanding' between machines and humans are presented in Gelepithis (2024, Chapter 4, pp. 179–217).

6.2 Implementation

The *process* of implementing goals is obviously drastically different between machines and humans.

The embodied brain shares with artificial circuits the ability to store and process information, but in addition, the brain creates subjective experiences. An individual, while achieving a particular goal, feels in a unique way an embodied fulfilment.

This fundamental difference was perfectly captured by Kasparov, who, describing his defeat by IBM's artificial creation, wrote that at least the Deep Blue 'was not enjoying beating me'. The elements of 'emotional intelligence' and 'self-awareness' suggested by the definition of intelligence in the 2015 meeting on intelligence mentioned earlier are not contained in the definition of intelligence as 'accomplishing complex goals'.[1]

---

[1]Surprisingly, although it is stated in Tegmark (2017, 184) that 'If we can one day figure out what properties an information-processing system needs in order to have a subjective experience [...]', on page 50 of the same book, the contradictory claim is made that emotional intelligence and self-awareness are contained in the definition of 'accomplishing complex goals'.

It is true that computers can replicate some of the features of cognition, including pattern recognition and prediction. However, the mechanical implementation of these processes is not accompanied by any first-person experience. In 1995, the philosopher David Chalmers called the elucidation of the neuronal mechanism giving rise to this *first-person experience* 'the hard problem' of consciousness; deciphering the neuronal basis of *implementing a given goal* was called by Chalmers the 'easy problem'.

It should be remembered that humans' underlying emotional intelligence originates from primordial desires related to the vital *evolutionary goals* of self-preservation (avoiding death) and reproduction (enjoying sex). This means that this type of intelligence is extremely broad and important (Damasio, 2012). In addition, organisms have the ability to be informed *directly* from their environment, as opposed to needing to be programmed by an external agent in order to be able to receive a particular type of information.

6.3 Creativity

The above remarks summarise, in a sense, self-evident differences between artificial and human intelligence. Beyond these differences, the suggestion that human thought can be simply reduced to the property of 'accomplishing complex goals' is, in my opinion, fundamentally wrong. This becomes clear by looking at a most important component of intelligence, namely creativity, which was, incidentally, included in the definitions proposed in the 2015 conference mentioned earlier. I believe that:

> *creativity is not defined in terms of achieving a specific goal. On the contrary, it is measured by the distance of the final unexpected achievement from a starting vague idea. Indeed, the defining property of creative individuals is their capability of establishing far remote associations and generating completely unexpected relations between different areas. These areas had appeared until that point so distinct that no one had thought of posing the goal of establishing a connection between them. As stated repeatedly in Fokas (2024) and in this book, the process of establishing such remote associations is mostly unconscious and therefore much more difficult to be 'programmed'. In this sense, it can be claimed that the origin of human creativity is non-literal and non-algorithmic, it is largely metaphorical, imaginative, and transcendental.*

Einstein expressed clearly the non-literal nature of creativity; he stated that, 'The words or the language as written or spoken do not seem to play any role in the mechanism of my thought' (Einstein, 1985, 32). Regarding the essence of human creativity, it seems to me that:

> *the less the final achievement is predefined and the more it is free from preconditions, the less the creative process is affected by misconceptions and the current way of thinking about a given problem. Hence, the higher the chances for a breakthrough.*

For example, the creation by Picasso of *Olga in an armchair* (1917) was the result of the specific goal of the great artist to depict Olga Koklova, who soon afterwards became his first wife. What was the goal associated with *Guernica* (1937)? The vague motivation of depicting the atrocities of the Spanish Civil War gave rise to this specific artistic meta-representation that arose directly from Picasso's unconscious. By comparing these two works, it becomes clear that in the arts, the less a goal is *a priori* defined, the higher the value of the artistic creation. The same is true in mathematics and the sciences.

Regarding mathematics, in what follows I discuss two examples from my own work. As discussed in Chapter 21, EEG gives rise to the important 'inverse mathematical problem' of computing the neuronal current from the knowledge of the measured electric potential. A general algorithm for determining the current was introduced in Fokas (2024). The numerical implementation of this algorithm requires the computation of a certain auxiliary function (Dassios and Fokas, 2020; Fokas *et al.*, 2004). This well-defined goal can be achieved via the training of a two-layer neural network, which provides yet another illustration of the importance of machine learning (Hashemzadeh *et al.*, 2020). Achieving this specific goal is conceptually very different from my work on the Lindelöf hypothesis (Fokas, 2019), where a novel approach was introduced to this historical problem.[2] The key step of this approach is the derivation of new identities satisfied by the Riemann zeta function. The genesis of these unexpected identities was not the result of a specific goal, but the outcome of unconscious processes motivated by my vague idea of imbedding the Riemann zeta function in a larger mathematical framework.

---

[2]Lindelöf's hypothesis is closely related to the Riemann hypothesis, which is the most famous open problem in the history of mathematics.

Regarding the sciences, particularly physics, Richard Feynman clearly expressed the fact that breakthroughs are not the result of well-defined goals. He wrote, 'Whatever way comes out, it's nature, and she is going to come out the way she is! Therefore, when we go to investigate it, we shouldn't pre-decide what it is we're going to find'.

Creativity is further discussed in my forthcoming book, *From Ancient Greece to Viennese Modernism: Lessons in Creativity*, and in Boden (2003).

6.4 From Platonism to AI

AI provides the apotheosis of mathematics, of algorithms, and in general of reason . In this sense, it represents the strongest possible endorsement of the key elements of Platonism. As noted in the epilogue of Fokas (2024), the overreliance of Plato on reason led him to elevate theoretical constructions and, in particular, his Forms *above* reality. For example, according to Plato, the stars are less important than their orbits. He argued that stars can be observed directly, so they are susceptible to the misleading information gathered by the 'unreliable senses', whereas their orbits can be inferred via the 'perfection of theoretical reasoning'. He wrote that 'the stars […] are far inferior […] to the orbits that carry them, which are perceptible to reason and thought […]'. Plato, pushing the importance of the rational to its ultimate limit, reached, in my opinion, the erroneous conclusion that the *only* 'true reality' consists of certain rational metarepresentations. Indeed, according to him, his disembodied Forms were the 'essence of things', whereas actual sensory experiences were simply 'shadows of reality'.

Incidentally, a first look at quantum mechanics appears to endorse this Platonic position. As discussed in detail in Chapter 14, solutions of the so-called Schrödinger equation provide the only way not only to quantify but also to *define* certain quantum quantities. However, there is a difference between the *understanding* of a phenomenon and its *existence*. Namely, the crucial role of the mathematical formalism for understanding physics at micro-scales does not negate the existence of the reality of this strange world, with respect to our usual experiences.

I believe that in the same way that Plato elevated certain abstract notions above reality, some scholars have elevated *reductionism* and *computability* above the embodied brain and the astounding processes associated with it. These scholars claim that any phenomenon can be fully understood by its reduction to appropriate constituent elements, which can then be simulated via powerful computers. For example, the MIT expert on robotics, Rodney Brooks, after presenting the 1955 quote of John McCarthy,

> "the conjecture that every aspect of learning or any other feature of intelligence can in principle be so precisely described that a machine can be made to simulate it",

he states, 'As a materialist reductionist, I agree with that' (Brooks, 2019). However, contradicting this position, he later admits that the current computational approaches are inadequate. Actually, the entire above article by Brooks is about possible ways to overcome the limitations of AI. For example, Brooks states, 'if we are pushing things into the information metaphor, are we missing things?' And after reviewing certain behavioural experiments involving flatworms, he notes that neurons, in contrast to current AI computations, are adaptive: 'it's some sort of adaptation and our computation is not locally adaptive'. Later, after correctly noting that the discoveries related to computations and calculus were 'natural', in contrast to the extraordinary 'conceptual jumps' needed for quantum mechanics and general relativity, he states, 'The question is whether there is something like that out there that could potentially give us a better [information]metaphor'.

It seems that proponents of reductionism and computability are biased by their knowledge of *physics, where there exist fundamental physical laws, which take the form of elegant mathematical equations.* Reductionists start with the assumption that matter arises from combining building blocks. Then, they argue that our knowledge of the fundamental physical laws allows us to gain a full understanding of these constituent parts. Furthermore, they claim that the scrutiny of these parts can yield understanding of the whole. Several physicists have questioned the validity of this claim. For example, Philip Anderson (Nobel Prize in Physics, 1977) in 'More is Different' (Anderson, 1972), states:

> "the ability to reduce everything to simple fundamental laws does not imply the ability to start from those laws and reconstruct the universe. Actually, the more the particle physicists tell us about the fundamental laws, the less relevant seem [these laws] to be to the very real problems of the rest of science, much less to those of society".

The reductionist approach relies on the notion of *locality,* i.e. the assumption that, in the neighbourhood of a given element, only local interactions are important. However, this assumption is not always valid. For example, several phenomena in quantum mechanics are characterised by long-distance interactions.

I believe that the application of reductionism in biology is even more problematic than in physics. First, it is argued in Fokas (2024) that, in contrast to the existence of fundamental physical theories, the 'theory of biological evolution' is not a rigorous theory but merely provides a useful framework for understanding biology. In particular, physical theories are predictive, whereas the theory of evolution is not. *The modus operandi of biological evolution appears to be 'trial and error'. Moreover, new solutions are usually sought within the constraints of what has already been constructed.* Evolution's trial-and-error approach explains the abundance in biology of *redundancy*, namely the fact that the same task can be accomplished, with varying degrees of efficiency, via completely different mechanisms. Evolution invents new improvements without necessarily abandoning the use of older inventions. *This gives rise to overlapping systems of high redundancy and immense, 'illogical' complexity.* An example of this process is provided by the plethora of completely different mechanisms used for neuronal plasticity and employed for learning, as discussed in Chapters 9 and 10 of Fokas (2024) as well as in the earlier chapter.

This crucial feature of biological evolution provides a strong argument against the position that organisms are the result of a 'grand design'. Moreover, it apparently eliminates the possibility that fundamental biological processes follow *global fundamental laws.* Since mathematics reveals its power precisely in relation to such laws, these arguments suggest that perhaps mathematics, in particular, and modelling, in general, cannot have the pivotal impact in biology that they have had in physics. Importantly, biological interactions do not generally satisfy the assumption of locality.[3]

The phenomenon of 'emergence properties' has been observed in many physical and biological circumstances. Namely, new patterns or

---

[3] Another difference with physics is that biology does not follow the model of 'scientific revolutions', as described by Thomas Kuhn. This was emphasised by the leading evolutionary biologist Ernst Mayr (1904–2005) in *The Growth of Biological Thought* (Mayr, 1982).

properties appear spontaneously in complex systems that are not present in their individual parts. For example, in many dynamical systems, order can arise from chaos or chaos can appear from ordered behaviour after a sufficient number of individual components are combined. Emergentists correctly advocate that complex systems cannot always be described in terms of their constituent parts.

The above remarks suggest that the dogmatic apotheosis of reductionism and computability should perhaps be replaced by a more general framework. Such a framework was introduced in Fokas (2024) and is based on notions which reflect basic neuronal mechanisms, including associations, continuity, generalisation, abstraction, and interconnectedness.

6.5 Godel's Incompleteness Theorem and Unconscious Processes

A decisive argument in favour of the superiority of human thought is provided by the fact that consciousness is much richer than both mathematics and computation. As argued in Fokas (2024), this is a direct consequence of Gödel's *incompleteness theorem*, which states that there exist infinitely many true statements which cannot be formally verified via mathematics, whereas some of these statements are intuitively obvious. Humans seem able to recognise the truth of certain mathematical statements by stepping outside the 'formal approach' used in mathematics and adopting a 'common sense' way of thinking.

Philosopher John Lucas and physicist Sir Roger Penrose have presented convincing arguments that Gödel's theorems demonstrate that human understanding cannot be fully captured by computational algorithms.

Importantly, Turing proved that this very serious limitation cannot be overcome via mechanical computations. Interestingly, Doukas Kapantais (2016) has shown that an 'idealised human machine' *can* compute a specific function (called the Ackerman function), while this function fails to be computed by a Turing machine.[4]

---

[4]The failure of the Turing machine is due to its inability to compute functions whose domain is numbers and whose values are functions.

Finally, and perhaps most importantly, artificial algorithms are based on logical constructions relying on the conscious function of parts of the cerebral cortex. However, researchers in AI ignore that, in addition to the crucially important fact that every conscious experience is preceded by an unconscious process, importantly, the passage from the unconscious to consciousness is accompanied by a loss of information.

Regarding the relation between AI and the unconscious, the following statement by the leading neuroscientist Christof Koch is, in my opinion, both surprising and misleading: 'Unconscious processing is far less mysterious than consciousness; it is after all what computers do' (Koch, 2012, 89). This statement is surprising because, as was argued in Fokas (2024), unconscious and conscious processes form a continuum. This implies that it is impossible to resolve the mysteries of consciousness without simultaneously elucidating the unconscious. It is misleading because, although Koch is of course correct that computers mimic only the first steps of perception, which is indeed implemented in the unconscious, even during this first stage the embodied brain is fully engaged in a variety of ways (which include emotions), whereas the computer is *not*. Actually, AI is based on computational tools, and hence, as stated earlier, it only mimics aspects of the brain's conscious processes; AI cannot address underlying unconscious mechanisms. In this sense, computer algorithms are closer to conscious processes than to unconscious ones. Hence, the question of 'what computers do' is more pertinent to consciousness!

In summary, *human thought, in contrast to artificial algorithms, is created via the interaction of conscious and unconscious processes in the dynamic environment of the embodied brain and is crucially affected by hormones, neuromodulators, and a variety of unconscious homeostatic mechanisms. The process of our thinking is critically affected by our continuously changing culture and our ability to construct metarepresentations.*

The above arguments suggest that human thought is much broader than mathematics, the sciences, and technology; therefore, I do not expect that it can be replaced by AI. This point of view is consistent with positions expressed by several other scholars; see for example Polson and Scott (2019), Fry (2019), and Stewart (2024). However, apparently, deep learning possesses a remarkable novel feature, as discussed in the following, which surprisingly has not been emphasised.

6.6 A New Type of Intelligence

Even strong proponents of the unlimited potential of deep learning appreciate several of its current limitations in comparison to human thought. In particular, neural networks cannot handle unanticipated problems, and in general, they cannot discover novel solutions unless they have been 'trained' with similar problems. For example, since datasets for training self-driving cars contain relatively few examples of extraordinary situations, what will be the reaction when a deer leaps across the car? Additional limitations include: the inability to represent in the computer a complete 'world model', as well as everyday knowledge; the narrow capabilities of AI, namely that current machines cannot implement different tasks; and the inability of AI to transfer knowledge from one domain to another.

The relatively narrow focus of current AI systems is exemplified by the fact that GPT-4 can create beautiful text but is useless when it comes to driving a car.

Regarding the question of transferring knowledge, Sifakis raises an even more fundamental issue: he claims that the computer *transforms* but *cannot create* information. According to this well-known computer scientist, creating information is the prerogative of the brain. By 'information', Sifakis means 'semantic information', namely that which carries meaning, in the sense analysed by Turing. This should be contrasted with other types of information, such as 'syntactic information' and 'transferring information'; the latter determines what type of channel is needed for propagating a specific amount of information with a given accuracy (this was studied by the mathematician and electrical engineer Claude Shannon). In my opinion, this position needs clarification. Evidently, the computer cannot generate information in an autonomous manner as done by the human brain. However, after it is programmed, a computer *can* generate not only new information but also knowledge. For example, GPT can give useful financial advice, and AlphaFold has greatly enhanced our biological knowledge by determining the geometrical structure of millions of proteins.

Interestingly, deep learning is characterised by a paradox: *starting with no knowledge creates new knowledge*. Consider, for example, the problem of building a robot to drive a bicycle. Using AI, this problem can be approached in two different ways. The first approach is based on modelling the movement of an object using the laws of classical mechanics. The formulation of this model requires knowledge of the laws of motion. Moreover, the construction of the associated algorithm demands mathematical computations of high complexity. The second approach, based on deep learning, does not involve modelling and therefore does not require knowledge of Newtonian mechanics. The role of the model is now replaced by the neural network itself and its parameters. Despite this lack of any input knowledge, the associated neural network, after sufficient 'training', can drive the bicycle at least as well as the first robot!

The unexpected success of deep learning provides a clear illustration of the ingenuity of nature. A child develops various skills, including linguistic ability and the cognitive capacity to manipulate complicated concepts, not in an analytic way by understanding the mathematical formulation of basic physical laws and by following specific algorithms, but in a holistic manner via the process of *imitation*. Deep learning is successful precisely because it is closer to the way that the brain learns than the earlier logic-based forms of AI. Similarly, elements of deep reinforcement learning mimic the important psychological processes of positive reinforcement used by the brain.

It is emphasised in Fokas (2024) that a crucial component of learning is the continuous elaboration of *associations*. These associations are of fundamental importance for deciphering relations between different entities; in this way, they facilitate the process of understanding. Remarkably, deep learning, by analysing huge sets of data, has the capacity to establish opaque associations which are *hidden to the human mind*. By exploring these associations, the new AI systems identify patterns and relationships in data structures which were previously invisible. Harnessing these relations, it is then able to solve complicated problems. I find this extremely interesting because it defines *a new type of intelligence*.

Apparently, AI perceived a reality which is hidden to the human brain. How is this achieved? Unfortunately, no one knows; it does not seem possible that, by analysing a neural network step by step, we will be able to understand, let alone predict, the outcome. Actually, the more elaborate the structure of a neural network, the less probable it is that we will ever understand how this new type of intelligence operates. Clearly, it is impossible to decipher the *modus operandi* of a neural network involving trillions of parameters. The higher the level of a neural network's complexity, the opaquer it becomes.

This new type of intelligence is tremendously exciting but at the same time worrisome. It is exciting because it opens new ways of unprecedented achievements. On the other hand, until now, scientists and engineers have been able to explain and predict how a system worked, and they have been able to consider the output of machines as natural extensions of their input. But this is not the case anymore.[5]

The inability to understand this new intelligence is highly unfortunate. First, we will increasingly have to rely on systems that operate in an unpredictable manner. If we are unable to trace the origin of each individual decision made by an AI system, should we implement its recommendations? If we do not, how can we ensure that we are not eliminating a solution to a given problem which is much superior to our recommendation? More importantly, will we be able to detect truly catastrophic recommendations or reverse wrong choices? Second, we will not be able to achieve deep insight using the remarkable results of deep learning. For example, using the algorithm developed by Floudas and his group on protein folding (Subramani and Floudas, 2012), it was possible to make predictions with only 85% accuracy for a special class of proteins called sandwich proteins. However, this work was based on the discovery of specific geometrical rules (Fokas *et al.*, 2005) which enhanced our insight into the topological interactions of these particular proteins. What is the biological insight gained by the astounding success of AlphaFold?

Our inability to understand how this 'new intelligence' operates is illustrated by the following paradox: the newest and most powerful technologies, called reasoning systems, developed by several companies, including OpenAI, Google, and the Chinese start-up DeepSeek, for unknown reasons, generate more errors than the earlier, less powerful, algorithms. For example, according to tests performed by OpenAI itself, its new o4 system 'hallucinated', i.e. made up answers, 48% of the time when running the benchmark test called PersonQA; this involves answering questions about public figures. The earlier version, o3, hallucinated 33%, and the first version, o1, only about 16%. The corresponding numbers for the test SimpleQA, which asks more general questions, were 79%, 51%, and 44%. It is clear that such systems cannot be used for important purposes, such as preparing court documents, managing sensitive business data, and making medical decisions.

It is truly astonishing that deep learning exhibits remarkable 'creativity'. For example, as discussed in Chapter 1, the Deep Q-Network

---

[5] In addition, because AI systems can perceive subtle characteristics that humans do not, occasionally this leads to errors. For example, an animal was perceived by a system based on deep learning as a gun.

'understood' that a useful alternative to knocking out rows is knocking out columns, and AlphaGo 'invented' the brilliant 'move 37'. AlphaZero, by 'recognising' patterns of moves across vast sets of possibilities which human minds cannot anticipate, 'concluded' that sacrificing the queen was the most likely move to lead to victory.

After asking ChatGPT to write a poem about the Fokas method, it wrote:

"In the realm of math, where equations abide, a method arose, a true breakthrough. Fokas, the creator, bequeathed us his art to tackle equations that tear us apart. Partial differential equations, with many variables changing, all interlaced, heat, sound, and fluid, the real-world they show. Their secrets uncovered, through Fokas' tableau. A toolbox so clever, transforming equations to new forms, gentler and keener. The answers we seek, in this new realm we find, a journey through math, with clear, focused mind. As solutions emerge, the journey rewinds, to original form, the answers now bind. Versatile and elegant, this method prevails, in unravelling mysteries, Fokas unveils".

It would be marvellous if this 'new type of intelligence' could be deciphered since it would open the way for far-reaching discoveries.

The more transformative the impact of AI on our civilisation, the greater the imperative that scientists and engineers involved with deep learning resist the ongoing hype and remain humble. A mature researcher, although aware that the nature of highly complex physical and artificial phenomena, such as how the computer 'understands' important relationships via the analysis of 'big data', can never be fully knowledgeable continues to live a life dedicated to searching. This will inevitably lead to further progress and occasionally to breakthroughs. In this way, they will feel not only deep *eudaimonia* but also awe for the inventiveness of nature as expressed in its greatest creation, which remains the human brain.

Incidentally, Democritus, the Stoics, Cicero, and many other philosophers have advocated avoiding overexcitement that can obscure logic and judgement. Particularly, Democritus introduced the term *athambia* (lack of amazement), later embraced by Cicero. However, in my opinion, it is impossible not to be amazed by nature and the abilities of the human brain. Moreover, the feelings accompanying this awe provide a great motivation for the endless search to uncover as many as possible of nature's secrets, as well as its astonishing creations.

References

Anderson, P. W. (1972). More is different: Broken symmetry and the nature of the hierarchical structure of science. *Science, 177*(4047), 393–396.

Boden, M. (2003). *The Creative Mind.* Oxfordshire: Routledge.

Brooks, R. A. (May 2019). The Cul-de-Sac of the computational metaphor. *Edge.*

Damasio, A. (2012). *Self Comes to Mind.* Vancouver: Vintage Books.

Dassios, G. and Fokas, A. S. (2020). *Electroencephalography and Magnetoencephalography: An Analytical-Numerical Approach,* Vol. 7. Walter de Gruyter GmbH & Co KG.

Davis, E., Morgenstern, L. and C. Ortiz. (n.d.). The Winograd Schema Challenge. https://cs.nyu.edu/faculty/davise/papers/WinogradSchemas/WS.html (Accessed 2019).

Einstein, A. (1985). A letter to Jacques Hadamard. In Ghiselin, B. (ed.), *The Creative Process: A Symposium.* Berekeley: Iniversity of California Press.

Fokas, A. S. (2019). A novel approach to the Lindelöf hypothesis. *Transactions of Mathematics and Its Applications, 3*(1), tnz006.

Fokas, A. S. (2024). *Ways of Comprehending: The Grand Illusion and the Essence of Being Human.* Singapore: World Scientific.

Fokas, A. S., Kurylev, Y., and V. Marinakis. (2004). The unique determination of neuronal currents in the brain via magnetoencephalography. *Inverse Problems, 20*(4), 1067.

Fokas, A. S., Papatheodorou, T. S., Kister, A. E., and Gelfand, I. M. (2005). A geometric construction determines all permissible strand arrangements of sandwich proteins. *Proceedings of the National Academy of Sciences, 102*(44), 15851–15853.

Fokas, A. S. (in preparation). *The Golden Ages of Athens and Vienna.* Singapore: World Scientific.

Fry, H. (2019). *Hello World, How to Be Human in the Age of the Machine.* New York: Doubleday.

Gelepithis, P. (2024). *Unification of Artificial Intelligence and Psychology: Volume One - Foundations.* London and New York: Palgrave Macmillan.

Hashemzadeh, P., Fokas, A. S., Saratov, A., Demange, J., and Chec, L. (2020). A surrogate model of the forward problem in EEG imaging (submitted).

Kapantaïs, D. (2016). A refutation of the Church-Turing thesis according to some interpretation of what the thesis says. In *Computing and Philosophy: Selected Papers from IACAP 2014,* pp. 45–62.

Koch, C. (2012). *Consciousness: Confessions of a Romantic Reductionist.* Cambridge: MIT Press.

Kurzweil, R. (2005). *The Singularity Is Near: When Humans Transcend Biology.* New York: Penguin.

Mayr, E. (1982). *The Growth of Biological Thought: Diversity, Evolution, and Inheritance.* Cambridge: Harvard University Press.

Polson, N. and Scott, J. (2019). *AIQ, How Artificial Intelligence Works and How We Can Harness Its Power for a Better World.* London: Bantam Press.

Sifakis, J. (2020). *Understanding and Changing the World* (in Greek). Athens, Greece: Armos.

Stewart, W. (2024). *The Human Biological Advantage over AI.* Springer: AI & Society.

Subramani, A. and Floudas, C. A. (2012). β-sheet topology prediction with high precision and recall for β and mixed α/β proteins. *PloS One,* 7(3), e32461.

Tegmark, M. (2017). *Life 3.0: Being Human in the Age of Artificial Intelligence.* New York: Knopf.

Part II

Medical Breakthroughs

The past couple of decades have been marked by tremendous progress in medicine. In particular, the use of technology continues to have a decisive impact on the monitoring and diagnosis of various disorders. For example, wearable digital health technologies can be used for recording pulse rate, oxygen saturation, and rhythm assessment; this is particularly helpful for identifying the rather common arrythmia, atrial fibrillation (Spatz *et al.*, 2024). In addition, the use of cardiac implantable electronic devices represents a major breakthrough in the management of heart rhythm disorders. Such devices include pacemakers for patients suffering from bradycardia and other rhythm abnormalities, and implantable cardioverter defibrillators capable of preventing sudden death from ventricular arrythmias (Al-Khatib, 2024).

Additional examples of wearable devices include the following: a wristband that can detect a possible seizure by combining information from sensors monitoring heart rate, muscle contraction, movement, oxygen saturation, and skin conductance, following which it wirelessly notifies a smartphone (Donner *et al.*, 2024); a wristwatch that measures the duration spent in the different phases of sleep, which is useful for monitoring patients suffering from depression.

The combination of deepening understanding of key physiological processes and the continuous progress in technological innovations leads to major discoveries with direct clinical impact. For example, the availability of rapid and long-acting insulin formulations, continuous glucose monitoring, and the availability of a hybrid closed-loop system for delivering insulin has led to the current situation where insulin therapy is closer to mimicking the endogenous insulin profile. Currently, the emphasis is on engineering 'glucose-responsive insulin', i.e. insulin which is active when blood glucose level increases and inactive when the level is in the normal range (Cefalu and Arreza-Rubin, 2024).

In this part of the book, we review some of the major developments in immunology, cancer, gene therapy, and interventional approaches to treating patients suffering from an acute neurological stroke. The last topic is discussed together with analogous approaches used in the treatment of a heart attack; this illustrates the importance of the concepts of analogical thinking and unification, as discussed in Fokas (2024). Additional examples of a unified approach to medical disorders are also discussed, including the possible role of inflammation in major depression.

Metchnikoff's ideas regarding the role of the immune system in maintaining homeostasis were discussed in Chapter 5 of Fokas (2024). These ideas provide the basis of the so-called 'innate immunity', which was rediscovered, extended, and placed in the proper molecular biological framework in the late 1990s. Until then, the prevailing concept regarding immunity was the so-called 'adaptive immunity', namely immunosurveillance against 'foreign invaders'. The new paradigm shift led to truly astounding developments, including the discovery of various proteins, such as interferons and interleukins, which can be considered the 'hormones of the immune system'. These recent breakthroughs, as well as the discovery of monoclonal antibodies and their huge impact on the treatment of various autoimmune diseases, are discussed in Chapters 7 and 8.

Chapters 9 and 10 review the recent astounding progress in the treatment of several types of cancer. In particular, the effect of specific 'biological therapies' is discussed, and the different approaches to enhancing the immune response against cancerous cells are explained. These include the following: the use of specific monoclonal antibodies in general and antibodies against 'programmed death' receptors in particular; the engineering of CAR-T lymphocytes; and the employment of 'checkpoint inhibitors.' The current armament against cancer does not have the same impact on all solid tumours. For example, the prognosis for melanoma and lung adenocarcinoma is now much better than before the introduction of the new therapies, but progress in the treatment of colon and ovarian cancers has lagged behind. The impact of the new therapies on the treatment of several cancers is briefly discussed in Chapter 10.

Chapter 11 emphasises the vital role of the concepts of *unification* and *analogical thinking* for the understanding and treatment of several common medical syndromes. This follows from the fact that many different physiological and pathological processes express themselves in a similar manner. An important example is provided by the manifestation of *inflammation* and the underlying aetiological impact of *immunology*. The possible role of immune mechanisms in the occurrence of major depression is also discussed. An illustrative example of the fact that pathological processes of common origin require similar medical interventions is the emergence, in the past few years, of a similar approach to the treatment of *neurological strokes* as that used for treating *heart attacks*. Protein misfolding provides a striking manifestation of the fact that a common aetiological cause, namely topological defects in the three-dimensional structure of proteins, gives rise to a variety of different disorders.

Chapter 12 reviews the recent breakthrough regarding our ability to modify DNA in a precise manner. This is an exciting and, at the same time, extremely dangerous development. On the one hand, it is already possible to introduce a gene to patients lacking a specific protein, which can produce a sufficient amount of the missing polypeptide. On the other hand, there is the risk of using this powerful new development for all the wrong reasons, including efforts towards 'eugenics'. We might create new strains of viruses or bacteria that could wipe out all of humanity. Establishing regulations mitigating such risks is, literally, a matter of life or death.

It is worth noting that despite the enormous progress made on a number of different medical fronts, several well-defined questions, surprisingly, remain open. For example, in the early 1980s, results from clinical trials reported improved survival with long-term beta-blocker treatment following a heart attack. However, recently, the use of beta-blockers has been questioned. Actually, a study involving 5,020 patients with myocardial infarction but without heart failure (more precisely, with left ventricular ejection fraction of 50% or more) showed that beta-blockers therapy did not have any benefits (Yndigegn *et al.*, 2024); unfortunately, this was an open-label trial with no placebo group, and hence the question regarding the usefulness of beta-blockers remains open.

The US has played a crucial role in most of the above-mentioned astonishing developments. Unfortunately, the entire medical-scientific infrastructure supporting these astounding advances is currently under threat. This includes the National Institutes of Health (NIH), which has been one of the main engines driving biomedical breakthroughs. In addition, numerous executive orders, signed by President Trump during the first 100 days of his second presidency, have led to the dismantling of US foreign assistance and global health efforts. The new policies, in addition to already having severe adverse effects on vulnerable populations worldwide, will undoubtedly have serious implications for people in the US.

In 1948, the World Health Organization (WHO) was created with the influential participation of US representatives. Global collaborative efforts, many under the auspices of the WHO, have had an extremely beneficial effect on health; from the eradication of smallpox and the near eradication of polio to the recognition of the relationship between infectious diseases and malnutrition. The US has now withdrawn from the WHO! In 1961, John F. Kennedy established the U.S. Agency for International Development (USAID). The impact of USAID has been dramatic, from stemming pandemic-causing infectious diseases, including severe respiratory syndrome, Zika, Ebola, and COVID-19, to the discovery and implementation of oral hydration therapy, which saved the lives of millions of children. USAID has now been eliminated. Cutting USAID, in addition to pushing millions of people into poverty, is estimated to lead to at least 369,000 child deaths each year (Osendarp *et al.*, 2025).

The US has been the world leader in addressing emerging medical challenges. For example, the President's Emergency Plan for AIDS Relief (PEPFAR), which was announced early in the HIV epidemic, and the NIH's role in promoting global scientific collaborations, led to the early identification of the associated virus and the development of life-saving treatments. It has been estimated that PEPFAR has saved 26 million lives (Duggan and Bhutta, 2025).

The massive cuts to US federal funding and the federal workforce related to health and disease prevention have compromised the US' ability to respond to outbreaks, have significantly affected veteran care, and, importantly, will have a long-term negative impact on medical research, from basic scientific studies to clinical trials. Half of the Centers for Disease Control and Prevention (CDC) have been closed, and 1,500 CDC staff have been fired. These reductions undermine the CDC's mission, namely, its ability to protect America from any health threats. For example, the CDC was unable to respond when Wisconsin sought help related to lead poisoning; this is, apparently, the first time the CDC has not been able to respond to a state's request for epidemiological assistance. The CDC is directly involved with PEPFAR. The Trump administration's fiscal 2026 budget proposes to dramatically reduce the funding for PEPFAR, as well as programmes for malaria prevention, global tuberculosis control, and the control of various tropical diseases. In addition, it proposes to close environmental health programmes (Frieden, 2025). These changes have already compromised food and drug safety and will dramatically reduce America's global efforts to save lives related to HIV, tuberculosis, malaria, and many other global threats. In addition, there will be reduced information to guide education and to implement regulations of emerging environmental threats, such as the dangerous effects of nanoparticles and microplastics.

In May 2025, US Secretary of Health and Human Services Robert F. Kennedy announced that COVID-19 vaccines had been removed from the CDC's recommended immunisation schedules for healthy children and pregnant women. This was the first time that such an announcement was not made by the CDC Advisory Committee on Immunization Practices. A couple of weeks later, Kennedy announced that all 17 members of this committee had been removed; they were replaced by vaccine sceptics. Since 1964, this committee has been responsible for evaluating evidence for FDA-approved vaccines. In this regard, it is noted that despite the fact that hundreds of Texans, mostly young unvaccinated children, contacted measles in the spring of 2025 and two children died, Kennedy refused to unequivocally endorse the measles-mumps-rubella vaccine.

If the current policies of the US are not reversed, their impact, worldwide, will be truly devasted.

References

Al-Khatib, S. M. (2024). Cardiac implantable electronic devices. *New England Journal of Medicine, 390*, 442–54.

Cefalu, W. T. and Arreza-Rubin, G. (2024). Engineering an insulin complex to treat diabetes. *New England Journal of Medicine, 390*, 2214–16.

Donner, E., *et al.* (2024). Wearable digital hearth technology for epilepsy. *New England Journal of Medicine, 390*, 736–45.

Duggan, C. and Bhutta, Z. (2025). Putting America First—Undermining health for populations at home and abroad. *New England Journal of Medicine, 392*, 1769–1771.

Fokas, A. (2024). *Ways of Comprehending: The Grand Illusion and the Essence of Being Human.* World Scientific.

Osendarp, S., *et al.* (2025). The full lethal impact of massive cuts to international food aid. *Nature, 640*, 35–7.

Spatz, E., Ginsburg, G., Rumsfeld, J., and Turakhia, M. (2024). Wearable digital health technologies for monitoring in cardiovascular medicine. *New England Journal of Medicine, 390*, 346–356.

Yndigegn, T., *et al.* (2024). Beta-blockers after myocardial infarction with preserved ejection fraction. *New England Journal of Medicine, 390*, 1372–81.

Chapter 7

Breakthroughs in Immunology

7.1 Adaptive Immunity

As noted in Chapter 5 of Fokas (2024), Metchnikoff's ideas were ignored for many years, and the prevailing concept regarding immunity was the so-called 'specific or adaptive immunity' (Germenis, 2019). According to this dogma, particular types of white blood cells, called *T-lymphocytes*, *B-lymphocytes*, and *macrophages*, possess specialised receptors. These receptors have the ability to recognise particular proteins, called *antigens*, which are located on the surface of various foreign cells and on pathogens. Following intricate interactions between these three types of cells, the B-lymphocytes finally produce *antibodies* that attack cells carrying the antigens. In addition, certain subtypes of T cells can also attack pathogens directly.

The pillar of this concept is the 'clonal theory', proposed in the mid-1950s by Frank Macfarlane Burnet. This theory described a highly elaborate process of how specific lymphocytes can recognise specific antigens, and how this leads to the *clonal expansion* of specific B-lymphocytes. Through this process, daughter B-lymphocytes arise from a parent cell, and thus many copies of a given B cell are produced that are specific for a given antigen. T cells are born in the thymus. The importance of the thymus in the immune system was discovered in 1961 by the French-Australian researcher Jacques Miller, who surgically removed the thymus from one-day-old mice and found that this led to a deficiency of a lymphocyte population; these cells were later called T lymphocytes, after the organ of their origin (Miller, 1961a, 1961b). Until this discovery, the thymus was considered by many to be an 'evolutionary accident' of no functional importance.[1]

A particular type of stem cells, called CD34+ stem cells, migrate to the thymus, where they differentiate and acquire their unique identity, in the sense that each T cell expresses its own unique receptor. The abundance of different T cells is calculated on the evidence that each T cell expresses its own unique T-cell receptor (TCR), composed of two chains, called TRCα and TRCβ. Each of these chains is generated through the process of recombining multiple genetic elements. This combinational diversity leads to an astonishingly large number of receptors, of the order of 10 billion. The importance of T cells becomes evident through the multiple disorders associated with HIV infection, which compromises their function.

A crucial component of the clonal theory is the postulate that the immune system can distinguish between self and foreign antigens. Indeed, soon after their genesis in the thymus, T cells undergo 'negative selection'. This is a filtering step that removes T cells that have a strong binding affinity to self-proteins. In this way, mature lymphocytes develop 'immune self-tolerance', i.e. the capacity to react only to foreign antigens and to ignore host tissues. In general, the term 'immune tolerance' refers to the prevention of an immune response against a particular antigen. Immune self-tolerance is the capacity of the immune system to be tolerant of self-antigens, so it does not attack the body's own cells, tissues, and organs. When this tolerance is compromised, disorders such as autoimmune diseases or food allergies may occur. The first step in achieving self-tolerance, called 'central tolerance', takes place in the thymus. A particular type of T cells, called regulatory T (Treg) cells, recognise self-peptides; when activated, Treg cells inhibit those T cells that are activated by self-peptides. This complex dynamic interaction of self–nonself discrimination is the basis of immune self-tolerance. Many self-reactive T cells escape thymic negative selection. This necessitates peripheral mechanisms to ensure that self-tolerance is maintained. 'Peripheral self-tolerance' will be discussed later in this chapter.

Remarkably, until the end of the 1980s, immunology had concentrated exclusively on lymphocyte-mediated immunity against foreign

---

[1]Galen of Pergamon claimed that the thymus is 'the seat of the soul'. On the other hand, in 1961, the Nobel Laureate Sir Peter Medawar wrote that the thymus 'is an evolutionary accident of no very great significance' (Taylor, 2023).

antigens. According to Cohen, this rather restrictive theory 'describes a model of autistic lymphocytes that can only hear the voice of their receptor' (Cohen *et al.*, 2004).

7.2 Innate Immunity

In more recent years, the earlier incomplete understanding of immunity has been greatly enhanced with the development of the 'danger model', advocated by Polly Matzinger. This model is based on Charles Janeway's insight that the fundamental purpose of the immune system is to regulate 'processes that can cause damage, as opposed to attacking only what is foreign' (Matzinger, 1994). This new concept, which was initially either ignored or largely rejected, precipitated a paradigm shift and led to the study of 'natural or innate immunity'. This is the study of *generic mechanisms used by the organism for identifying and reacting to 'danger signals.'*

The modern view of the immune response emphasises its dynamic nature and its capacity to deal with *both endogenous and exogenous dangers*. Immune mechanisms have two key functions: first, to maintain *normal-tissue homeostasis*, and second, *immunosurveillance* against pathogens and cancer cells. Importantly, it is now recognised that dysmetabolic conditions, including diabetes and obesity, disturb normal-tissue homeostasis and hence can also elicit immune responses. Specifically, lipotoxicity, i.e. the abnormal accumulation in obese individuals of harmful lipid mediators in fat cells, liver, and muscles, results in cellular stress and tissue dysfunction; this leads to subclinical inflammation. In this case, a homeostatic perturbation initially induced by lipotoxicity yields inflammation, which may be perpetuated, leading to a vicious cycle. Similarly, hyperglycaemia can lead to glucose toxicity and tissue damage, which, in turn, can also lead to inflammation.

In homeostasis, an important source of 'danger signals' is cellular debris, which needs to be removed. Interestingly, well before the emergence of the 'danger theory', the Greek-born immunologist Stratis Avrameas, along with others, demonstrated the existence of autoreactive B cells, which produce autoantibodies (Avrameas *et al.*, 2018). This particular type of antibody is *poly-reactive*, meaning that it recognises a variety of antigens, including self-antigens. Under physiological conditions, the receptors of the autoantibodies bind to self-antigens, forming autoantigen–autoantibody complexes, which are subsequently eliminated from circulation, primarily via phagocytosis. Fundamental homeostatic biological activities, such as cellular repair and enzymatic catalysis, are monitored by a variety of different auto-poly-reactive immune receptors. In addition, in disease states, including demyelination, stroke, and certain pulmonary diseases, autoantibodies lead to the rapid sequestration of the self-antigens of the defective cells.

Auto-poly-reactivity endows the immune system with the readily available capacity to recognise and interact with self and environmental constituents. This innate mechanism gives rise to the formation of a vast network that plays a vital role in immune homeostasis. This highly dynamic network is continuously affected by stimuli arising from the internal and external milieu of the organism. In particular, this network is crucially affected by the human microbiota (Fokas and Ktistakis, in preparation). Studies regarding this network, which clearly contradicted the dogma of clonal theory, were for many years largely ignored.

The role of the immune system regarding its vital function for normal homeostasis is being continuously elucidated. An example is provided by the recently understood role of macrophages in engulfing and removing decaying or spent mitochondria. This is particularly important for high-energy, durable cells, which are mitochondrially enriched. Such cells include retinal rods and cones, skeletal muscle cells, and cardiomyocytes. Remarkably, each cardiomyocyte normally touches, on average, five macrophages. Genetically modified mice without such macrophages have abnormally functioning mitochondria, which leads to cardiac dysfunction (Nicolás-Ávila *et al.*, 2020).

As stated earlier, in addition to homeostasis, the second important role of the immune system is immunosurveillance. In particular, the immune system is highly effective against the most important pathogens, namely bacteria, viruses, and parasites; in addition, it is also effective against cancer cells. For both functions of the immune system, there is a delicate balance between, on the one hand, the need to develop potent effector cells to eliminate tissue debris as well as to combat dangerous pathogens and cancer cells and, on the other hand, the need to develop immune self-tolerance so that autoimmune inflammation is avoided.

An example of the ongoing elucidation of mechanisms used in innate immunosurveillance is the understanding by Zhijian Chen and his group of the ability of our cells to detect foreign DNA; this is of crucial importance for initiating an immune response against viruses (Sun *et al.*, 2013;

Wu *et al.*, 2013). Foreign DNA activates a specific enzyme called cyclic GMP-AMP synthase (cGAS). This enzyme binds to a protein called stimulator of interferon gene (STING), and this leads to the production and secretion of type I interferon, which is a specific protein (to be discussed later in this chapter) that orchestrates antiviral immunity. It is conjectured that the presence of viral DNA in the cytosol, in contrast to our DNA which resides in the nucleus, is the reason that cGAS responds only to the viral DNA without being activated by the abundant DNA naturally present in our cells.

The capacity of antibodies and phagocytic cells to clear cell debris as well as pathogens and cancer cells from the body and to enhance inflammation is significantly augmented by the *complement system*. This is an evolutionarily conserved key player in humoral innate immunity, consisting of approximately 30 proteins, most of which are produced in the liver.

These proteins circulate in a precursor form and are activated either by an infection or by an inflammatory process. The complement proteins help 'label' pathogens and damaged cells throughout the body, affixing them with specific protein tags, which finally results in the formation of large macromolecules. In this way, the 'key effectors' of the immune system, namely the macrophages, can identify the tagged cells as 'cellular trash' and eliminate them. Using this multicomponent mechanism, the complement system modulates tissue homeostasis and also contributes to immune surveillance. The activation of the complement cascade is inhibited by a group of proteins called *complement inhibitors*.[2] The complement system includes proinflammatory components (such as C_3, C_4, C_9, and factor B), as well as negative regulators (such as C_1 inhibitor).

Importantly, during development, the complement system participates in the fundamental physiological mechanism of synaptic pruning; this is achieved via the activation of microglia and is affected by the protein C1q, which is excreted by neurons (Underwood, 2016). It has been hypothesised that if the complement system causes microglia to prune too many or too few synapses, then various developmental and neurodegenerative disorders can occur, including Alzheimer's disease (Underwood, 2016); this suggests that blocking C1q could be beneficial for both

---

[2]The complement is activated via three different routes. (i) the classical pathway is activated through antigen–antibody complexes; (ii) a second pathway is activated through lipopolysaccharides on the microbial cell surface; and (iii) the lectin pathway is activated by certain residues on the pathogen surface.

neurodegenerative and immune diseases. Apparently, the complement system is involved in a variety of autoimmune diseases which affect the following: peripheral neurons, including the Guillain–Barre syndrome, paraproteinemic neuropathies, myasthenia gravis, and Lambert–Eaton myasthenia gravis; and the central nervous system, including multiple sclerosis, neuromyelitis optica spectrum disorder, autoimmune encephalopathies, and Alzheimer's disease.[3]

7.3 Pattern Recognition Receptors

A detailed presentation of the exceedingly complicated components and interactions of the immune system is beyond the scope of this volume. In what follows, some key recent developments will be reviewed, starting with Janeway's attempt to answer the following basic question: 'how does an immune reaction start'? In 1989, he claimed that there was a 'tremendous gap' in the understanding of this question and speculated that the starting point of an immune reaction is the recognition by immune cells that a particular molecule or cell is not simply 'foreign' but in general a 'threat to the body' (Janeway, 1989). Furthermore, he hypothesised that this recognition is achieved via the existence of certain specialised receptors, called *pattern recognition receptors*. These receptors, supposedly, have the ability to *interlock with dangerous cells*. Janeway's paper was ignored for several years, with only a few exceptions, including Matzinger and Ruslan Medzhitov, who at the time was a student at Moscow University (Madzhitov, 2013). Janeway and Medzhitov met in 1994; subsequently, Medzhitov joined Janeway's laboratory, and they began the search to identify the genes in human immune cells which encode pattern-recognising receptors. The first such gene, called *toll*, was discovered by Jules Hoffmann in 1996 in the fruit fly (Lemaitre *et al.*, 1996).

Its human analogue, called TLR4, was identified soon afterwards by Janeway and Medzhitov (Madzhitov *et al.*, 1997). Importantly, in 1998,

---

[3] The high complexity of the complement system is consistent with the development of a variety of potential therapeutic agents, including the following: Compstatin and high-dose intravenous immunoglobulin, which act against the critical protein of the complement system known as C_3; eculizumab, ravulizumab, zilucoplan, and tesidolumab, which act against the protein C_5 (Dalakas *et al.*, 2020). The bioscience company Annexon is developing several antibodies that can bind and block the action of complement proteins. Amyndas Pharmaceuticals is also developing complement therapeutics.

Bruce Beutler established that the TLR4 gene encodes for a protein that is indeed able to interlock with a component from the outer wall of bacteria (Poltorak *et al.*, 1998). This part of the wall consists of *lipopolysaccharides* (LPS). Following these seminal discoveries, a variety of similar genes, called *toll-like receptors*, were identified. For example, the receptors encoded by *TLR5* and *TLR10* interlock with molecules found in parasites, whereas *TLRs* 3, 7, 8, and 9 are important for detecting certain types of viruses in the endosomes. This is complemented by another type of genes, called *retinoid acid-inducible genes*, that encode for receptors found inside the cells and which identify invading viruses in the cytoplasm.

The above remarkable developments led to the understanding that innate immunity provides the *first* line of defence against pathogens. For example, a fungal or bacterial infection developing in a wound is immediately dealt with by an innate immune response and is often resolved in a few days. T- and B-lymphocytes are mobilised only if the infection is resistant to this innate response. There are cases where the infection is rapidly 'aborted' by the innate immune system and there is no time for antigen-presenting cells to be mobilised and to present antigens. Interestingly, there are cases where B- and T-lymphocytes, although not important for fighting an infection, do get activated so that adaptive protection is developed during reinfection. As correctly suggested by Metchnikoff, innate immunity is mostly implemented by phagocytosis with the aid of a variety of white blood cells. These cells include *macrophages, neutrophils*, and *monocytes*, as well as a recently discovered category of cells called *dendritic* cells.

7.4 The Interaction of Innate and Adaptive Immunity

According to the notions of continuity and unification promoted in Fokas (2024) and in this volume, there must be some specific mechanism ensuring that innate and adaptive immunity form a continuous, unifying process.

This is indeed the case. This unifying process is facilitated by macrophages and other white cells, and most importantly by dendritic cells. The latter cells, which were discovered by Ralph Steinman in 1973, possess a multitude of pattern-recognising receptors and, in this way, can identify and engulf different types of 'dangerous cells' (Steinman, 2007). Then, they move to one of the main components of the *lymphatic system*, namely a nearby lymph node or the spleen. There, by employing special proteins protruding from their surface, they present a fragment of the engulfed molecule to T cells. If any of these T cells have the 'right receptor' to recognise the presented molecule, then one of the *two signals* needed for the activation of these T cells is fulfilled. The second signal involves the appearance of a *co-stimulating protein* on the dendritic cell's surface. Such proteins are held within the dendritic cell and are shuttled to its surface *only* after the pattern-recognising receptor of the dendritic cell has locked into a *germ*. In other words, the second signal informs the lymphocyte that the engulfed cell is a *dangerous* cell. The receptors on the T cells capable of recognising the second signal are specific molecules (which include CD28, CD80, and CD86). If a T-cell receives two signals, then it multiplies between a hundred and a thousand times. This is the reason why the spleen and lymph nodes swell during infections. Finally, these cells enter the bloodstream and are ready to implement adaptive immunity. Remarkably, without the appearance of the second signal, the lymphocyte becomes *tolerant*. In this way, dendritic cells contribute towards the development of immune self-tolerance. This novel mechanism of developing tolerance occurs in the peripheral system, as opposed to the earlier discussed 'central tolerance', which takes place in the central lymphoid tissue, namely the thymus.

Remarkably, there exist additional controls on the surface of activated T cells, called 'negative regulators' or 'checkpoints'; these include CTLA-4 and programmed death-1 (PD-1), which will be discussed in the following chapter. The associated mechanisms are as important for controlling T-cell activity as the costimulatory pathway. The activation of checkpoints leads to tolerance induction. As expected, inhibition of checkpoints exacerbates autoimmunity, whereas agonists of checkpoints are now used in the treatment of autoimmune diseases (Bluestone and Anderson, 2020).

Yet another mechanism for inducing peripheral tolerance is provided by the action of specialised cells that can suppress pathological immune responses targeting self-tissue. In addition to Treg cells mentioned earlier, such cells include type 1 regulatory T cells (Tr1) and type 3 helper T cells (Th3), which produce transforming growth factor β.

Incidentally, Steinman had the fortune to be in the same Rockefeller University building as George Palade, who was awarded the Nobel Prize in Medicine in 1974 for his innovations in electron microscopy, as well as for the discovery of ribosomes. Using Palade's electron microscope, Steinman could see the many branch-like protrusions emanating from a dendritic cell (*dendron* in Greek means 'tree'). Steinman shared with Jules

Hoffmann and Bruce Beutler the Nobel Prize in Medicine in 2011, despite the fact that, unknown to the Nobel Committee, Steinman died three days before the Nobel Prize was awarded to him.[4] It is worth noting that this Nobel Prize caused a controversy. Twenty-four eminent immunologists published a letter in the leading journal *Nature* stating that the Nobel Committee 'should also have acknowledged the seminal contributions of Janeway and Medzhitov' (Janeway had already died due to lymphoma).

The proteins on the surface of lymphocytes are encoded by a handful of genes belonging to the so-called *Major Histocompatibility Complex* (MHC). Approximately 1% of the genome varies from person to person, and interestingly, the MHC genes vary the most among different individuals. This is part of the reason why there are variations in the individual responses to infections.

The understanding that innate and adaptive immunity form a continuum finally provided an answer to the question raised by Janeway regarding a 'dirty little secret'. This referred to the fact that vaccines work well only when certain substances, called 'adjuvants', are present. It is now clear that adjuvants are needed in order to activate the innate response. Indeed, one of the commonly used experimental adjuvants is LPS, which, as noted earlier, is important for the activation of dendritic cells. Accordingly, in 2009, a molecule similar to LPS was approved for use as an adjuvant in a vaccine against the human papillomavirus; this virus can cause cervical cancer as well as cancer in the lymph nodes.

Incidentally, the mode of action of the extensively used 'aluminium salts adjuvants' remains unknown. In addition, several other important questions regarding vaccines remain open. For example, perhaps as a result of an age-related decline in innate and adaptive immune responses, numerous studies have shown that vaccine efficacy decreases with age. However, paradoxically, some vaccines perform remarkably well in elderly populations. For example, the Shingrix vaccine for shingles has 90% efficacy in people over 70.

---

[4]Despite the fact that it was thought that Steinman's chance for surviving pancreatic cancer more than a year was less than 5%, he survived for 4.5 years. Perhaps, this was the result of the use of a variety of experimental treatments proposed by several of his collaborators and colleagues. These treatments were based on attempts to stimulate his dendritic cells in the hope of attacking his cancer cells.

7.5 The 'Hormones' of the Immune System

After sensing a microbial moiety—a cancer cell, tissue damage, or dys-metabolism—the response of the immune system can be expressed in different forms, including fever, increase of leukocyte counts, cardiovascular reactions, endocrine responses, and reorientation of metabolism. The last one includes the increased production of a plethora of specialised soluble proteins, which can be considered 'the hormones of the immune system' (Davies, 2018). Their goal is to regulate the appropriate immune response depending on the type of offending agent—for example, whether it is a bacterial or a viral infection—as well as to connect the immune system with other vital body systems. Collectively, these proteins are known as *cytokines*. Major functional categories of cytokines include the following: *chemokines*, which recruit immune cells to the site of infection or injury; proinflammatory proteins, which promote inflammation and antimicrobial activity; *colony-stimulating factors*, which promote haematopoiesis; anti-inflammatory cytokines, which ensure the return to homeostasis after the danger has passed; and *interferons*, which are of crucial importance for antiviral action. The last of these, which were the first cytokines to be discovered, cause the body to produce substances that 'interfere' with the ability of a virus to be infectious. When a pattern-recognising receptor of any cell locks onto a germ, it activates a set of genes called *interferon-stimulating genes*. This process, in addition to triggering the production of cytokines, also initiates the production of other important molecules, including prostaglandin E_2 which mediates inflammation, fever, and pain and which is blocked by aspirin.

The interferon-stimulating genes encode proteins which decisively contribute to the fight against bacteria and other germs, especially viruses, by making cells resistant to infection and inhibiting viral replication and spread. For example, one such protein, called *tetherin*, attaches to some viruses as they try to leave one cell to infect another. Interestingly, 10% of the influenza virus genes are designated to fight the effects of interferon. Another example is provided by interferon-γ, which amplifies an ongoing immune response.

There exists a different important family of cytokines called *interleukins*. As their name indicates, they are important for interconnecting and regulating the action of *leucocytes*, which is the technical name for white blood cells (*leukos* in Greek means 'white'). There exist different types of interleukins, denoted by IL-1, IL-2, etc., currently up to IL-37.[5] For example, following an infection, IL-1 dramatically increases the lifespan of neutrophils. Interestingly, after the elimination of an infection, IL-10 acts as an anti-inflammatory agent, preventing the perpetuation of immune reactions which may target self-tissue. The action of IL-10 and other molecules stimulates the type 3 helper T-cells mentioned earlier, which further contributes to the development of immune tolerance. Interleukin-12 and interleukin-23, two important cytokines produced by myeloid cells, stimulate the activation of T-helper lymphocytes.

The presence of autoantibodies neutralising interferons or interleukins results in specific infections. For example, the presence of autoantibodies against interferon-γ results in susceptibility to various agents, including *Mycobacterium tuberculosis*; autoantibodies against interleukin-6 can give rise to severe *Staphylococcus aureus* infection; and autoantibodies against interleukins 17 and 22 are correlated with chronic mucocutaneous candidiasis. Autoantibodies targeting type I interferon have been described in severe diseases associated with COVID-19 (Bastard *et al.*, 2020). Similarly, cryptococcal and nocardia infections are associated with auto-antibodies against granulocyte-macrophage colony-stimulating factor (Netea and van de Veerdonk, 2024).

Interestingly, although thymic atrophy begins in infancy and is markedly accelerated in puberty and then in adulthood, a study comparing 1146 adults who had undergone thymectomy with controls, established that the former group were twice as likely to die in the first five years after surgery; the incidence of cancer was twice as high in the same period. The levels of several cytokines, including interleukin-23 and interleukin-33, were 10 times higher in patients with thymectomy, suggesting that the thymus continues to play an important role in modulating the immune response (Kooshesh *et al.*, 2023). This conclusion is further strengthened by the observation that patients suffering from thymomas, which are tumours originating from thymic epithelial cells, exhibit immune dysregu-lation. In particular, such patients may have autoimmunity, as well as increased susceptibility to infections. Indeed, most patients with thymo-mas have infectious complications, which include the bacteria *Streptococcus*

---

[5]There are four main groups of interleukins known today: (i) those that cause inflamma-tion, including IL-1, 6, 23, and TNF; (ii) those produced by T-helper-1 cells, such as IL-2; (iii) those produced by T-helper-2 cells, such as IL-4 and 10; and (iv) those produced by T-17, such as IL-17.

pneumoniae, *Campylobacter jejuni*, *Pseudomonas*, and *Klebsiella* species, as well as opportunistic infections. The latter infections are associated with autoantibodies against interleukin-23 (Cheng *et al.*, 2024).

Incidentally, the discovery of IL-1 is related to the following litigation incident. In 1984, at a meeting in the Bavarian Alps, Philip Auron announced that his team had isolated the gene for one of the forms of IL-1. He was immediately challenged by Christopher Henney, the cofounder of the biotech company Immunex, who claimed that his company had already analysed this gene and its sequence was different from the one shown by Auron. Henney's results, published soon after in *Nature*, contained the sequence of two IL-1 genes, called alpha and beta, and in contradiction to his earlier remark, one of these sequences was identical to that presented by Auron. Later, Henney was accused of plagiarising an earlier paper by Auron's team that had actually been rejected by *Nature*. The crucial supportive evidence for this allegation was the fact that Auron's rejected paper contained an error, and the same identical error was contained in the gene sequence of the patent application submitted by Immunex. Following a litigation lasting 12 years, Immunex paid US$21 million to the company Cistron, formed by Auron, which by then was bankrupt (Davies, 2018, Chapter 4).

The interactions of cytokines themselves, as well as the interactions of cytokines with a plethora of other proteins and with several organs, are fiendishly complex. In particular, primary inflammatory cytokines (such as IL-1, IL-6, and TNF) induce the production of secondary mediators (such as chemokines, colony stimulator factors, prostaglandins, and nitric oxide); these mediators amplify leukocyte recruitment and enhance local innate immunity, which in turn stimulates the adaptive response. In addition, inflammatory cytokines act on the hypothalamus-pituitary-adrenal axis, resulting in the production of adrenocorticotropic and glucocorticoid hormones. The latter, together with the production of anti-inflammatory cytokines (such as IL-10, IL-1Rα, and transforming growth factor β), are part of negative regulators of inflammation hormones.

Importantly, IL-6 is a potent inducer of the production of the so-called *acute-phase proteins*. The most well-known among them is CRP, which was originally described in connection with *Streptococcus pneumoniae*. Now, almost a century after the discovery of CRP, approximately 200 acute-phase proteins are known. CRP belongs to a class of evolutionarily conserved proteins called pentraxins (CRP is also called PTX1). Other important proteins in this class are the serum amylase component P, denoted by SAP (also known as PRX2), and PTX3. CRP levels in the plasma can increase by as much as 1,000 times in response to an acute stimulus, especially to IL-6. Very high values of CRP, SAP, and PRX3 often indicate infection with bacteria, fungus, or viruses (in particular, high values of PTX3 are associated with inflammation of the vascular tree). Prolonged activation of SAP may lead to amyloidosis.

The SAA family is another important family of acute-phase proteins. In particular, high values of SAA1 and SAA2 are often associated with infection, acute respiratory syndromes, rheumatoid arthritis, and cancer. Several acute-phase proteins are involved in iron homeostasis; some of them bind circulating iron, thus preventing pathogens from utilising it. One of these proteins, ferritin, increases during chronic inflammation, in contrast to low values of ferritin observed in iron-deficiency anaemia.[6] Several acute-phase proteins play a direct role in coagulation and tissue repair. In addition, some of the proteins in this family, such as fibrinogen and fibronectin, bind microbes, facilitating their clearance by phagocytes.

Most of the acute-phase proteins are produced in the liver, but macrophages, endothelial cells, adipose tissue, and even skeletal muscle can also produce such proteins. The increased production of acute-phase proteins is accompanied by the decreased production of several other proteins, including albumin, which contributes to muscle wasting occurring in chronic infections.

References

Avrameas, S., Alexopoulos, H., and Moutsopoulos, H. M. (2018). Natural auto-antibodies: An undersugn hero of the immune system and autoimmune disorders - A point of view. *Frontiers in Immunology*, *9*, 1320.

Bastard, P., *et al.* (2020). Autoantibodies against type I IFNs in patients with life-threatening COVID-19. *Science*, *370*(6515), eabd4585.

Bluestone, J. A. and Anderson, M. (2020). Tolerance in the age of immunotherapy. *New England Journal of Medicine*, *383*, 1156–1166.

Cheng, A., *et al.* (2024). Anti–Interleukin-23 autoantibodies in adult-onset immunodeficiency. *New England Journal of Medicine*, *390*(12), 1105–1117.

Cohen, I. R., Hershberg, U., and Solomon, S. (2004). Antigen-receptor degeneracy and immunological paradigms. *Molecular Immunology*, *40*, 993–996.

---

[6]Other circulating peptide hormones important in iron metabolism are hepcidin, haptoglobin, and hemopexin.

Dalakas, M. C., Alexopoulos, H., and Spaeth, P. J. (2020). Complement in neurological disorders and emerging therapies with anti-complement biologics. *Nature Review Neurology, 16*, 601–617.

Davies, D. M. (2018). *The Beautiful Cure: The Revolution in Immunology and What It Means for Your Health.* Chicago: The University of Chicago Press.

Fokas, A. S. (2024). *Ways of Comprehending: The Grand Illusion and the Essence of Being Human.* Singapore: World Scientific.

Fokas, A. S. and Ktistakis, N. (in preparation). Ageing and the quest for rejuvenation; diet, epigenetics, microbiota.

Germenis, A. E. (2019). *Immunity: Metaphor and Reality* (in Greek). Vouton: Crete University Press.

Janeway, C. A. (1989). Approaching the asymptote? Evolution and Revolution in immunology. *Cold Spring Harbor Symposia on Quantitative Biology, 54*, 1–13.

Kooshesh, K. A., Foy, B. H., Sykes, D. B., Gustafsson, K., and Scadden, D. T. (2023). Health consequences of thymus removal in adults. *New England Journal of Medicine, 389*(5), 406–417.

Lemaitre, B., Nicolas, E., Michaut, L., Reichhart, J. M., and Hoffmann, J. A. (1996). The dorsoventral regulatory gene cassette spätzle/Toll/cactus controls the potent antifungal response in Drosophila adults. *Cell, 86*(6), 973–983.

Madzhitov, R. (2013). Pattern recognition theory and the launch of modern innate immunity. *The Journal of Immunology, 191*, 4473–4474.

Madzhitov, R., Preston-Hurlburt, P., and Janeway, C. A. (1997). A human homologue of the drosophila toll protein signals activation of adaptive immunity. *Nature, 388*, 394–397.

Matzinger, P. (1994). Tolerance, danger, and the extended family. *Annual Review of Immunology, 12*, 991–1045.

Miller, J. F. A. P. (1961a). Analysis of the thymus influence in leukaemogenesis. *Nature, 191*(4785), 248–249.

Miller, J. F. (1961b). Immunological function of the thymus. *Lancet, 278*(7205): 748–749.

Netea, M. G. and van de Veerdonk, F. L. (2024). Anti-interleukin-23 autoantibodies and severe infections. *New England Journal of Medicine, 390*(12), 1143–1146.

Nicolás-Ávila, J. A., *et al.* (2020). A network of macrophages supports mitochondrial homeostasis in the heart. *Cell, 183*(1), 94–109.

Poltorak, A., *et al.* (1998). Defective LPS singling in C3H/Hej and C57BL/10ScCrmice: Mutations in the Tlr4 gene. *Science, 282*, 2085–2088.

Steinman, R. M. (2007). Dendritic cells understanding immunogenicity. *European Journal of Immunology, 37*(Suppl 1), S53–S60.

Sun, L., Wu, J., Du, F., Chen, X., and Chen, Z. J. (2013). Cyclic GMP-AMP synthase is a cytosolic DNA sensor that activates the type I interferon pathway. *Science, 339*(6121), 786–791.

Taylor, N. (2023). The thymus—Not a graveyard after all, even in adults? *New England Journal of Medicine, 389*(5), 470–471.

Underwood, E. (2016). Wired. *Science, 353*, 762–765.

Wu, J., *et al.* (2013). Cyclic GMP-AMP is an endogenous second messenger in innate immune signaling by cytosolic DNA. *Science, 339*(6121), 826–830.

Chapter 8

Clinical Implications of Immunological Breakthroughs

8.1 Monoclonal Antibodies

An ingenious approach to the treatment of autoimmune diseases was initiated by the immunologist Sir Marc Feldmann and the clinician Sir Ravinder Maini. They observed the occurrence of high quantities of the cytokine *tumour necrosis factor* (TNF) in the inflamed joints of patients suffering from rheumatoid arthritis (RA), which is a prototypical autoimmune disease. In this extensively studied disorder, immune cells accumulate in joints, where they cause the destruction of cartilage and bone (Moutsopoulos and Zambeli, 2021). It was well known at the time that the TNF, which was discovered in 1975, is produced by many cells, including white blood cells. This factor promotes systemic inflammation, fever, apoptotic cell death, and cachexia. Feldmann and Maini proposed that the activity of the TNF could be inhibited by employing *monoclonal antibodies*. In particular, they speculated that if a mouse is exposed to human TNF, it would create antibodies against this *foreign* molecule. If these antibodies could be harvested, they could then be used to inhibit the TNF of RA. Indeed, in proof-of-concept studies, Feldmann and Maini used an antibody to neutralise TNF in *ex vivo* synovial cell cultures from RA patients. They found that this inhibited the production of IL-1, a major mediator of bone damage, and concluded that 'anti-TNFα agents may be useful in the treatment of rheumatoid arthritis' (Brennan *et al.*, 1989). Anti-human TNF antibodies were constructed in 1988 by Jan Vilcek.

For this purpose, he immunised mice with the human TNF protein, which was produced from the TNF gene that had been isolated earlier by the biotech company Genentech. As expected, the B-lymphocytes of the immunised mice produced antibodies against the human TNF protein. In order to generate immortal versions of these cells, so that Vilcek could have a stable supply of monoclonal antibodies, he fused these B cells with myeloma tumour cells (Vilcek, 2009). This 'immortalisation' technique was introduced earlier at the MRC Molecular Biology Institute at Cambridge by César Milstein and Georges Köhler (Nobel Prize in Medicine, 1984). As a result of this discovery, Vilcek, who was motivated to study cytokines after listening to a lecture by Alick Isaacs in his native Czechoslovakia in 1957, became a millionaire.[1]

The above-mentioned cells are called 'monoclonal' because they originate from a single B cell. Taking into consideration that, on average, an individual has 10 billion B cells, each of us can generate 10 billion different monoclonal antibodies. An important step towards proving the central role of TNF antibodies in RA was taken in 1991 by George Kollias' group in Greece, which established that mice that were genetically engineered to produce human TNF developed inflammation in their joints analogous to what occurs in RA (Keffer *et al.*, 1991). In 1992, Feldman, Maini, and collaborators showed that the use of the anti-TNF antibodies reduced inflammation and cartilage destruction in afflicted mice (Williams *et al.*, 1992). Subsequent clinical studies in humans established that the use of a humanised anti-TNF monoclonal antibody (*infliximab*) or a form of a soluble TNF receptor-Fc antibody fusion protein (*etanercept*) was remarkably more efficacious than any of the treatments for RA available at the time.

The anti-TNF antibody was eventually approved for the treatment of RA in 1998. It is now used extensively, often in combination with other medications, especially with cortisone and methotrexate. This treatment is also used for several other autoimmune diseases, including Crohn's disease, ulcerative colitis, psoriasis, and ankylosing spondylitis. Subsequently, a plethora of monoclonal antibodies were produced for diverse targets, including cytokines such as IL-1 (*anakinra*) and IL-6 (*tocilizumab*) and cells such as B lymphocytes (*rituximab*). Some monoclonal antibodies are

---

[1] In 2005, Vilcek donated US$105 million to New York University, the largest gift any New York State health institute had received at that time.

effective against certain types of cancer; this will be discussed in the following two chapters. In this way, a new transformative approach to the treatment of autoimmune diseases and cancer was initiated.[2]

Incidentally, regarding the medications methotrexate and cortisone, it is noted that the former dampens the immune response via inhibiting folate. Cortisone is similar to the endogenous hormone cortisol, which is released (along with epinephrine) from the adrenal glands in response to stress. The effectiveness of this synthetic analogue of cortisol for the treatment of RA was already established in 1929 by Philip S. Hench. Cortisone was identified in 1948 at the Mayo Clinic by the chemists Edward Calvin Kendall and Harold L. Mason. Kendall, Hench, and Tadeus Reichstein were awarded the Nobel Prize in Medicine in 1950 for their study of adrenal gland hormones; Kendall also discovered the hormone thyroxine, which is released by the thyroid gland.

The brief summary of some of the recent major developments in immunology presented in the earlier chapter makes it clear that the most striking feature of the immune system is its *complexity*. Indeed, the deeper one investigates immunity, the higher levels of complexity one discovers. For example, it is now known that various types of T cells exist. In particular, *Natural Killer T cells* activate *Natural Killer lymphocytes* that attack germs directly, whereas other T cells are responsible for regulating the immune response. Remarkably, in 2001, following the persistent efforts of the Japanese scientist Shimon Sakaguchi, six different research teams announced the identification of human *suppressor T cells*, which have the capacity to dampen the immune response. Similarly, in 2003, it was discovered that the activity of a particular gene, called *Foxp3*, could convert a normal T cell into a *regulatory T cell* (Hori *et al.*, 2003). Such cells are particularly important in the gut, where they regulate the symbiosis between the immune system and the microbiota.

---

[2]The biotech company Centocor made a substantial investment in the use of anti-TNF antibodies for the treatment of sepsis. After it became clear in 1992 that this approach was ineffective, within a few months, its share price fell from US$50 to US$6. The use of anti-TNF antibodies for the treatment of several autoimmune diseases rescued this company. It was finally sold in 1999 to Johnson & Johnson for US$4.9 billion.

8.2 Activation of the Immune System and Clinical Implications

As mentioned in the earlier chapter, T cells that possess the receptor CTLA-4 contribute to tolerance; this new type of T cell was discovered by Jim Allison (Walunas *et al.*, 1994). Since these cells *inhibit* the immune response, the use of monoclonal antibodies against CTLA-4 receptors induces the *activation* of the immune system.[3] The Japanese scientist Tasuku Honjo (who shared with Allison the Nobel Prize in Medicine in 2018) established that a similar effect can be achieved by blocking the PD-1 receptor, which appears on a different type of T cell (Okazaki and Honjo, 2007). Incidentally, the name 'PD-1' is a misnomer, originating from Honjo's earlier research on proteins encoded by genes that cause cells to die. As will be discussed in the following chapter, inhibition of checkpoint pathways has revolutionised cancer immunotherapy; it has turned once deadly cancers, such as melanoma and non-small-cell lung cancer, into treatable diseases. Taking into consideration that currently more than 20 similar receptors are known which are capable of inhibiting the immune system, it is natural to expect that many similar weapons against cancer will soon be developed (Long, 2008). In this context, it is noted that, in 2016, the entrepreneur Sean Parker donated US$250 million for the creation of the Parker Institute, aiming to carry out 'a Manhattan Project for curing cancer with the immune system'.

The U.S. Food and Drug Administration has approved the CTLA-4 antagonists abatacept and belatacept for the treatment of RA (Bluestone and Anderson, 2020). The PD-1 receptor agonist, peresolimab, provides an additional weapon in the armamentarium against this disease. It has the potential to 'reset the immune response or to provide immune tolerance' (Gravallese and Thomas, 2023).

There is no doubt that the recent astonishing developments in immunology have greatly enhanced our understanding of a multitude of mechanisms responsible for the development of immune tolerance. Importantly, this has suggested several ways to artificially enhance this tolerance selectively for the treatment of autoimmune diseases. For example, the

---

[3] Many other scientists have made decisive contributions in these developments. In particular, CTLA-4 receptors were first identified in 1987 at Pierre Goldstein's laboratory in Marseilles. Monoclonal antibodies to these receptors were constructed at Jeff Bluestone's laboratory in the University of Chicago.

monoclonal antibody teplizumab induces the inactivation of effector T cells without affecting the regulatory T cells.

The continuous process of deepening of our insight into the immune response is illustrated by the recent progress in the understanding of RA. This prevalent disease usually presents with symmetric, polyarticular pain and swelling of small joints in the hands and feet. The most common form of this disorder is seropositive RA, characterised by the presence of anti-citrullinated protein antibodies (ACPAs) and other less specific autoantibodies, which are known as rheumatoid factors. Autoantibodies can be detected in the blood a median of 4.5 years before the clinical onset of the disease. RA can appear at any age, with a peak incidence in the third through the fifth decades of life; it is 2–3 times more common in women than in men. Like many other autoimmune diseases, it results from the interaction of genetic factors and specific environmental exposures, including the occurrence of bacterial infections. The most prominent risk factors are cigarette smoking and certain genetic sequences in the HLA-DRβ domain (Gravallese and Firestein, 2023). In addition to proinflammatory processes, inadequate production of anti-inflammatory cytokines (such as IL-1Rα and IL-10) is also important for the transition from the preclinical to the clinical form of RA. Synovitis is a hallmark of the disease, with the influx of various inflammatory cells into the synovial fluid.

It has been known that the synovial fluid of patients with RA contains several types of particular cells, called *fibroblasts*. It has now been shown that approximately two weeks before a flare-up of the disease, fibroblast-like cells, called 'prime', appear in the blood of patients suffering from RA. The number of these cells subsequently decreases. This suggests that the prime cells may move into the inflamed synovium, representing a circulatory precursor of inflammatory synovial fibroblasts (Orange *et al.*, 2020). The ability to predict the periods of flare-ups should have major therapeutic implications.

At the moment, the initial treatment of RA remains low-dose methotrexate. This treatment is adequate for 25–40% of cases. If this fails, the triple therapy of methotrexate, sulfasalazine, and hydroxychloroquine can achieve reasonable control. Inadequate response requires the addition of other agents, including a biological agent such as one of those mentioned earlier (initially anti-TNF), or a Janus kinase (JAK) inhibitor; JAK is a small intracellular enzyme which plays a key role in cell signalling.

In addition to RA, the recent breakthroughs in immunology have had a transformative impact on several other autoimmune diseases and beyond. Some examples follow.

Since cytokines are instrumental in fighting germs, it is not surprising that specific cytokines can be used for the treatment of specific infections. For example, interferon-α is used for the treatment of hepatitis B and hepatitis C. Since some cytokines dampen the immune response, it is natural to expect that they can be useful for disorders caused by over-activation of the immune system. This is indeed the case where, for example, interferon-β is useful in the treatment of multiple sclerosis.

Systemic lupus erythematosus (SLE) is an autoimmune disease caused by aberrations affecting both the innate and adaptive immune systems. SLE can affect different body systems, including the joints, skin, kidneys, heart, lungs, and the central nervous system. This disorder was earlier managed with nonsteroidal anti-inflammatory medications, hydroxychloroquine, low-dose glucocorticoids, methotrexate, and azathioprine. More severe cases were managed with mycophenolate or cyclophosphamide. The introduction of rituximab and other specialised antibodies, such as belimumab which inhibits specific B-lymphocyte activating factors, has reduced the long-term sequelae of SLE. The recent understanding of the role of dendritic-cell differentiation in SLE pathogenesis explained the positive effect of interferon observed earlier; furthermore, it paved the way for the use of medications blocking the interferon receptors (such as the monoclonal antibody anifrolumab) (Wallace, 2022).

Inflammatory bowel disease is an immune-mediated intestinal disease characterised by inflammation. It usually takes the form of either ulcerative colitis, which affects the mucosa of the colon, or Crohn's disease, which can affect any part of the gastrointestinal tract, although it is usually localised in the ileum and the colon. Most patients with Crohn's disease present with fatigue, abdominal pain, and diarrhoea; the most morbid complication is the development of perianal fistulas, occurring in approximately 20% of cases. More than 240 genetic loci are implicated in the pathogenesis of Crohn's disease. It has been established that the appearance of this disorder requires changes in the microbiota followed by an immune response. It is believed that IL-23 activates type 17 helper T cells, which produce various cytokines, including TNF, IL-6, IL-17A, and IL-22. Anti-TNF therapy, via infliximab and, later, adalimumab, as well as treatment with inhibitors of IL-23 and of IL-12, has substantially increased remission periods.[4] In addition, the use of upadacitinib, an inhibitor of JAK, showed promising results (Abreu, 2023).

Any part of the body can suffer an autoimmune attack. For example, in Takayasu's arteritis the aorta of patients under 50 years of age develops granulomatous infiltrates. For many years, this disorder was managed with high-dose glucocorticoids and typical immunosuppressants, including methotrexate, azathioprine, and mycophenolate. Although the pathogenesis of this disease remains unknown, the empirical use of TNF and IL-6 inhibitors turned out to be more efficacious than the use of immunosuppressants. Incidentally, in the autoimmune attack of large- and medium-sized arteries excluding the aorta, which is a disorder called giant-cell arteritis, combining glucocorticoid with upadacitinib (a Janus kinase inhibitor), which blocks the signalling of several cytokines, including interleukin-6 and interferon-γ, was more efficacious than using glucocorticoid alone (Blockmans *et al.*, 2025).

The deeper our understanding of the mechanisms responsible for the regulation of the immune system, the higher the likelihood that new effective treatments will emerge for autoimmune diseases. For example, naïve T cells, following their encounter with pathogen-derived antigens, become effector T cells attacking the relevant pathogens. Some of these cells remain in the tissue, becoming memory T cells, which can provide rapid protection in case of reinfection. Regulatory T cells also remain in the tissue in order to promote healing and, importantly, to regulate potential effector T-cell reactivation. Autoimmunity can arise as a result of the inhibition of regulatory T cells. It was shown by Pompura *et al.* (2021) that the adipose tissue of patients suffering from multiple sclerosis has a lower concentration of oleic acid. Interestingly, exposure of the regulatory T cells of these patients to oleic acid partially remedied the suppressive function of these cells, which is of crucial importance for preventing autoimmunity. This provides further indirect evidence for the usefulness of olive oil, whose oleic acid content is approximately 80%. Incidentally, the

---

[4]Such inhibitors include ustekinumab, risankizumab, and guselkumab. IL-23 consists of a subunit, called p40, which is also shared with IL-12, and a unique subunit, called p19. Risankizumab binds to p19, thus inhibiting only IL-23, whereas ustekinumab binds to p40, thus inhibiting both IL-23 and IL-12. In a clinical study that directly compared risankizumab and ustekinumab, the former was shown to be superior to the latter with respect to endoscopic remission at week 48, suggesting that, perhaps, targeting IL-23 is more effective than targeting IL-12.

demonstration that the depletion of B cells, following the use of specific monoclonal antibodies, was beneficial to patients suffering from multiple sclerosis implies that B lymphocytes are also involved in the pathogenesis of this disease (Hauser *et al.*, 2020). Incidentally, gradual progressive neurologic impairment can occur throughout the course of multiple sclerosis. This impairment, which is called 'disability accrual', is independent of relapse activity and is thought to be caused by chronic neuroinflammation within the CNS involving myeloid cells. Tolebrutinib, an oral, brain-penetrating Bruton tyrosine kinase inhibitor that targets myeloid cells, lowered the risk of disability progression (Fox *et al.*, 2025).

Amyotrophic lateral sclerosis, which affected, among others, my Cambridge colleague Stephen Hawking, is a neurodegenerative disorder characterised by loss of motor neurons in the motor cortex, brainstem, and spinal cord. One of the hallmarks of this disease is the transport of certain proteins from the cytoplasm to the mitochondria, which causes mitochondrial damage via the process of reactive oxygenation. This induces the activation of STING (mentioned in the earlier chapter) and other genes, followed by the release of several proinflammatory cytokines. Ongoing efforts target various steps in this complicated process (Van Damme and Robberecht, 2021).

Neuromyelitis optica spectrum disorder is an autoimmune disease which affects the optic nerve and the spinal cord, as well as the brain in rare cases. It often presents with sudden visual loss or paralysis. Satralizumab, a monoclonal antibody targeting the IL-6 receptor, when added to the usual treatment of immunosuppression, lowers the risk of relapse (Yamamura *et al.*, 2019).

Fifteen patients with serious systemic autoimmune diseases refractory to conventional treatment were infused with CAR T cells targeting CD19. At a median follow-up of 15 months, ranging from 4 to 29 months, all patients were in remission or had a major reduction in their symptoms, their autoantibodies had disappeared, and they had discontinued their immunosuppressive and anti-inflammatory therapy (including glucocorticoids) (Isaacs, 2024). This appears to be another revolutionary therapy offered by immunology, which is still in its early days but expected to be very effective in the future.

In the rare lung disease called autoimmune pulmonary alveolar proteinosis, there exists excessive accumulation of surfactant within the alveolar airspace. A new treatment has now emerged: once-daily inhaled molgramostim, which is a recombinant human granulocyte-macrophage colony-stimulating factor (Seymour, 2025).

Importantly, further progress continues to be made with the treatment of well-known disorders. For example, for sarcoidosis, in which granulomas can appear in the lungs and other organs, the standard treatment has been an initial dose of 40 mg prednisone per day, which is later tapered to 5 mg per day for up to week 24. It is now established that methotrexate provides an alternative initial therapy (Baughman, 2025). Similarly, over the last 25 years, rituximab has been used for the treatment of the autoimmune disease myasthenia gravis, which is extensively discussed in Fokas (2024). Taking into consideration that in this disease autoimmune antibodies attack the neuromuscular junction and that rituximab depletes the B-cell lineage, this is a logical treatment. However, there is a subgroup of patients with antibodies against acetylcholine receptors who do not respond well to rituximab. For such patients, inebilizumab offers an advantage (Wolfe and Shelly, 2025). Interestingly, there is a new treatment for coeliac disease, which is a prevalent disorder caused by autoimmune activation triggered by gluten. The use of gluten protein (gliadin) encapsulated in specific nanoparticles induces gluten tolerance (Kelly *et al.*, 2021).

In conclusion, it should be emphasised that monoclonal antibodies have revolutionised the treatment of a plethora of other diseases, from malaria to migraines. Regarding malaria, current drug-based strategies, such as seasonal malaria chemoprotection in children using sulfadoxine-pyrimethamine and amodiaquine, are difficult to implement (as they require parents to adhere to a multidose oral drug regimen and strict coordination with the healthcare system); in addition, these strategies have a limited duration of protection. For children in sub-Saharan Africa and other areas with a moderate-to-high incidence of *Plasmodium falciparum* malaria, the World Health Organization (WHO) has recommended the widespread use of the vaccine RTS, S/AS0. However, the four-year efficiency of this vaccine is only 32–39%. In children in Mali, the intravenous administration of a monoclonal antibody (CIS43LS), which acts against the sporozoites of *Plasmodium falciparum*, had an efficacy of 88% over the six-month malarial transition period (Kayentao *et al.*, 2022). For a similar population, monoclonal antibody (L9LS), which has an extended half-life and offers the advantage of subcutaneous administration, provided an efficacy of 77.4%; this is close to the WHO target of

80% efficacy against clinical disease at 3–4 months follow-up (Kayentao *et al.*, 2024).

Regarding migraines, the new gepant drugs consist of monoclonal antibodies acting against a protein called calcitonin gene peptide (CGRP). Apparently this protein is involved in the activation of sensory neurons in the meninges associated with the early stages of a migraine attack.

References

Abreu, M. T. (2023). Jak1 inhibition to treat Crohn's disease. *New England Journal of Medicine*, *388*, 2005–2009.

Baughman, R. P. and Lower, E. E. (2025). *New England Journal of Medicine*, *393*, 303–305.

Blockmans, D., Penn, S. K. *et al.* (2025). A phase 3 trial of upadacitinib for giant cell arteritis. *New England Journal of Medicine*, *392*, 2013–2024.

Bluestone, J. A. and Anderson, M. (2020). Tolerance in the age of immunotherapy. *New England Journal of Medicine*, *383*(12), 1156–1166.

Brennan, F. M., Chantry, D., Jackson, A., Maini, R., and Feldmann, M. (1989). Inhibitory effect of TNF alpha antibodies on synovial cell interleukin-1 production in rheumatoid arthritis. *Lancet*, *2*, 244–247.

Fox, R. J., Bar-Or, A., *et al.* (2025). Tolebrutinib in non-relapsing secondary progressive multiple sclerosis, *New England Journal of Medicine*, *392*, 1883–1892.

Gravallese, E. M. and Firestein, G. S. (2023). Rheumatoid arthritis—Common origins, divergent mechanisms. *New England Journal of Medicine*, *388*, 529–542.

Gravallese, E. M. and Thomas, R. (2023). Reinforcing the checkpoint in rheumatoid arthritis. *New England Journal of Medicine*, *388*, 1905–1907.

Hauser, S. L., *et al.* (2020). Ofatumumab versus teriflunomide in multiple sclerosis. *New England Journal of Medicine*, *383*, 546–557.

Hori, S., Nakamura, T., and Sakaduchi, S. (2003). Control of regulatory T cell development of the transcription factor Foxp3. *Science*, *299*, 1057–1061.

Isaacs, J. D. (2024). CAR T-cells-a new horizon for autoimmunity. *New England Journal of Medicine*, *390*, 758–759.

Kayentao, K., *et al.* (2022). Safety and efficacy of a monoclonal antibody against malaria in Mali. *New England Journal of Medicine*, *387*, 1833–1842.

Kayentao, K., *et al.* (2024). Subcutaneous administration of a monoclonal antibody to prevent malaria. *New England Journal of Medicine*, *390*, 1549–1559.

Keffer, J., *et al.* (1991). Transgenic mice expressing human tumour necrosis factor: A predictive genetic model of arthritis. *EMBO Journal*, *10*, 4025–4031.

Kelly, P., Murray, J. A., Leffler, D. A., *et al.* (2021). TAK-101 nanoparticles induce gluten-specific tolerance in celiac disease: A randomized, double-blind, placebo-controlled study. *Gastroenterology*, *161*, 66–80.

Long, E. O. (2008). Negative singling by inhibitory receptors: The NK cell paradigm. *Immunological Review*, *224*, 70–84.

Moutsopoulos, H. M. and Zambeli, E. (2021). *Immunology-Rheumatology in Questions*. 2nd ed. Springer-Nature.

Okazaki, T. and Honjo, T. (2007). PD-1 and PD-2 ligands: From discovery to clinical application. *International Immunology*, *19*, 813–824.

Orange, D. E., *et al.* (2020). RNA identification of PRIME cells predicting rheumatoid arthritis flares. *New England Journal of Medicine*, *383*(3), 218–228.

Pompura, S. L., *et al.* (2021). Oleic acid restores suppressive defects in tissue-resident FOXP3 Tregs from patients with multiple sclerosis. *The Journal of Clinical Investigation*, *131*.

Seymour, J. F. (2025). GM-CSF in autoimmune pulmonary alveolar proteinosis. *New England Journal of Medicine*, *393*, 812–815.

Van Damme, P. and Robberecht, W. (2021). STING-induced inflammation—A novel therapeutic target in ALS? *New England Journal of Medicine*, *384*, 765–767.

Vilcek, J. (2009). From IFN to TNF: A journey into realms of lore. *Nature Immunology*, *10*, 555–557.

Wallace, D. J. (2022). Progress toward better treatment of lupus. *New England Journal of Medicine*, *387*, 939–940.

Walunas, T. L., *et al.* (1994). CTLA-4 can function as a negative regulator of T cell activation. *Immunity*, *1*, 405–413.

Williams, R. O., Feldmann, M., and Maini, R. N. (1992). Anti-tumor necrosis factor ameliorate joint disease in muting collagen induced arthritis. *Proceedings of the National Academy of Sciences of the United States of America*, *89*, 9784–9788.

Wolfe, G. I. U. and Shelly, S. (2025). Myasthenia Gravis-redemption for B-Cell depletion. *New England Journal of Medicine*, *392*, 2382–2384.

Yamamura, T., Kleiter, I., Fujihara, K., Palace, J., and Greenberg, B. (2019). Trial of satralizumab in neuromyelitis optica spectrum disorder. *New England Journal of Medicine*, *381*, 2114–2124.

Chapter 9

Breakthroughs in the Fight against Cancer

Cancer is an ancient disease. Interestingly, a tumour was found in a 240 million-year-old femur fossil of an extinct animal called Pappochelys, which is a shell-less relative of modern turtles, discovered in Germany in 2019 (Haridy *et al.*, 2019). Regarding cancer in humans, the mummified remains of a Peruvian Inca, approximately 2,400 years old, show clear indications of a melanoma.

The first weapon against cancer was surgery with or without radio-therapy (which is discussed in Part III of the book). But by the 1950s, it was clear that this treatment could only cure a third of cancers.

The next major step in the fight against cancer was the development of effective drugs. At the turn of the 20th century, Paul Ehrlich attempted to develop chemicals to treat cancer and introduced the term 'chemo-therapy'. The first successful such chemicals were nitrogen mustard, used in 1943 at Yale University against lymphomas, and a folic acid antagonist, used in 1947 by Sidney Farber against leukaemia in children. Nitrogen mustard harms DNA through a process called alkylation, where an alkyl group moves from one molecule to another; it is derived from mustard gas, a chemical used in warfare that many countries stored during the Second World War.[1] Folic acid, which was synthesised in 1937, is found

---

[1]An air raid in Bari, Italy, led to the accidental release of this gas, which affected several hundred individuals. Their medical examination showed a reduction in their lymphocytes, consistent with the main effect of nitrogen mustard.

in large quantities in green leafy vegetables and is vital for the function of the bone marrow. Folate deficiency produces a bone marrow picture reminiscent of the effects of nitrogen mustard. The folic acid antagonist introduced by Farber was later replaced by an anti-folate immunosuppressant medication, methotrexate, mentioned in the earlier chapters.

The largest post-war programme of cancer drug development was conducted at the Sloan-Kettering Institute in New York in the US before the National Cancer Institute got involved. This biomedical facility housed almost the entire programme and staff of the US' Chemical Warfare Service. Discoveries in this institute and further developments had an unexpected major positive effect. Namely, childhood leukaemia and several types of lymphoma, testicular cancer, and cancer of the uterus (choriocarcinoma) became largely curable!

The fundamental principle of chemotherapy is to target cancer cells by exploiting differences in the growth patterns of cancer versus normal cells, especially the ability of cancer cells to divide without control. The importance of cell division becomes evident by noting that an adult has approximately 100 trillion cells and that there is an enormous number of continuously dividing cells replacing the dying cells. Before a cell divides, it grows in size, and then it duplicates and separates its chromosomes so that they can be distributed evenly between the two daughter cells. Aberrations in this highly complicated process, which is meticulously coordinated during the 'cell cycle', can lead to cancer. In 2001, L. Hartwell, P. Nurse, and T. Hunt were awarded the Nobel Prize in Medicine for elucidating aspects of how the cell cycle is controlled. Chemotherapy targets the cell division process and hence affects all dividing cells, in addition to cancer cells. Normal cells that divide extensively, like blood cells, are more affected by chemotherapy. Interestingly, for most cancers, only a small set of cells progresses with high rates through the entire cell cycle. For this and several other reasons, progress due to general chemotherapy soon reached its limits. Fortunately, molecular biology provided a way forward.

Cancer arises from a clone of cells that, after escaping the built-in constraints governing cellular function, is able to continue proliferating and evade immune surveillance. Advances in molecular biology have revealed that long-accumulating mutations cause cancer. The changes that occur help cancer cells bypass the normal rules that keep healthy cells functioning properly, especially the impact of certain genes known as 'tumour suppressing genes' that help maintain the integrity of the genome. In addition, the cancer cells use specific mechanisms to evade natural immune surveillance. Such behaviour gives rise to uncontrolled proliferation, tissue invasion, and total restructuring of the local tissue microenvironment.

There is growing evidence that, among the thousands of mutations acquired by a cancer cell, only a handful, called 'driver mutations', can instruct the cell to function as an autonomous clone. Driver mutations usually take the following forms. First, the mutation involves the replacement of one DNA base with another. Second, there occur insertions or deletions of a small number of DNA bases. Third, large chromosomal regions, or even entire chromosomes, may be gained or lost. Fourth, rearrangements occur, including the fusion of one gene to another or the juxtaposition of one gene with the regulatory apparatus of another; the particular case when a chromosome breaks into two parts and the two fragmented pieces reattach to different chromosomes is referred to as 'translocation' (Nangalia and Campbell, 2019).

Remarkably, the vast majority of driver mutations occur within 300–600 protein codings.[2] Furthermore, these mutations take generic forms, which provide natural targets for specific therapies. These mutations create distinctive 'signatures in the cancer genome', which can often reveal the cause of the particular cancer. More than 30 such signatures have been identified. For example, a particular such pattern can be used to specify whether lung cancer has been caused by tobacco or how ultraviolet light leads to melanoma. The larger the number of driver mutations, the worse the prognosis; in addition, there exist particular mutations, such as the TP53 mutation, that are associated with poor prognosis.

Genes that induce normal cells to become cancerous are called oncogenes, and the normal versions of these genes are called proto-oncogenes. As noted earlier, it is possible for a proto-oncogene to become an oncogene via a mutation in just a single base. For example, a single nucleotide change in the DNA in a family of proto-oncogenes, called RAS, gives rise to an alteration in the corresponding RAS protein, which can have devastating consequences: the one amino acid change in the protein makes it hyperactive, thereby driving the cell into a cancerous state. Oncogenic mutations in the RAS proto-oncogenes give rise to the oncogenes KRAS, NRAS, and HRAS, which are the most prevalent oncogenes in human cancers. Among these three oncogenes, KRAS is the

---

[2] There exist approximately 21,000 protein-coding genes. The protein-coding regions make up only 1% of the human genome.

most commonly mutated (85% of all RAS mutations). Remarkably, KRAS, which arises from the RAS proto-oncogene via a single mutation, is responsible for a quarter of lung cancers, half of colon cancers, and 90% of all pancreatic cancers (the occurrence of KRAS in human tumours was discovered in 1981).

It turns out that several of the proto-oncogenes are associated with viruses. The elucidation of this fact has a long and illustrious history.[3] It begins in 1910 with Peyton Rous, who received for this work a Nobel Prize in Medicine more than half a century later (Rous was mentioned in Fokas, 2024) in relation to Alan Hodgkin). He extracted cells from a cancer tumour in a hen and injected them into a healthy chicken, which then developed cancer; Rous concluded that the tumour contained an infectious virus that could transmit cancer. It turns out that this is indeed the case; the associated virus, called Rous Sarcoma Virus (RSV), is a retrovirus. Namely, it is a special type of virus that uses RNA as its genomic material. When a cell is infected by a retrovirus, the retroviral RNA is converted into DNA, which is then inserted into the DNA of the host cell. The machinery used by the host cell to convert RNA to DNA involves the enzyme reverse transcriptase; the underlying process was elucidated by D. Baltimore, R. Dulbecco, and H. M. Temin (for this work, they shared the Nobel Prize in Medicine in 1975) (Temin and Mizutami, 1970; Baltimore, 1970).

What is the origin of the oncogenes carried by retroviruses? J. M. Bishop and H. Varmus understood that the oncogene in a retrovirus does not represent a genuine viral gene but a cellular gene which the virus had picked up during its replication in one of the cells it had infected in the past (for this work, Bishop and Varmus shared the 1989 Nobel Prize in Medicine) (Bishop, 2003; Varmus, 1990). In the particular case of RAS, this proto-oncogene was picked up by a retrovirus during the process of infecting a rat, causing sarcoma. The name KRAS is due to the identification of this virus, in the 1960s, by Werner Kirsten. The proto-oncogenes found in infected chickens, mice, rats, monkeys, or cats were named after the viruses they came from; for instance, the proto-oncogene MYC is linked to the avian myelocytomatosis virus. Over 20 proto-oncogenes

---

[3]The pre-history starts with the great geneticist Thomas Morgan, who, under the influence of the prominent geneticist Theodor Boveri (*On the Origin of Malignant Tumours*, 1914), proposed that genes in chromosomes are arranged in a linear form (a theory which is expanded in his book, *Mechanism of Mendelian Heredity* (1915)).

have been found. It turns out that these proto-oncogenes play an important role in cell growth and cell division.

At that time, it was thought that retroviruses were involved in avian and rodent cancers and not in humans. Are the associated proto-oncogenes relevant to human cancers? In 1977 and 1978, D. Spandidos suggested that human cancer cells have certain genes that control how these cells behave, and he also proposed a way to find and copy these genes. A few years later, the groups of R. Weinberg at MIT, M. Wigler at Cold Spring Harbour Laboratory, and M. Barbacid at the NIH were able to clone the first human oncogene, HRAS1, from cells of a bladder carcinoma. This gene is homologous with the corresponding RAS gene of the Harvey sarcoma virus.

The elucidation of the key role of proto-oncogenes contributed decisively to the more profound understanding of the genetic basis of cancer. In particular, it created a model for how human tumours form, which involves turning on oncogenes (such as RAS and MYC) one after another and turning off tumour suppressors (such as APC and P53). Moreover, these developments paved the way for the birth of the field of precision oncology.

9.1 Precision Therapies

It turns out that there exist 'critical nodes' in the signalling pathways of cancer cells. Biological therapies have identified several such nodes as their targets. This type of specific intervention has had a transformative effect on the treatment of cancer. In this regard, a pioneering result was the introduction, in 2007, of imatinib, which was discovered by A. Matter and developed by B. Druker and C. Sawyers. This drug targets the unique molecular abnormality causing a particular type of leukaemia called chronic myelogenous leukaemia (CML). This cancer is the result of the generation of the abnormal gene BCR-ABL, formed via translocation, giving rise to the so-called 'Philadelphia chromosome', discovered in 1961 by P. Nowell. This gene causes increased production of the enzyme tyrosine kinase, which is inhibited by imatinib. A 10-year follow-up study, published in 2017, showed that 83% of the patients treated with imatinib had a complete response. This shows that, for most patients, imatinib can convert this particular leukaemia into a manageable chronic disease. This development provided a paradigmatic shift in the approach to cancer treatment. It initiated the introduction of therapies targeting specific molecular abnormalities. These therapies are designed to identify drugs that inhibit specific molecular alterations responsible for certain cancers. Therapeutic approaches based on such drugs are called 'biological therapies' and form the foundation of 'precision medicine'.

Another major success of a drug targeting a molecular abnormality is the treatment of chronic lymphocytic leukaemia (CLL). The condition is the most common chronic leukaemia in adults and is usually diagnosed via a high lymphocyte count in the circulating blood. As discussed in Chapter 7, by the early 1950s, it had become clear that the lymphocytes are vital components of the immune system. Apart from blood, the bone marrow and lymphoid tissue, such as the lymph nodes and the spleen, also contain lymphocytes. As discussed in Chapter 7, lymphocytes are of different subtypes, including B cells, which secrete antibodies, and T cells, which directly or indirectly contribute to removing unwanted cells and eliminating foreign microbes. CLL is caused by a clonal expansion of B-lymphocytes. This expansion is driven by abnormal B-cell receptor signalling. This aberration is sustained by the overexpression of a protein preventing cell death, called B-cell lymphoma 2 (BCL2). Medications acting against these two targets—receptor signalling and BCL2—have had a transformative impact on the treatment of CLL, replacing chemotherapy. In particular, the combination of ibrutinib, which inhibits an essential component of the B-cell receptor signalling apparatus called Bruton tyrosine kinase, and of venetoclax, which is a selective and potent BCL2 inhibitor, has led to a complete recovery for most patients.

The successes against AML and CLL are illustrative of how the implementation of precision medicine is changing the treatment of cancer. Since alterations in specific genes promote carcinogenesis, the essence of these therapies is the understanding that specific molecular alterations can be the targets of specific therapies. It is now possible to identify several particular carcinogenic alterations and, more importantly, to target some of them with several inhibitors and antibodies. For example, the amplification of the human epidermal receptor-2, which can cause breast cancer, can be targeted with, among other drugs, trastuzumab. The mutation of the epidermal growth factor receptor (EGFR), which can cause non-small-cell lung cancer, can be targeted with, among others, gefitinib. The mutation of the B-Raf proto-oncogene, which can cause melanomas and specific forms of lung cancer, can be targeted with several drugs, including dabrafenib and vemurafenib.

Unfortunately, not all patients exhibit pharmacologically targetable DNA alterations; also, some possess such alterations but do not respond to therapy. Additionally, tumours are diverse; factors such as unstable genes, changes in how genes are turned on or off, and varying gene activity due to environmental changes—strongly influenced by treatment—play a key role in the survival of different cancer cells, causing this diversity. In particular, the formation of new mutations results in the development of resistance to a given therapy. In his uplifting book, David Agus describes the modern approach to cancer treatment used for Steve Jobs: after the development of resistance to his metastatic cancer treatment, the DNA of his newly mutated tumour was re-examined, and new targets were identified; this procedure was repeated several times.

Using double or triple therapy offers hope for further progress. For example, an inhibitor of the proto-oncogene BRAF fails for colorectal cancer, characterised by the presence of the BRAF V600E mutation. It turns out that this failure is due to a rapid activation of a feedback mechanism through the action of EGFR. Progress has been achieved by adding an anti-EGFR monoclonal antibody to the BRAF inhibitor.[4]

9.2 Immunotherapy

Another remarkable implication of molecular biology involves stimulating the immune system to attack cancer cells. The modern understanding of immunology discussed in the earlier chapters, together with key advances in molecular biology reviewed in Fokas (2024), have had a significant impact on medicine in general and on cancer treatment in particular. Mutations within a cancer cell result in the generation of new proteins called neoantigens, which can be recognised by the immune system as 'nonself'. Understanding the highly complex immune response has led to efforts to target neoantigens with antibodies and T cells. In particular, the development of specific antibodies such as rituximab in the mid-1990s was crucial for treating many autoimmune diseases and certain types of cancer. For example, rituximab was successful for the treatment of B-cell lymphoma, a cancer caused by the proliferation of abnormal

---

[4]Actually, the best results so far have been obtained by employing a triple therapy involving the use of the BRAF inhibitor encorafenib, the anti-EGFR monoclonal antibody cetuximab, and the mitogen-protein kinase inhibitor binimetinib (Kopetz *et al.*, 2019).

B-lymphocytes. Also, learning that the cytokine interleukin-2 helps T-lymphocytes grow led to the development of several drugs, such as ipilimumab, to change how the immune system works. This work has had a significant impact on both autoimmune diseases and the treatment of several types of cancer, including metastatic melanoma and renal tumours.

Considering that, as a result of genetic and epigenetic changes, cancer cells differ substantially from normal cells, it is surprising that cancer cells are not directly eliminated by T cells. This is due to various factors, which include the following: an overwhelming neoantigen load; chronic T-cell receptor stimulation, which drives T cells into a hyporesponsive state (known as T-cell exhaustion); and the generation of a variety of signals from the microenvironment and, in particular, the use of specific proteins to activate the 'checkpoints', as discussed in Chapter 7. A major break-through in cancer immunotherapy, for which J. Allison was awarded the Nobel Prize in Medicine in 2018, is the so-called 'checkpoint inhibition therapy'. This is based on the manipulation of 'checkpoint regulators'; there now exist drugs called 'checkpoint inhibitors', blocking the action of these proteins, which allows the immune system to attack cancer cells. In the past decade, following the approval of ipilimumab for the treatment of melanoma, regulatory agencies across the world have approved almost 100 additional uses for a dozen checkpoint inhibitors. Remarkably, using these medications, long-term remission can be achieved in patients with several cancers, including melanoma, some cases of non-small-cell lung cancer, Hodgkin's lymphoma, and the so-called 'mismatch repair-deficient' tumours of various histological origins, such as particular types of gastrointestinal, endometrial, breast, prostate, bladder, and thyroid cancers. The repair-deficient tumours involve mutations in genes which are important for correcting mistakes made when DNA is copied in a cell; the neoantigens of these tumours are highly specific, so they are ideal candidates for immune checkpoint inhibitors. For example, 12 patients with such stage II or III rectal cancers were treated with dostarlimab, a PD-1 inhibitor; they all achieved complete remission, and at a median follow-up of 12 months, none exhibited cancer regrowth (Cercek *et al.*, 2022).

Unfortunately, checkpoint inhibitors may have toxic effects, provoking autoimmunity and several adverse reactions in the lungs, liver, and skin, with an approximately 1% fatality rate (Johnson *et al.*, 2023).

Cancer cells use a mechanism to dampen the T-cell response against them by activating one of the Programmed Death (PD) receptors mentioned in the earlier chapters. Patients with several advanced cancers have achieved durable tumour regression by administering specific antibodies against a particular type of these receptors, known as PD-L1. For example, urothelial carcinoma cells are typically positive for PD-L1; the addition of standard supportive care for patients with advanced or metastatic urothelial carcinoma of specific antibodies against PD-L1, such as avelumab, significantly prolonged survival.

Importantly, combining several of the new treatments has led to truly remarkable results. For example, for patients with advanced melanoma, the combination of ipilimumab and the anti-PD 1 agent nivolumab led to spectacular results (which is further discussed in the following chapter).

It is worth noting that medical imaging, which was reviewed in Fokas (2024), is important, not only in the diagnosis but also in the treatment of cancer. For example, using specific ligands such as 89Zr-Atezolizumab, PET can identify PD-L1 locations and help direct immunotherapy.

In the 2010s, the development of genetically engineered T cells that can target cancer cells marked another immunological breakthrough. Several cancerous cells, including those associated with B-cell leukaemia and lymphomas, are characterised by specific neoantigens on their surfaces. Such a marker is called CD19. By removing T cells from the blood of a patient with this type of cancer and using molecular biology techniques, it is possible to insert a new gene into these cells in vitro so that they can express specific receptors that recognise CD19. Large numbers of these engineered T cells are grown in the laboratory and then administered to patients by infusion, targeting the cancer cells expressing CD19. The engineering of such cells, which are called Chimeric Antigen Receptor (CAR) T cells, has revolutionised the treatment of patients with advanced blood cancers (June and Sadelain, 2018). Since 2017 and as of 2024, more than 10 distinct CAR T-cell products have been approved for non-Hodgkin lymphoma, acute lymphoblastic leukaemia, and multiple myeloma.

The current generation of autologous CAR T-cell therapies, i.e. therapies where the patient's own T cells are used, often faces production delays, which limits their applicability in rapidly progressing diseases, such as acute myeloid leukaemia (which continues to have a dismal five-year survival rate below 20%). Progress has now been made to bypass this difficulty by using allogeneic therapy, i.e. using T cells from other individuals. Interestingly, in this treatment, graft-versus-host disease, i.e. rejection of these 'foreign' cells, did not occur despite the fact that no immunosuppression was used (Hu *et al.*, 2024). This therapy can also be used for the elimination of cancer cells which escaped after a given therapy. For example, allogeneic CAR T cells were used in a girl suffering from relapsing T-cell acute lymphoblastic leukaemia to eliminate residual tumour cells, allowing her to undergo a bone marrow transplant (Chiesa *et al.*, 2023).

Since CAR T-cell therapy is based on genomic manipulation, there exists the possibility of oncogenesis. Until the end of 2023, 22 cases of T-cell cancers that occurred after the treatment with CAR T cells were reported to the FDA.

The new understanding of the importance of the epigenome has led to new therapeutic targets for cancer, especially for certain types of acute myeloid leukaemia and for relapsing or refractory non-Hodgkin lymphoma. These new developments will be discussed in a future work (Fokas and Ktistakis, in preparation). Here, it is only mentioned that, as of 2020, there are nine epigenetic drugs available in the US for standard-of-care treatment (Bates, 2020).[5]

It is worth noting that DNA abnormalities can be identified by several commercial platforms, such as 'Foundation One' and the platform developed by the Mayo Clinic. Moreover, there exist specific platforms that can quantify the number of mutations carried by tumour cells and hence predict the response to immunotherapy; a higher number of mutations is associated with an increased number of neoantigens, which provides a larger number of targets for immunotherapy.

References

Baltimore, D. (1970). Viral RNA-dependent DNA polymerase: RNA-dependent DNA polymerase in virions of RNA tumour viruses. *Nature*, *226*(5252), 1209–1211.

Bates, S. E. (2020). Epigenetic therapies for cancer. *New England Journal of Medicine*, *383*(7), 650–663.

Bishop, J. M. (2003). *How to Win the Nobel Prize, An Unexpected Life in Science.* Cambridge: Harvard University Press.

Cercek, A., *et al.* (2022). PD-1 blockade in mismatch repair–deficient, locally advanced rectal cancer. *New England Journal of Medicine*, *386*, 2363–2376.

---

[5]These include two DNMT inhibitors (azacytidine and decitabine), four HDAC inhibitors (vorinostat, romidepsin, belinostat, and panobinostat), and one KMT inhibitor (tazemetostat). Footnote 2 introduces the terminology of DNMT, HDAC, and KMT.

Chiesa, R., *et al.* (2023). Base-edited CAR7 T cells for relapsed T-cell acute lymphoblastic leukemia. *New England Journal of Medicine, 389*, 899–910.

Fokas, A. S. (2024). *Ways of Comprehending: The Grand Illusion and the Essence of Being Human.* Singapore: World Scientific.

Fokas, A. S. and Ktistakis, N. (in preparation). Ageing and the quest for rejuvenation; diet, epigenetics, microbiota.

Haridy, Y., *et al.* (2019). Triassic Cancer-osteosarcoma in a 240-million-year-old stem turtle. *JAMA Oncology, 5*, 425–426.

Hu, Y., *et al.* (2024). Sequential CD7 CAR T-cell therapy and allogeneic HSCT without GVHD prophylaxis. *New England Journal of Medicine, 390*, 1467–1480.

Johnson, P. C., Gainor, J. F., Sullivan, R. J., Longo, D. L., and Chabner, B. (2023). Immune checkpoint inhibitors-the need for innovation. *New England Journal of Medicine, 388*, 1529–1532.

June, C. H. and Sadelain, M. (2018). Chimeric antigen receptor therapy. *New England Journal of Medicine, 379*(1), 64–73.

Kopetz, S., *et al.* (2019). Encorafenib, binimetinib, and cetuximab in BRAF V600E–mutated colorectal cancer. *New England Journal of Medicine, 381*(17), 1632–1643.

Nangalia, J. and Campbell, P. J. (2019). Genome sequencing during patient's journey through cancer. *New England Journal of Medicine, 381*, 2145–2156.

Temin, H. M. and Mizutami, S. (1970). RNA-dependent DNA polymerase in virions of Rous sarcoma virus. *Biotechnology, 24*, 51–56.

Varmus, H. E. (1990). Nobel lecture, Retroviruses and oncogenes. *Bioscience Reports, 10*, 413–430.

Chapter 10

Novel Approaches to Cancer Treatment

Cancer development depends on genetic predisposition and environmental factors. In some cancers, such as cervical, liver, and stomach (gastric) cancers, infections are of key importance. Progress has been made in quantifying the interaction of these contributing factors. The quantification of factors contributing to the development of gastric cancer provides an example. This malignancy, which is the fourth most common cause of cancer-related deaths worldwide, is usually diagnosed late after metastases have occurred and has a poor five-year survival rate of 32%. Risk factors include smoking, as well as a diet low in fruits and vegetables and high in salted, smoked, or preserved foods. Epstein–Barr virus infection and histopathological lesions (such as chronic atrophic gastritis and intestinal metaplasia) are also important risk factors.

It has long been established that infections with the bacterium Helicobacter pylori, which can also cause gastric ulcers, are a major risk factor for gastric cancer. What is the relationship between this infection and genetic predisposition? It is well known that pathogenic variants of specific genes are risk factors for certain cancers. For instance, having harmful changes in the ATM, BRCA2, BRCA1, and PALB2 genes increases the risk of developing breast, ovarian, prostate, and pancreatic cancers. It turns out that certain pathogenic variants of nine other specific genes are more common among patients with gastric cancer than the general population. Interestingly, Helicobacter pylori infection significantly increases cancer risk in individuals carrying these nine genes. Namely, noninfected carriers have a lifetime risk of 5% of developing gastric cancer, but this risk jumps to 45.5% for infected carriers. Infected individuals who are not carriers of these nine pathogenic genes have a lifetime risk of 14.4% (Usui *et al.*, 2023). These estimates illustrate the synergistic effect of genetic predisposition and specific environmental drivers (in this case, infection with Helicobacter pylori) for the development of cancer. The elucidation of such synergies is of vital significance for the prevention, deeper understanding, and treatment of cancer.

Genetic predisposition accounts for 42% of the life risk of prostate cancer, with environmental factors accounting for the remaining portion. Red meat consumption correlates with prostate cancer risk, while vegetables may provide a protective effect. Antioxidants, such as vitamin E and selenium, may reduce the risk of developing cancer. In a rigorous study, high intake of cruciferous vegetables containing isothiocyanate sulforaphane was correlated with a diminished risk of prostate cancer. The effect of this chemoprotective substance is apparently due to its promotion of the expression of carcinogen-detoxification enzymes, which limit the damage caused by carcinogens.

Prostatitis may have an important role in initiating the pathway to cancer. Specifically, it may cause proliferative inflammatory atrophy, which in turn gives rise to cells of prostatic intraepithelial neoplasia and finally to cancer. This scenario may explain the predisposing role of the genes RNASEL and MSR1, which apparently makes the carrier vulnerable to certain infections. The genes PTEN and AR may modulate the progression of prostatic carcinogenesis, whereas the genes AR, CYP17, and SRD5A2 affect the action of androgens (Nelson *et al.*, 2003).

Eliminating smoking, adhering to a healthy diet, and, importantly, following cancer-screening recommendations remain crucial components of cancer prevention. In particular, adhering to strict recommendations of screening for colorectal, breast, cervical, and prostate cancers dramatically decreases deaths from these malignancies. For example, it has been calculated that 90% of colorectal cancer-related deaths could be prevented with early detection.

Regarding colon cancer, colonoscopy remains the gold standard. It is now recommended for individuals older than 45 years of age. 'Liquid biopsy' has emerged as an important non-invasive screening tool; it detects tumour-related DNA in the peripheral blood. In a clinical study, the measurement of circulating tumour DNA (ctDNA) for colorectal cancer had 83% sensitivity (i.e. 83% of individuals with early cancer were identified) and 90% specificity (i.e. 90% of individuals with a positive test had early cancer) (Chung *et al.*, 2024).

Guidelines for the screening of cervical and breast cancers are well established. On the other hand, some physicians questioned in the past the use of prostatic serum antigen (PSA) screening. The past 25 years have seen a decrease in prostate cancer mortality in all racial and ethnic groups. The reduction has been attributed largely to the improved therapies and the routine use of PSA screening. Hence, PSA screening is now accepted. One of the problems with the earlier use of PSA was that it led to a large number of unnecessary biopsies following high PSA values. Today, if a patient has an elevated PSA level, a (multiparametric) magnetic resonance imaging (MRI) is performed to determine whether to perform a biopsy. Lesions are classified on a scale ranging from 1 to 5 (using the Prostate Imaging Reporting and Data System); patients with scores of 3–5 are usually selected for biopsy. In addition to the measurement of PSA, which is the total PSA found in the blood, the measurement of free PSA, i.e. the PSA that is not attached to proteins, is also recommended. The higher the ratio of free PSA to total PSA, the lower the risk of prostate cancer. For borderline cases, in addition to the above score and PSA level, one may use data from PET, which can quantify 'prostate-specific membrane antigen', as well as information from genomic assessment (known as genomic classifiers) (Sarto, 2023).

When should testing for PSA begin, and what are the threshold values? Prostate cancer tends to appear earlier and is more aggressive in certain ethnic groups, including Black men. In these cases, screening should start around the age of 40. The threshold values are age-dependent; the current recommendations are 3.5 ng/mL for men in their 50s and 4.5 ng/mL for men in their 60s. Men with newly elevated PSA levels should have repeat testing a few months later to corroborate the measurement before they undertake further evaluation.

It is now possible to measure the low relative risk of prostate cancer associated with hundreds of specific genetic variants. This gives rise to the so-called polygenic risk score. Men in the highest decile of risk for prostate cancer have more than three times the risk of those at average risk, and men in the lowest decile have about one-third the risk of those at average risk. A screening programme based on the assessment of the polygenic risk score identified a number of clinically significant cases that otherwise would have been missed. There is the suggestion that this score should be used for men with a positive PSA result to help the physician to determine who should undergo MRI (Hunter, 2025).

Incidentally, in addition to its use as a screening tool, the measurement of ctDNA is also useful for the detection of 'minimal residual disease', which involves detecting cancer cells remaining after a given treatment. This information can be useful for deciding the best follow-up course of action. For example, ctDNA can be used to answer the following question: should chemotherapy be used after surgery for stage II colon cancer? It turns out that for such patients, the detection of ctDNA increased the risk of relapse by 18 times in comparison to patients with negative ctDNA (Montagut and Vidal, 2022).

A common complication of cancer is cachexia, a syndrome characterised by progressive weight loss, which can lead to depletion of adipose tissue and muscle, and ultimately to death. Cachexia is caused by reduced dietary intake and abnormal metabolism or catabolism. Several mechanisms that affect food intake reside in the brain stem. In particular, the GDF-15 protein is a type of cytokine (part of the transforming growth factor β family) that, when it attaches to receptors on nerve cells in a specific part of the brain stem called the nucleus of *the solitary tract*, causes a decrease in appetite and feelings of nausea. Many tumours overexpress this protein; also, several chemotherapies, especially cisplatin, induce the expression of GDF-15 in normal tissues. A major breakthrough in the fight against cachexia was the development of ponsegromab, a monoclonal antibody inhibiting GDF-15. Among patients with cancer cachexia and elevated GDF-15 levels, the use of ponsegromab resulted in increased weight gain and overall activity level, as well as the reduction of other cachexia symptoms (Groarke *et al.*, 2024).

Another possible complication of cancer is venous thromboembolism. The oral anticoagulant, apixaban, is effective for the prevention of recurrent venous thromboembolism but is associated with bleeding complications. Very recently, it was shown that extended treatment with reduced-dose apixaban is as effective for preventing thromboembolic episodes and is associated with a lower incidence of clinically significant bleeding complications (Mahe *et al.*, 2025).

Interestingly, since 2006, several prospective observational studies have suggested that physical activity after the occurrence of several types of cancer yields survival benefits. Now, for the case of colonic cancers, such benefits have been rigorously established: a three-year structured exercise programme initiated soon after adjuvant chemotherapy resulted in significantly longer disease-free survival and findings consistent with increased overall survival (Courneya, 2025). Perhaps, this is the result of the effects of exercise on metabolic growth factors, inflammation, and immune function.

10.1 Breast, Prostate, and Other Common Cancers

Current clinical practice combines the traditional use of pathology with the tools of molecular biology. Traditional information from pathology, such as the 'grade', which quantifies the 'departure' of the tumour from normal tissue and the rate of its proliferation, is now supplemented by specific data obtained from gene-expression assays of tumour samples. For example, breast cancer treatment is designed on the basis of the following features: tumour size and grade; occurrence of metastasis in the axillary nodes; expression of oestrogen and progesterone receptors, which identifies patients that will benefit from endocrine therapy; overexpression of the human epidermal growth receptor-2 protein, which identifies patients who will benefit from biological therapy targeting these receptors; and additional information regarding the genetic make-up of the tumour cells. According to these features, hormone therapy is combined with novel, specific inhibitors of signalling pathways responsible for the proliferation and survival of cells.

The prognosis of breast cancer patients is calculated taking into consideration the above features and 21 specific genes (Sparano *et al.*, 2015). With a score of less than 10, patients are not subjected to chemotherapy, while patients with a score higher than 25 are treated with chemotherapy. For patients in the range of 11–25, the question of whether chemotherapy would be useful was open for a long time, but it is now known that it is not needed. This new approach to breast cancer treatment and prognosis illustrates the importance of precision medicine. It uses specific drugs based on individualised information, which includes genome-based data.

Incidentally, traditional chemotherapeutic drugs kill dividing cells by stopping DNA replication (the so-called S-phase) or mitosis (the so-called M-phase). There now exist new drugs which arrest progression through the so-called G-phase: they promote transient cell cycle withdrawal into a quiescent state; this is a state where cells rest and do not replicate. Such drugs are now used for oestrogen-receptor-positive, human epidermal growth factor-negative, advanced breast cancer.

There are breast cancers that lack expression of oestrogen receptors, progesterone receptors, and human epidermal growth factor receptor 2. These cancers are referred to as 'triple-negative breast cancers', i.e. they are defined by what they lack. Recently, these cancers have found a new characterisation: it turns out that they exhibit higher tumour expression of PD-L1 receptors and more tumour-infiltrating lymphocytes than other breast cancers. This makes them good candidates for treatment with immune checkpoint inhibitors. Indeed, in a trial involving women with stage II or III triple-negative breast cancer, when pembrolizumab, which blocks the programmed death receptors on T cells, was added to chemotherapy, the overall survival improved from 82% to 87% (Schmid *et al.*, 2020).

Targeting specific genetic aberrations continues to yield better outcomes. For example, the drug imlunestrant inhibits the oestrogen receptors encoded by a specific mutation of the oestrogen receptor gene (SER1). The addition of this drug to standard therapy for patients with advanced breast cancer whose cancer carried this gene significantly prolonged the progression-free survival (Jhaverri *et al.*, 2024).

Another use of molecular biology is to help with the decision regarding whether to avoid radiotherapy after breast-conserving therapy. It has been established that patients at least 55 years of age who underwent breast-conserving surgery for grade 1 or 2 localised (tumour size <2 cm, no lymph node involvement) cancer did not need radiotherapy, provided that they had the 'luminal A' subtype, namely that their tumour had more than 1% oestrogen receptors, more than 20% progesterone receptors, and no human epidermal growth type 2 receptors (Whelan *et al.*, 2023).

Also, regional nodal irradiation was not beneficial in patients who had negative axillary nodes following chemotherapy (Mamounas *et al.*, 2025). Incidentally, it has now been established that vasomotor symptoms in women receiving endocrine therapy for hormone receptor-positive breast cancer can be treated effectively with elinzanetant, which is a medication that antagonises the neurokinin system (Cardoso *et al.*, 2025).

The overall rate of survival of breast cancer began to increase by the early 1990s, and this trend continues up to now. The same is true for prostate cancer. The effect of oestrogens on breast cancer was documented as early as 1896 in a publication by G. T. Beatson in the prestigious medical journal *Lancet*. Charles Huggins was awarded the Nobel Prize in Medicine in 1966 for pioneering the hormonal treatment of men with prostate cancer. Breast and prostate cancer, as well as several other tumours, are susceptible to hormonal manipulation due to the fact that the growth of their normal-tissue counterparts is regulated by hormones.

In the UK, between 1999 and 2009, 82,429 men between 50 and 69 years of age underwent a prostate biopsy as a result of elevated PSA; 2,664 were diagnosed with localised prostate cancer. Among these men, 1,643 enrolled in a trial to evaluate the effectiveness of three different approaches: approximately a third were randomly assigned to receive active monitoring, a third to undergo prostatectomy, and a third to receive radiotherapy. In the prostatectomy group, doctors suggested radiotherapy if the cancer was found at the edges of the surgery, if the tumour had spread beyond its capsule, or if the PSA level after surgery was above 0.2 ng/mL. In the active-monitoring group, an increase of at least 50% in the PSA level during a 12-month period or any concern raised by the patient or clinician triggered a review. After an average follow-up of 15 years (between 11 and 21 years), the results showed that about 3% of patients died from prostate cancer and about 22% died from other causes; importantly, these rates were similar across all three groups, although patients in the active-monitoring group were slightly more likely to have developed metastases compared to those in the other two groups. These results provide long-term evidence that active surveillance is appropriate for many patients with localised cancer; in this way, many patients can avoid possible postoperative complications. Interestingly, more than half of the active-surveillance patients crossed over to prostatectomy during long-term follow-up (often based on rising PSA levels). However, the authors of this study suggest that some patients' decision to abandon sur-veillance was driven more by anxiety than by compelling clinical reasons (Hamdy *et al.*, 2023).

Prostate surgery for a localised disease is guided by an MRI. For can-cer cells in the lower or back part of the prostate, one option is high-intensity focused ultrasound; this treatment uses ultrasound waves to accurately target and destroy cancer tissue while protecting nearby healthy areas. In 2022, the National Health Service of the UK approved the use of the NanoKnife system for prostate cancer treatment; this is a minimally invasive approach that uses needles to deliver electrical energy to target and destroy cancerous prostate tissue, sparing surrounding healthy structures.

For about 75 years, the primary treatment for men with metastatic, hormone-sensitive prostate cancer has been androgen-deprivation therapy through either surgical or medical castration. About 90% of men respond to such therapy with an improvement in symptoms and a reduction in PSA; the median response duration is approximately 18 months. In the period 2004–2022, eight therapeutic agents were approved by the Food and Drug Administration for the treatment of men with advanced prostate cancer. Four are androgen inhibitors (abiraterone, enzalutamide, apalutamide, and darolutamide); two are chemotherapies that suppress the function of cancer microtubules (docetaxel and cabazitaxel); one is an α-emitting radiopharmaceutical agent (radium-223); and one is an autologous cell–based immunotherapy (sipuleucel-T). It is now known that overall survival is significantly longer if darolutamide is added to docetaxel and androgen-deprivation therapy (Smith *et al.*, 2022); the median duration of survival with this new treatment is approximately 2.5 years.

Additional research has shown that the progression of advanced prostate cancer is driven by androgens produced within the tumour; this is referred to as 'castration-resistant prostate cancer'. Two drugs, abiraterone and enzalutamide, can inhibit residual androgen stimulation of tumour tissue; abiraterone inhibits an enzyme that is critical to the production of androgens and glucocorticoids, whereas enzalutamide binds the androgen receptor and inhibits the transcription of androgen-dependent genes. Despite the different mechanisms of action, these drugs have led to comparable and substantial improvements in survival. Moreover, for the cases in which the tumour cells carry the oncogenes BRCA1 or BRCA2, the drug olaparib is a better option than abiraterone or enzalutamide (the mechanism of action of olaparib will be discussed later in this section in connection with the treatment of ovarian cancer) (Mateo *et al.*, 2023).

In addition to their impact on breast and prostate cancer, biological treatments have had a transformative effect on several cancers. In what follows, some such cases are briefly discussed.

The treatment of patients with advanced melanoma who were treated with a combination of ipilimumab and the anti-programmed death 1 agent nivolumab was followed up for a minimum of five years; remarkably, they had a median overall survival of approximately five years, as opposed to an overall survival of approximately 40 months or 20 months when treated only with nivolumab or ipilimumab, respectively (Larkin *et al.*, 2019). As of 2025, the only approved anti-CTLA-4 agent is ipilimumab. A direct comparison of this agent with the two anti-PD-1 antibodies, nivolumab and pembrolizumab, showed that among patients who were free of disease at three years, 10-year melanoma-specific survival was 97% with nivolumab, 96% with nivolumab plus ipilimumab, and 88% with ipilimumab (Wolchok *et al.*, 2025).

Since 1975, the treatment of choice for Hodgkin's lymphoma has been the ABVD regimen (consisting of doxorubicin, bleomycin, vinblastine, and dacarbazine). The presence of the so-called Reed–Sternberg cells, which express the CD30 receptor, is the hallmark of Hodgkin's lymphoma. This suggests that the use of the anti-CD30 antibody brentuximab vedotin (BV) should be useful for the treatment of Hodgkin's disease. This is true: the ABVD treatment was directly compared to the BV+AVD treatment, where bleomycin was swapped for BV; after an average follow-up of six years, the chance of not having the disease progress was 75% with the first treatment and 82.3% with BV+AVD (Ansell *et al.*, 2022).

For CML (mentioned in an earlier chapter), only 5–7% of patients do not respond to imatinib. Furthermore, in 40–45% of the patients, it is possible to discontinue treatment after 3–5 years (although for some, it may be necessary to restart therapy later in life). A new tyrosine kinase inhibitor called asciminib may be useful for patients who do not respond to ipilimumab or those who require restarting treatment (Abruzzese *et al.*, 2024).

For CLL (mentioned in an earlier chapter), the armament of effective drugs is even more extensive than for CML. Two second-generation Bruton tyrosine kinase (BTK) inhibitors, called acalabrutinib and zanubrutinib, are at least as effective as ibrutinib and have better safety profiles. A B-cell lymphoma 2 (BCL2) antagonist, venetoclax, alone or in combination with the monoclonal antibody obinutuzumab, is useful for relapsed CLL. Using these drugs, the median remission lasts in excess of five years (Woyach and Byrd, 2023).

B-cell lymphoma is a spectrum of aggressive B-cell cancers with heterogeneous clinical progression. First-line chemotherapy cures most patients, but those with relapsed or refractory disease have a dire prognosis. CAR T-cell therapy that targets CD 19 achieves durable remission in up to 40% of the latter group of patients.

In multiple myeloma, where plasma cells become cancerous, bone marrow transplant is the treatment of choice. In patients who are ineligible for transplantation, the addition of the anti-CD38 monoclonal antibody isatuximab to the standard triple therapy is quite beneficial. Namely, the addition of isatuximab used as initial therapy in patients 18–80 years of age increased the 'progression-free' survival at five years from 45% to 63% (Facon *et al.*, 2024); the term 'progression-free survival' means that the patient is not getting worse as a result of cancer progression.

There is a precursor state of multiple myeloma referred to as 'smouldering myeloma'. Until recently, such patients were actively monitored, with no treatment being available. A recent study found that using daratu-mumab, a type of monoclonal antibody that targets CD38, greatly reduced the chance of developing active multiple myeloma. Overall, patients treated with this monotherapy had a higher overall survival in comparison to the patients who were actively monitored (Dimopoulos *et al.*, 2024).

Patients with renal cancer are treated with surgery, which can be cura-tive. However, among patients with high risk factors, the disease recurs in up to 50% of cases. Among these patients, those treated with pembroli-zumab had a 38% lower risk of death than those given a placebo (Voss and Motzer, 2024).

As of 2020, the five-year survival rates in the UK are between 80% and 100% for breast, prostate, melanoma, various types of leukaemia, Hodgkin's lymphoma, and testicular, thyroid, and uterine cancer. Unfortunately, lung, stomach, liver, pancreas, ovarian, and brain cancers continue to have a poor prognosis. However, even in these cases, after a drought for decades, there is now progress. Some examples follow.

Lung cancer remains the leading cause of cancer-related deaths worldwide. Lung tumours are classified as small-cell (also called neuro-endocrine) and non-small-cell cancers. About 15% of all lung tumours are of the former type, which typically metastasises by the time of diagnosis. Most of the non-small-cell tumours are adenocarcinomas, squamous-cell carcinomas, and large-cell carcinomas. All tumours are associated with smoking, except adenocarcinoma. For non-small-cell lung cancer (NSCLC) of stage II or III, the addition of pembrolizumab to standard chemotherapy (with either cisplatin and gemcitabine or cisplatin and pemetrexed) before or after surgery was quite beneficial; it increased the 24-month 'event-free survival' from 40.6% to 62.4% (Lovly, 2023). The terminology 'event-free survival' means that no complications occur due to the original cancer.

For patients with NSCLC of stage IB to IIIA a new approach to treat-ment consists of nivolumab plus chemotherapy followed by surgery and 1 year of nivolumab monotherapy. Adding nivolumab to chemotherapy improved the overall survival from 55% to 65.4%. Importantly, patients who had a complete pathological response, namely, patients with no evi-dence of cancer after surgery, had a 5-year overall survival of 95.3% (Calvo, 2025).

About a third of lung adenocarcinoma and most other NSCLC cases contain the KRAS oncogene. Patients who did not respond to chemo-therapy and an immune checkpoint inhibitor showed some improvement when treated with oral adagrasib, which works against the KRAS onco-gene. Patients with NSCLC who have mutations in the epidermal growth factor receptor (EGFR) can be treated with osimertinib, achieving an overall survival of 38.6 months (Ramalingam *et al.*, 2020). Some lung cancers carry translocations of the anaplastic lymphoma kinase (ALK) gene; it was shown in 2010 that 57% of such patients had a response to treatment with crizotinib, a tyrosine kinase inhibitor. Since then, four more similar inhibitors (alectinib, brigatinib, ensartinib, and lorlatinib) were found to be superior to crizotinib.

More than 1% of tumour cells express PD-L1 receptors in about 50% of patients with advanced oesophageal squamous-cell carcinoma. Hence, it is not surprising that treatment with the anti-PD-1 monoclonal antibody nivolumab results in significantly longer overall survival than chemo-therapy. This approach is now approved for advanced oesophageal squamous-cell carcinoma, irrespective of PD-L1 expression status (Doki *et al.*, 2022).

Pancreatic ductal adenocarcinoma is the deadliest of all common cancers. By the time symptoms manifest, the liver has occult micro-metastases, rendering chemotherapy ineffective. Approximately 41% of patients with pancreatic ductal adenocarcinoma harbour the mutation G12D, which is a single amino acid mutation in the KRAS family; also, a very small percentage of patients—around 1%—carry the G12C mutation. The results of clinical trials of small molecules that inhibit KRAS, such as sotorasib, have shown only moderate responses. Apparently, this is due to the complexity of the processes affected by the KRAS gene. In particular, a signal-related kinase (ERK) is of crucial importance for the action of this gene. The elucidation of factors affecting the expression of KRAS, like ERK, is expected to lead to the discovery of better ways to inhibit the consequences of KRAS (Klomp *et al.*, 2024a, 2024b). There is an encour-aging report that one patient with the gene KRAS carrying the mutation G12D responded well after being treated with CAR T-cell therapy.

Hepatocellular carcinoma is linked to inflammation; most cases of this type of liver cancer are caused by infections with hepatitis B and C, but the number of cases from non-alcoholic fatty liver disease is increas-ing quickly. The combination of atezolizumab and bevacizumab, which focuses on the programmed death receptor 1 immune checkpoint and vascular endothelial growth factor, has greatly increased survival rates. Depending on the disease's cause, appropriate therapy will likely lead to further therapeutic progress. Interestingly, a paradoxical effect of obesity has been observed: although obesity can lead to immune dysfunction and non-alcoholic fatty liver disease, in patients with hatatocellular carcinoma it is associated with a greater antitumour efficacy and survival (Kellry and Greten, 2021).

In patients with locally advanced or metastatic urothelial cancer, the combination of pembrolizumab and the antibody enfortumab vedotin demonstrated a significant benefit. Namely, compared with standard first-line chemotherapy (gemcitabine plus cisplatin or gemcitabine plus carbo-platin), the biological therapy increased both the progression-free survival and the overall survival (Niegisch, 2024).

Mutations of the gene BRCA cause some ovarian tumours. In most of these cases, using a type of medicine called PARP inhibitors (such as niraparib, veliparib, and olaparib) can be helpful; these medicines block a certain enzyme (poly ADP ribose polymerase) that helps repair DNA in damaged cells. Patients who do not have the BRCA gene might still gain from this treatment if their cancer cells have a reduced ability to repair DNA, which is shown by a high score in a certain indicator called homol-ogous recombination deficiency (Longo, 2019).

Glioblastoma is the most formidable brain tumour; current treatment involving surgery, radiotherapy, and alkylating chemotherapy has limited efficacy. Most glioblastoma tumours have specific epidermal growth fac-tor receptors (EGFRs), as well as a modified version of these receptors called EGFR-vIII. Three patients whose tumours processed these recep-tors were treated with CAR T cells activated by EGFR and EGFR-vIII; two of them had a transient response (Choi *et al.*, 2024). Efforts are made to increase and prolong the efficacy of this new approach. Another treat-ment involves the use of virotherapy. A specific virus was introduced via a specially designed cannula into the pons of 12 children suffering from high-grade gliomas. After tumour cells were infected, specific receptors on the cancer cells were activated, which, via a complicated signalling process, led to the production of an inflammatory innate and adaptive immune response, including the release of interferon types I and III. The typical dismal survival of 8–12 months was prolonged to a median of 17.8 months (Perez-Larraya *et al.*, 2022).

It should be emphasised that there is a particular type of glioma, called IDH-mutant glioma (due to the mutation of isocitrate dehydroge-nase), which in the US affects about 2,500 individuals annually. This cancer has a good prognosis; radiotherapy for grade 2 results in a progression-free survival of more than four years; adding chemotherapy increases survival to more than eight years.[1]

The so-called RET proto-oncogene encodes a protein forming a transmembrane receptor used by a family of growth factors. Problems with the function of this gene, which can occur due to changes or joining with other genes, have been found in a small number of thyroid, non-small-cell lung, and colorectal cancers. Specific RET inhibitors, such as selpercatinib, have been shown to increase the survival of patients suffering from cancers which harbour RET alterations (Kurzrock, 2020).[2]

10.2 Anticipated Progress

From the above brief summary of various approaches to cancer treatment, it can be predicted with absolute certainty that further progress will be made.

Our understanding of tumourigenesis continues to deepen. For example, it has been hypothesised that oncogenes activate the DNA damage response (DDR) pathway. This activation triggers various cellular responses, which results in the elimination of cancer cells from the earliest stages of carcinogenesis. However, the accumulation of DNA damage overwhelms the cells' error-free repair capacity, leading to error-prone processes. This situation gives rise to genomic instability and cancer progression. A key prediction of this model is that as cancer develops, senescent cells become active. Oncogene-induced senescence is part of the tumourigenesis barrier imposed by DNA damage checkpoints. Bartkova (2006), Halazonetis *et al.* (2008), and Gorgoulis *et al.* (2018), made significant progress in deciphering how the mechanisms of DNA damage and protein stress responses could be used for elucidating cancer development and ultimately yielding a better treatment. Cellular senescence constitutes a stress response mechanism in reaction to various types of stimuli. Senescent cells exhibit cell-cycle arrest and altered function. It was

---

[1] A drug inhibiting IDH mutant enzymes, called vorasidenib, which was used 1–5 years after initial treatment, had a modest effect. Perhaps, its effect could be more pronounced if it is used early on (Schiff, 2023).

[2] Specifically, for patients with medullary thyroid carcinoma, a comparison of selpercatinib with the multikinase inhibitors cabozantinib and vandetanib showed that selpercatinib was superior for prolonging progression-free survival (Hadoux *et al.*, 2023).

previously believed that this cell-cycle arrest was permanent; however, new findings in cancer research suggest the 'escape-from-senescence' concept: under certain conditions, senescent cells can start dividing again and become very aggressive, which is linked to cancer recurrence. It is conjectured that conventional cancer treatments, which are incapable of eliminating senescence, may benefit from including senolytic agents that selectively eliminate senescent cells (Zampetidis *et al.*, 2021; Hanahan, 2022).

In addition to the deepening understanding of general principles dictating tumourigenesis, the enhanced knowledge of specific genomic alterations that appear in different types of cancer will continue to result in the development of highly specific drugs targeting these alterations. In this regard, it is noted that it is now possible (via a genomic assessment of specific repeat loci) to quantify the microsatellite instability of a given cancer. Also, molecular classification can identify whether for a given cancer with a certain DNA repair process, called 'mismatch repair deficiency' (dMMR), is defective or if it works fine, known as 'mismatch repair proficiency' (pMMR). Such information will become increasingly important in guiding treatment. For example, in a mismatch repair–proficient, microsatellite-stable, endometrial cancer, the addition of pembrolizumab to standard chemotherapy (paclitaxel plus carboplatin) is beneficial (Eskander *et al.*, 2023).

Ideally, the specific neoantigens associated with a given tumour should be defined, and the specific subset of T cells responding to these neoantigens should be stimulated to attack them. Efforts towards such an approach, which for example, would avoid the toxicity associated with checkpoint inhibitors, are ongoing; see, for example, Krishna *et al.* (2020), where the tumour-specific neoantigens of various solid cancers are analysed.

An alternative, more direct approach is to engineer T cells that resist exhaustion and, more generally, the dysfunctional effects exerted by the tumour and its environment. It has been shown that the gene that encodes a specific receptor of the TNF-receptor family (called LTBR) had a very positive impact, not only on the ability of T cells to proliferate but also on maximising cytotoxicity (via the production of interleukin-2 and interferon-γ) (Legut *et al.*, 2022).

For patients who have relapsed after anti-CD19 CAR T-cell therapy or are resistant to this treatment, a promising strategy involves developing

CAR T cells that secrete specific proinflammatory cytokines. This is based on the hypothesis that cytokine secretion enhances the antitumour activity of CAR and tumour-infiltrating T cells. Indeed, CAR T cells, when appropriately modified to secrete interleukin-18, showed promising efficacy in patients with lymphoma after the failure of standard anti-CD19 CAR T-cell therapy (Svoboda *et al.*, 2025).

It was discovered in the 1960s that certain childhood tumours contain circular DNA structures located in the nuclei of cells, outside chromosomes. The occurrence of such extrachromosomal DNA (ecDNA) has been attracting a lot of interest recently. In an analysis of a large number of tumour samples, ecDNA was detected in 17% of cases, most commonly in tumours with poor prognosis, including sarcoma, glioblastoma, and gastroesophageal adenocarcinoma (Baily *et al.*, 2024). Specific strategies are being developed targeting ecDNAs, including the development of drugs against proteins coded by these circular DNA structures.

In summary, the last couple of decades have witnessed unprecedented success in the fight against cancer. There is no doubt that this stupendous, life-saving process will continue.

References

Abruzzese, E., *et al.* (2024). CML 25 years later – Poised for another breakthrough? *New England Journal of Medicine, 391*, 955–957.

Ansell, S. M., Radford, J., Connors, J. M., Długosz-Danecka, M., Kim, W.-S., Gallamini, A., Ramchandren, R., *et al.* (2022). *New England Journal of Medicine, 387*, 310–320.

Baily, C., *et al.* (2024). Origins and impact of extrachromosomal DNAs in cancer cells. *Nature, 635*, 193–200.

Bartkova, J. (2006). Oncogene-induced senescence is part of the tumorigenesis barrier imposed by DNA damage checkpoints. *Nature, 444*, 633–637.

Calvo, V. (2025). Redefining lung cancer therapy-A long-awaited shift in strategy. *New England Journal of Medicine, 393*, 809–810.

Cardoso, F., Parke, S., Brennan, D. J., *et al.* (2025). Elinzanetant for vasomotor symptoms from endoctine therapy for breast cancer.

Choi, B. D., *et al.* (2024). Interventricular CAR v3-TEAM-E T cells in recurrent glioblastoma. *New England Journal of Medicine, 390*, 1290–1298.

Chung, D. C., *et al.* (2024). A cell-free DNA blood-based test for colorectal cancer screening. *New England Journal of Medicine, 390*, 9723–9783.

Courneya, K. S., Vardy, J. L., *et al.* (2025). *New England Journal of Medicine, 393*, 13–25.

Dimopoulos, M. A., *et al.* (2024). Daratumumab or active monitoring for high-risk smouldering multiple myeloma? *New England Journal of Medicine, 392*, 1777–1788.

Doki, Y., *et al.* (2022). Nivolumab combination therapy in advanced esophageal squamous-cell carcinoma. *New England Journal of Medicine, 386*, 449–462.

Eskander, R. N., *et al.* (2023). Pembrolizumab plus chemotherapy in advanced endometrial cancer Ramez. *New England Journal of Medicine, 388*, 2159–2170.

Facon, T., *et al.* (2024). Isatuximab bortezomib, lenalidomide, and dexamethasone for multiple myeloma. *New England Journal of Medicine, 391*, 1597–1609.

Gorgoulis, V. G., Pefani, D.-E., Pateras, I. S., *et al.* (2018). Integrating the DNA damage and protein stress responses during cancer development and treatment. *Journal of Pathology, 246*, 12–40.

Groarke, J. D., *et al.* (2024). Ponsegromab for the treatment of cachexia. *New England Journal of Medicine, 391*, 2291–2303.

Hadoux, J., *et al.* (2023). Phase 3 trial of a comparison of selpercatinib in advanced RET-mutant medullary cancer. *New England Journal of Medicine, 389*, 1851–1861.

Halazonetis, T. D., Gorgoulis, V. G., and Bartek, J. (2008). An oncogene-induced DNA damage model for cancer development. *Science, 319*, 1352–1355.

Hamdy, F. C., *et al.* (2023). Fifteen-year outcomes after monitoring, surgery, or radiotherapy for prostate cancer. *New England Journal of Medicine, 388*, 1547–1558.

Hanahan, D. (2022). Hallmarks of cancer: New dimensions. *Cancer Discovery, 12*, 31–46.

Hunter, D. J. (2025). A polygenic risk score in practice. *New England Journal of Medicine, 392*, 1444–1446.

Jhaverri, K. L., *et al.* (2024). Imlunestrant with or without abemaciclib in advanced breast cancer. *New England Journal of Medicine, 392*, 1189–1202.

Kellry, R. K. and Greten, T. F. (2021). Hepatocellular carcinoma-origins and outcomes. *New England Journal of Medicine, 385*, 280–281.

Klomp, J.A., *et al.* (2024a). Defining the KRAS-and ERK-dependent transcription in KRAS-mutant cancers. *Science, 384*, eadk0775.

Klomp, J. E., *et al.* (2024b). Determining the ERK-regulated phosphoproteome driving KRAS-mutant cancer. *Science, 384*, eadk0850.

Krishna, S., *et al.* (2020). Stem-like CD8 T cells mediate response of adoptive cell immunotherapy against human cancer. *Science, 370*, 1328–1334.

Kurzrock, R. (2020). Selpercatinib aimed at RET-altered cancers. *New England Journal of Medicine, 383,* 868–869.

Larkin, J., Chiarion-Sileni, V., Gonzalez, R., Grob, J. J., Rutkowski, P., Lao, C. D., Cowey, C. L., *et al.* (2019). Five-year survival with combined nivolumab and ipilimumab in advanced melanoma. *New England Journal of Medicine, 381*(16), 1535–1546.

Legut, M., *et al.* (2022). A genome-wide screen for synthetic drivers of T cell proliferation. *Nature, 603,* 728–735.

Longo, D. L. (2019). Personalised medicine for primary treatment for serous ovarian cancer. *New England Journal of Medicine, 381,* 2471–2474.

Lovly, C. M. (2023). Perioperative immunotherapy – A key toward improved outcomes for early-stage lung cancer? *New England Journal of Medicine, 389,* 560–561.

Mahe, I., *et al.* (2025). Extended reduced-dose apixaban for cancer-associated venous thromboembolism. *New England Journal of Medicine, 392,* 1363–73.

Mamounas, E. P., Bandos, H., *et al.* (2025). Omitting regional nodal irradiation after response to neoadjuvant chemotherapy. *New England Journal of Medicine, 392,* 2113–2124.

Mateo, J., *et al.* (2023). Olaparib for the treatment of patients with metastatic castration-resistant prostate cancer and alteration in BRCA1 or BRCA2 in the PROfound trial. *Journal of Clinical Oncology, 42,* 571–583.

Montagut, C. and Vidal, J. (2022). Liquid biopsy for precision adjuvant chemotherapy in colon cancer. *New England Journal of Medicine, 386,* 2330–31.

Nelson, W. G., *et al.* (2003). Prostate cancer. *New England Journal of Medicine, 349,* 366–381.

Niegisch, G. (2024). Enfortumab Vedotin and Pembrolizumab—A new perspective on urothelial cancer. *New England Journal of Medicine, 390,* 944–946.

Perez-Larraya, J. D., *et al.* (2022). Oncolytic DNX-2401 virus for paediatric diffuse intrinsic pontine glioma. *New England Journal of Medicine, 388,* 2471–81.

Ramalingam, S. S., *et al.* (2020). Overall survival with osimertinib in untreated EGFR-mutated advanced NSCLC. *New England Journal of Medicine, 382,* 41–50.

Sarto, O. (2023). Localised prostate cancer —Then and now. *New England Journal of Medicine, 388.*

Schiff, D. (2023). Headway against brain tumours with molecular targeting of IDH-mutant gliomas. *New England Journal of Medicine, 389,* 653–654.

Schmid, P., *et al.* (2020). Pembrolizumab for early triple-negative breast cancer. *New England Journal of Medicine, 382,* 810–821.

Smith, M. R., *et al.* (2022). Darolutamide and survival in metastatic, hormone-sensitive prostate cancer. *New England Journal of Medicine, 386,* 1132–1142.

Sparano, J. A., Gray, R. J., Makower, D. F., Pritchard, K. I., Albain, K. S., Hayes, D. F., Geyer Jr, C. E., *et al.* (2015). Prospective validation of a 21-gene expression assay in breast cancer. *New England Journal of Medicine*, *373*, 2005–2014.

Svoboda, J., *et al.* (2025). Enhanced CAR T-cell therapy for lymphoma after previous failure. *New England Journal of Medicine*, *392*, 1824–1835.

Usui, Y., *et al.* (2023). Helicobacter pylori, homologous-recombinant genes, and gastric cancer. *New England Journal of Medicine*, *388*, 1181–90.

Voss, M. H. and Motzer, R. J. (2024). Adjuvant immunotherapy for kidney cancer — A new strategy with new challenges. *New England Journal of Medicine*, *390*, 1432–1433.

Whelan, T. J., *et al.* (2023). Omitting radiotherapy after breast-conserving surgery in luminal A breast cancer. *New England Journal of Medicine*, *389*, 612–618.

Wolchok, J. D., *et al.* (2025). Final 10-year outcome with nivolumab plus ipilimumab in advanced melanoma. *New England Journal of Medicine*, *392*, 11–22.

Woyach, J. A. and Byrd, J. C. (2023). Time-limited initial therapy for young fit patients with CLL. *New England Journal of Medicine*, *388*, 1812–1813.

Zampetidis, C. P., *et al.* (2021). A recurrent chromosomal inversion suffices for driving escape from oncogene-induced senescence via subTAD reorganisation. *Molecular Cell*, *81*, 4907–4923.

Chapter 11

A Unified Approach to Aetiology
and Treatment in Medicine

The recent impact of immunology in the treatment of both cancer and autoimmune diseases discussed in the last two chapters illustrates the emergence of a *unified approach* for the understanding and treatment of different diseases. Several additional examples of the importance of the concept of unification are discussed in this chapter.

11.1 Inflammation

As discussed in Chapter 7, immunological mechanisms involve the mobilisation of a variety of different cells and molecules in specific amounts and in a well-defined time sequence. The highly dynamic inter-actions of the components of the 'generalised immune system' give rise to the generic manifestation of inflammation, as well as to a variety of more specific processes, from healing, regeneration, and angiogenesis, to the formation of scar tissue and 'apoptosis', i.e. cell-death (Germenis, 2019). Inflammation occurs in almost every pathological process, including infection, atherosclerosis, and cancer. It is a protective response aiming to defend and restore physiological functions when homeostatic mechanisms are inadequate. Inflammatory signals can override homeostatic ones because they have higher priority. This implies that, occasionally, in an attempt to restore homeostasis, inflammation can reset basic homeostatic parameters outside their physiological range; this has detrimental effects resulting in chronic pathological states (Kotas and Medzhitov, 2015).

A simple blood test monitoring inflammation is the measurement of the C-reactive protein, whose levels increase if inflammation persists.

In many cases, inflammation may not be the key mechanism, but still it contributes significantly to the final syndrome. For example, this occurs in a migraine. The first phase of this prevalent neurological disorder begins with a 'depolarisation wave' that self-propagates across the cortex. This wave disrupts the normal ionic currents, leading to a reduction of blood flow to the cortex. The aura phase of the migraine is the consequence of the resulting cerebral *hypoperfusion*. In particular, the reduction of the blood supply to the visual cortex causes the appearance of *scintillations* and *scotomata*. This depolarisation is accompanied by the release of inflammatory mediators, including nitric oxide and prostanoids. These molecules cause dilation of intracranial arteries (Ashina, 2020). Pain is the result of the activation of the trigemino-vascular system caused by the dilated arteries. Earlier, migraine was treated with Cafergot, which is a combination of caffeine and ergotamine, whose main effect is the constriction of the cerebral arteries. The current treatment of choice is triptans, which activate serotonin receptors at blood vessels and nerve endings in the brain.

Incidentally, long-term treatment with certain anti-hypertensive medications, such as propranolol and calcium blockers, for an unknown reason, protects from migraine attacks. The beneficial side effect of these medications is worth noting since it is usually emphasised that drugs have unwanted side effects. Some additional examples follow. Statins, which are prescribed in order to reduce the low-density lipoprotein (the harmful form of cholesterol), apparently reduce inflammation. Bisphosphonates, which are given to combat osteoporosis, provide some protection against colon cancer (Agus, 2011). Diabetic patients taking metformin have a lower risk for a variety of cancers (Evans *et al.*, 2005).

Atherosclerosis

First, relevant terminology is needed. 'Coronary arteries' is the technical name for the main arteries supplying the heart. Most heart attacks are caused by the occlusion of one of the coronary arteries supplying blood to the 'myocardium', which is the technical name for the muscle tissue forming the cardiac chambers (from the Greek words '*myo*', meaning muscle, and '*kardia*', meaning heart).

Remarkably, in the past 50 years, there has been a 50% reduction in the US of cardiovascular mortality. This great achievement has its origin in the identification of the following major risk factors for cardiovascular disease: smoking, high cholesterol, hypertension, physical inactivity, and diabetes. This was followed by the development of concrete strategies to counteract these risk factors and their effects. The first coronary artery bypass grafting (CABG), i.e. the first open heart surgery, was performed in 1961; the first percutaneous coronary intervention (PCI), i.e. the first angioplasty, in 1977; and the first expandable coronary stent was implanted in 1986. Regarding the risk of hypertension, it is noted that, in 1981, a new family of hypertensive medications was discovered, called angiotensin coenzyme inhibitors (such as captopril); in 1988, Sir James Black received the Nobel Prize in Medicine for elucidating the mechanisms that led to the invention of beta-blockers (such as atenolol). The Nobel Prize work of Michael Brown and of Joseph Goldstein on the low-density lipoprotein receptors and of Akira Endo on the identification of a specific enzyme involved in cholesterol metabolism (called HMG-CoA) led to the introduction, in 1987, of the first statin (lovastatin).

The crucial role of inflammation in atherosclerosis is now widely accepted. It has even been suggested that atherosclerosis may involve autoimmune processes (Cinoku *et al.*, 2019). However, although treatments for the secondary prevention of heart attacks, such as aspirin and statins, do have anti-inflammatory effects, the identification of a medication targeting inflammation *directly* has been elusive. Finally, progress was achieved in this direction when it was shown that the IL-1β antagonist canakinumab reduced the risk for an acute coronary syndrome (Ridker *et al.*, 2017). Also, low-dose colchicine, which is a well-known anti-inflammatory agent used in the treatment of gout and pericarditis, can be beneficial (Bhatt *et al.*, 2019): patients were given colchicine starting 30 days after a myocardial infarction and were followed for almost a year; these patients had a significantly lower risk of ischaemic cardiovascular events than the patients who received placebo. In a later study, patients with chronic cardiac disease were followed for a median of 28.6 months; the risk of cardiovascular events was significantly lower among those who received 0.5 mg of colchicine once daily than those receiving placebo (Nidorf *et al.*, 2020).

Evidence of the correlation of inflammation and atherosclerosis is provided by the fact that patients with rheumatoid arthritis, which is characterised by chronic inflammation, show similar coronary microvascular dysfunction to that occurring in diabetic patients (as noted in Chapter 7, inflammation is one of the features of dysmetabolic syndromes, such as diabetes).

Patients with atherosclerotic coronary arteries may suffer from an acute cardiac syndrome caused by the rupture of an atherosclerotic plaque and the formation of a clot, referred to in medicine as 'thrombus'. It has been recently realised that often, somehow, atherosclerotic plaques desta-bilise *without* causing serious clinical effects. Apparently, this is the result of the dynamic healing process following the rupture. In more detail, a plaque consists of lipids, necrotic debris, and macrophages; the plaque is enclosed in a fibrous cap composed of collagen and smooth-muscle cells. The rupture of the plaque causes the exodus of the highly thrombo-genic necrotic core, causing the aggregation of platelets and the formation of a clot. However, the rupture of the fibrous cap is immediately followed by a cascade of enzymatic processes, which include the following. First, the release from the artery's endothelium of several molecules, such as tissue plasminogen activator, which have the ability to lyse the thrombus. Second, the release from white blood cells of molecules, such as elastase, capable of forming an elastic cap. Third, the release of various growth factors that stimulate the local proliferation of smooth-muscle cells, which synthetise collagen. Although this highly complicated healing pro-cess may prevent the rapid development of an occlusive thrombus, in a longer timeframe it has a negative impact. Indeed, it transforms a lipid-rich plaque to a more fibrous lesion, causing stenosis of the associated artery.

Remarkably, the precise dynamic balance between the rupture of the plaque and healing is dictated by the immune system. In this regard, it is noted that there exist different types of macrophages. In particular, in addition to the pro-inflammatory cells, M_1 macrophages, there also exist the so-called 'alternative M_2 macrophages'. These macrophages secrete anti-inflammatory cytokines which counterbalance the action of M_1 mac-rophages.[1] They also secrete several profibrotic factors which promote the production of collagen and smooth-muscle cells.

---

[1] The alternative M_2 macrophages are stimulated by interleukins, including IL-4 and IL-13. These macrophages secrete IL-10 and several other anti-inflammatory cytokines, as well as several profibrotic factors, such as fibronectin, insulin-like growth factor 1, and trans-forming growth factor β.

Importantly, the macrophages appearing in atherosclerotic diseases are not terminally differentiated but can switch from one type to another. Indeed, there are specific mediators which can modulate the conversion of M_1 macrophages to alternative M_2 macrophages. Actually, specific drugs are being tested with the aim of aiding this conversion (Vergallo and Crea, 2020). Interestingly, it has been suggested that both proper diet and statins enhance the healing process of atherosclerotic plaques. In particular, this intervention gives rise to an increase in the thickness of the fibrous cup enclosing the plaque, as well as to a reduction in its lipid content and macrophage infiltration. Also, there is an acceleration in the calcium formation, which further stabilises these plaques.

The best imaging technique for visualising the effect of atherosclerosis in a coronary artery is optical coherence tomography (OCT), which has a resolution 10 times greater than that of intra-vascular ultrasonography. OCT uses a source of infrared light, and by measuring the reflections of the light from multiple depths within the tissue under examination, reconstructs the anatomy of the tissue.[2] Incidentally, OCT has had a transformative impact in ophthalmology. Indeed, following the 1991 work of David Huang, Eric Swanson, and James Fujimoto, who were able to image the retina of an eye of a cadaver using OCT, this technique has become indispensable for the care of the eye (Toth, 2023). In particular, it is of crucial importance for following the progression and the response to treatment of age-related macular degeneration.

Neuroimmunology

The highly dynamic process following the rupture of a plaque provides yet another illustration of the general fact that:

Biology is characterised by dynamic instead of static processes.

Moreover, it emphasises further the astonishing complicated nature of the immune system, which, in terms of its complexity, can only be compared to the central nervous system. In this sense, it is not surprising that the central nervous and immune systems share several similarities.

---

[2] OCT, by combining the technologies of fibreoptic communications and Michelson interferometry, creates images of micrometre scale.

In particular, both are able to recognise a variety of internal and external stimuli and to respond accordingly. Moreover, both systems retain a memory of earlier experiences.

For decades, it was assumed that the immune and nervous systems operate independently. However, recent research established that these systems collaborate to maintain homeostasis and to protect the host against infections and other disease processes. The elucidation of a variety of interconnections and interactions between these two systems provides yet another manifestation of *unification*.

Regarding homeostasis it is noted that metabolic thermogenesis, regulation of vascular tone and of blood pressure, intestinal mobility, and the autonomic nervous system are all affected by a variety of immune cells, including macrophages. On the other hand, the peripheral nervous system plays an important role in regulating the immune response. In particular, the sympathetic nervous system is a key component and regulator of the stem cell niche (Hanoun *et al.*, 2015). For example, sympathetic nerves, in response to granulocyte colony-stimulating factor, release norepinephrine, which triggers haematopoietic stem cell mobilisation (Katayama *et al.*, 2006). Also, adrenergic cells modulate the flow of lymphocytes through arteries and affect vascular permeability; moreover, these cells have been identified as key regulators of leukocyte recruitment within tissues.

Regarding infections and various disease processes, several studies established that inflammatory cytokines, mast cell mediators, lipid mediators, microbial products, and various metabolites all lead to neuronal activation (Ordovas-Montanes *et al.*, 2015). On the other hand, many immunological processes, such as the recognition of bacterial products by pain receptors, are under neurological control. For example, pain recognition during infection with the bacterium *Staphylococcus aureus* is not the direct result of inflammation, but it is due to the interaction of specific pain receptors with bacterial peptides. Interestingly, vagal sensory nerves are of key importance for detecting localised inflammation and informing immune cells to elicit an appropriate immune response (Hansen *et al.*, 2001).

The dynamic interaction of the immune and nervous systems is characterised by high complexity. For example, in the spleen, released norepinephrine from nerve fibres stimulates a subset of splenic T cells to produce acetylcholine, which in turn activates macrophages, which supress the production of tumour necrosis factor-alpha(TNF-α) (Rosas-Balina *et al.*, 2011).

The interaction of the immune and nervous systems is facilitated by specific synapses. For example, in order to initiate their own proliferation, T cells must initiate activation of appropriate transcription factors. For this purpose, cytokine and cytotoxic agents are released in the synaptic cleft, where, within two minutes, certain tyrosine kinases are activated, initiating a complicated process that finally leads to transcriptional activation. This process, in addition to inducing T-lymphocyte proliferation, gives rise to permanent synaptic formation directly affecting the central nervous system (Dustin and Colman, 2002).

A dysfunction of the collaboration of the immune and nervous systems can lead to exacerbated inflammation and collateral damage. Therapeutic approaches are under investigation aiming to inhibit, moderate, or harness the neuro-immune axis. One striking example is the finding that the electric vagal nerve stimulation decreased inflammation and pain in two-thirds of patients suffering with rheumatoid arthritis who were resistant to methotrexate (Kanashiro *et al.*, 2018). Also, efforts are made to exploit the fact that in the early stages of this prevalent autoimmune disease, the sympathetic nervous system enhances inflammation (Moutsopoulos *et al.*, 2018).

Neuroinflammation is of crucial importance during the acute phase following traumatic brain injury. Damage to the central nervous system elicits an inflammatory response from microglia and macrophages, as well as peripheral immune cells, such as neutrophils, monocytes, and T cells. Specifically, microglia and resident macrophages immediately respond to injury, sensing the presence of adenosine triphosphate (ATP) and intracellular mediators released from damaged or dying cells, particularly interleukin-33. This has a beneficial impact, which includes the preservation of the blood–brain barrier, debris clearance, and immune regulation. For this reason, medications suppressing inflammation should be avoided. In 2005, a definitive clinical trial found that administration of methyl prednisone within eight hours of injury increased the risk of death at six months post a traumatic brain injury (Edwards *et al.*, 2005). Nevertheless, if macrophages or microglia remain in the inflammatory state for months or years, neuroinflammation becomes maladaptive. In this case, medications such as the antibiotic minocycline that can suppress microglia and macrophage activity may be useful (Russo and McGavern, 2016). The question of determining the optimal time that neuroinflammation is beneficial remains open.

A unified treatment to coronary arteries and cerebral arteries occlusions

If there is a substantial occlusion in a coronary artery, initially (as a result of the reduced blood supply) the affected part of the myocardium suffers from ischaemia, which means 'stopping blood' (the corresponding Latin and Greek words are *ischaemus* and *iskhaimos*, respectively). The immediate goal of the urgent medical intervention is to remove the clot causing the occlusion so that death of the part of the myocardium supplied by the occluded artery is prevented. The technical name for the local necrosis is 'infarction'; hence, a pharmacological or an invasive intervention aims to prevent 'myocardial infarction'.

For this purpose, the first effective treatment was the use of a thrombolytic medication, i.e. a medication to lyse the clot. The medication used today for this purpose is called *alteplase* (recombinant tissue plasminogen activator).[3] Later, it was realised that for well-localised clots, an alternative, more effective approach is to remove the clot by inserting a catheter in the occluded artery and placing a stent. This life-saving procedure, known as PCI, was first performed in Zurich by the German cardiologist Andreas Gruentzig. Although the general technique of angioplasty was invented in 1964 by Charles Dotter, who is considered the 'father of interventional radiology', the medical use of angioplasty continues to expand. For example, up to half of patients presenting with a common acute myocardial infarction syndrome, called STEMI, have *multivessel* coronary artery disease (patients are classified according to the occurrence of a particular abnormality observed in the electrocardiogram, called ST-segment elevation). The question remained open until very recently of whether revascularisation for such patients should be limited to the lesion that caused the infarction or revascularisation of the other lesions should also be performed. Finally, it has been shown that complete revascularisation decreases the risk of recurrent myocardial infarction and death (Køber and Engstrøm, 2019).[4]

---

[3] Streptokinase, which has similar properties with alteplase, is produced from the bacterium *Streptococcus a*. This belongs to the same genus with the bacterium causing rheumatic fever, which is the reason for the hypothesis that the latter infection protects against atherosclerosis.

[4] Also, it is now known that in patients with a blockade of their left coronary artery with low or intermediate anatomical complexity, PCI is as good as CABG for preventing myocardial infarction and death at five years (Stone *et al.*, 2019).

A similar pathological process occurs in strokes, where the occluded artery supplies blood to the brain. In analogy to the approach followed in a heart attack, patients are treated during the first 3–4.5 hours after the onset of a stroke with alteplase. Before initiating treatment, it is required to measure the blood glucose to exclude hypoglycaemia and to perform a CT to rule out an intracerebral haemorrhage. It was later shown that alteplase administration up to nine hours after the onset of stroke was still helpful in patients with salvageable brain tissue identified with the use of CT or MRI. More recently, a large study in China established that in patients with mild posterior circulation stroke, alteplase administered 4.5–24 hours after stroke onset resulted in better outcomes at 90 days (Yan *et al.*, 2025).

Starting in 2015, in analogy to the percutaneous coronary angioplasty, 'endovascular thrombectomy', i.e. removal of the clot via a catheter, has been used for discrete clots occluding large cerebral vessels (Berkhemer *et al.*, 2015). Such large accessible arteries are the intracranial carotid artery, the proximal middle cerebral artery, and the basilar artery. Later, it was shown that the interventional treatment is beneficial even up to 24 hours, provided that imaging studies combined with clinical examination can establish that there is ischaemic tissue that has not yet been infarcted (Hacke, 2018). Such tissue is called *penumbral*. In particular, two rigorous clinical studies compared endovascular thrombectomy with best medical care in patients with basilar-artery occlusion and a moderate to severe deficit; in one study patients were treated up to 12 hours after the suspected occlusion and in the second up to 24 hours. In both cases, there was a similar benefit of the endovascular group.

In 2018, it was shown that tenecteplase, which is a genetically engineered variant of alteplase, has more potent thrombolytic activity than alteplase. This medication has the additional advantage that it can be given as a bolus, in contrast to alteplase, which is infused over a period of one hour (Baird, 2018). Since failed medical thrombolysis is usually followed by thrombectomy, the faster administration of tenecteplase can reduce the time interval between thrombolysis and thrombectomy. In addition, apparently tenecteplase is beneficial even if given in the interval of 4.5–24 hours after an ischaemic stroke (Xiong *et al.*, 2024).

Reteplase, a recombinant plasminogen activator, given in a double bolus separated by 30 minutes, has been approved in many geographic regions for the treatment of acute myocardial infarction. It is now established that among patients with ischaemic stroke within 4.5 hours after the onset of symptoms, reteplase is more likely to result in an excellent functional outcome than alteplase (Li *et al.*, 2024).

Another example of similarities in treating a serious cardiac episode and a stroke is provided by the use of ticagrelor. This medication inhibits platelet aggregation and hence prevents the formation of a clot. It has been well established in rigorous clinical trials that the addition of this medication to aspirin enhances significantly the effect of aspirin on preventing coronary thrombotic events. Until recently, aspirin was the main medication used for preventing a stroke following a minor stroke or a transient ischaemic attack (TIA). A TIA is a brief episode of neurological dysfunction due to the reduction of blood flow to the brain. The risk of a stroke can be as high as 10% in the week following a TIA. Blood pressure control and the use of aspirin reduce this risk by more than 80%. Motivated by the experience in cardiology, ticagrelor was used along with aspirin within 24 hours of symptoms of a minor stroke or a TIA. It was shown that in comparison to aspirin and placebo, the inclusion of ticagrelor significantly reduced the 30-day risk of disabling stroke or death (Amarenco *et al.*, 2020).

The above interventions continue to improve. For example, until recently, thrombectomy was limited to strokes where the ischaemic volume was less than 50 ml. It was hypothesised that if there was a larger necrotic area, it would increase the risk of haemorrhage. However, recently, thrombectomy was performed in large ischaemic stokes, and surprisingly, the incidence of haemorrhage was not significantly higher in the endovascular therapy group than in the medical care group (Fayad, 2023). Interestingly, the outcome of a 'hybrid' therapy, namely, the treatment with intravenous tenecteplase followed by endovascular thrombectomy, in patients presenting within 4.4 hours of the onset of acute ischaemic stroke was better than the treatment with thrombectomy alone (Qiu *et al.*, 2025). Regarding antiplatelet medications, it is worth noting that, although the glycoprotein receptor inhibitor tirofiban is used for acute coronary syndromes, it was not used until recently in acute strokes, for which thrombolysis was contra-indicated. In a recent rigorous clinical trial, by comparing the outcome of patients injected intravenously with tirofiban for up to 48 hours after an acute stroke, followed by aspirin for 90 days, with patients receiving aspirin alone, the benefit of tirofiban was established.

Depression

In the early 1990s, several publications appeared in the medical literature investigating interactions of the immune system with the brain and the mind (Dantzer and Kelley, 1989; Smith, 1991; Maes, 1995). Since that time, the area of *neuro-immunology*, which investigates how the immune system interacts with the central nervous system, especially *immuno-psychiatry*, which focuses on the impact of the immune system on mental health, has attracted a lot of interest. In particular, the distinguished Cambridge psychiatrist, Edward Bullmore, is a strong proponent of the theory that inflammation can lead to mental disorders and especially to major depression (Bullmore, 2018).

A serious difficulty in obtaining direct proof for this theory is that, as discussed in detail in Chapter 16 of Fokas (2024), the current understanding of the aetiology of major depression is far from complete. Hence, it is not possible to make the diagnosis using functional imaging techniques (such as PET or fMRI) or specific blood biomarkers. The general belief that depression is related to the reduced action of the neurotransmitters norepinephrine and serotonin is based on the positive effect on many patients of medications which enhance the availability of these basic neurotransmitters rather than on their direct blood measurements. Such medications include tricyclic antidepressants, like amitriptyline, which block the reuptake of norepinephrine and serotonin, as well as serotonin reuptake inhibitors (SSRIs), such as Prozac. It is worth noting that serotonin is an evolutionarily old neurotransmitter, which, for example, plays a crucial role in the neural networks of the worm *Caenorhabditis elegans*. Thus, its effect on our brain could indeed be important. However, blood biomarkers of serotonin provide neither a reliable criterion for the diagnosis of major depression nor can be used to predict the response to SSRIs. Moreover, it is questionable whether blood levels are correlated with serotonin levels in the brain. Also, whereas the serotonin levels in the brain rise soon after the digestion of an SSRI, a clinical effect (if any) does not occur for about four weeks. Furthermore, although many SSRIs are supposed to have a similar mechanism of action, sometimes different SSRIs have a completely different effect on a given patient.

Despite the lack of direct proof of the hypothesis that inflammation can lead to major depression, there are several indirect facts supporting this conjecture. They include genetic and epidemiological data, animal experiments, observations regarding patients who suffer from chronic inflammation, and general theoretical considerations. Details follow.

The analysis of the DNA from about 130,000 patients suffering from depression and 330,000 healthy controls identified 44 genes associated with depression (Sullivan *et al.*, 2019). The most significant association was established for the gene *olfactomedin 4*, which controls the stomach's inflammatory response to dangerous bacteria. Interestingly, 80 individuals who possess a mutation of olfactomedin 4 had an enhanced inflammatory response to bacterial infections. This finding provides a possible evolutionary explanation for the prevalence of depression: individuals with this mutation have a higher risk of depression, but at the same time they have a survival advantage following a serious bacterial infection (which was of crucial importance before the era of antibiotics).

A study in 2014 involving 15,000 children in Bristol, UK, found that children who had a slight inflammation at the age of nine were significantly more likely to be depressed 10 years later (Khandaker *et al.*, 2014). Also, it is well established that markers of inflammation, including C-reactive protein (CRP), increase in many stressful situations, such as poverty, debt, and social isolation; these conditions increase the risk of the appearance of depression later in life. It is interesting to note that about 60% of the cells in adipose tissue are macrophages, which are of critical importance for eliciting an inflammatory response. Thus, it is not surprising that overweight people have higher blood levels of CRP than slimmer people and are also more likely to be depressed (Luppino *et al.*, 2010).

Macrophages use cytokines to broadcast a general message about the current status of the inflammatory response to a given threatening situation. These proteins, which, as discussed in Chapter 7, include the tumour necrotic factor (TNF), interferon, and several interleukins, cause inflammation. Injecting mice with cytokines, or with lipopolysaccharide, which is the outer membrane of bacteria known to elicit a strong immune response, leads to 'depressive behaviour'. Specifically, they become *behaviourally indifferent* to the choice between water and sugary water, apparently because they have lost the hedonistic drive to consume sugar. These experiments provide direct support that, in animals, inflammation gives rise to anhedonia (loss of pleasure), which is a key characteristic of depressed individuals.

Occasionally, as discussed in Chapter 7, following specific aberrations of the immune system, lymphocytes elicit a hostile immune response against their own self, giving rise to a variety of autoimmune processes.

In about 90% of patients suffering from rheumatoid arthritis, *fatigue* is their main problem; moreover, about 40% of these patients are *depressed*, and many complain of '*brain fog*', i.e. difficulty in thinking and planning in a clear way. Bullmore, correctly in my opinion, claims that these manifestations are as much symptoms of this disease as the typical occurrence of swollen and painful joints. Importantly, as discussed in Chapter 7, a major breakthrough in the treatment of rheumatoid arthritis was the discovery, in the late 1990s, of a specific antibody acting against the TNF. This new medication, with the commercial name Remicade, in addition to ameliorating several physical symptoms of rheumatoid arthritis, also led to the elimination of depressive manifestations; actually, it gave rise to a 'Remicade high'.

Incidentally, the stimulation of the vagus nerve reduces the secretion of cytokines. In particular, in a study involving patients suffering from rheumatoid arthritis, the daily electrical stimulation of the vagus nerve for 20 minutes caused rapid and substantial reductions in blood cytokine levels, which was accompanied by the patients reporting fewer painful joint symptoms. When the stimulation was stopped for 10 days, both cytokine levels and symptom scores increased; after the stimulation was reinstated, cytokines and symptoms decreased again (Bullmore, 2018).[5]

There exist several theoretical arguments for a possible explanation for the connection between inflammation and depression. For example, cytokines *can* cross the blood–brain barrier. This leads to the excitation of the microglial cells, which begin to produce cytokines by themselves. Nerve cells make serotonin using tryptophan. Following the release of cytokines by microglia, neurons can use tryptophan to make other end-products, such as kynurenine, which leads to the decreased production of serotonin. This may be related to the observation that patients who do not respond well to the treatment with SSRIs are more likely to have higher levels of inflammation (Maes *et al.*, 1997).

Regarding the relation of inflammation (independently of its cause) and depression, several studies have measured in depressed individuals and in controls the values of major inflammation indicators, such as CRP and cytokines. Collectively, these data show that the blood concentrations of CRP and some cytokines exhibit a small increase in depressed patients (Dowlati *et al.*, 2010). In my opinion, this indicates that, in general, the

---

[5] Vagus stimulation can be achieved by rubbing both auricles, while simultaneously taking a deep breath and holding it (this is known as the Hering–Breuer reflex).

correlation between inflammation and depression may be weak. However, I believe that there may exist a subset of individuals for whom this correlation is stronger. A systematic effort should be made to identify this subset since such individuals may benefit from a specific anti-inflammatory treatment.

Neuropsychiatric disorders associated with the dysfunction of the immune system continue to emerge. In particular, in 1998, 50 cases were described in Swedo *et al.* (2022) involving children between the age of three and the beginning of puberty. These patients presented with a sudden onset of obsessive-compulsive behaviour (OCD) and/or a variety of tics; these basic symptoms were often accompanied by attention deficit hyperactivity, separation anxiety, and emotional liability. The common feature of all these cases was a recent streptococcal infection. It is hypothesised that these manifestations are the result of *molecular mimicry*, in which antibodies produced against streptococcal proteins are able to cross-react with specific brain proteins, particularly in the basal ganglia. This so-called Paediatric Autoimmune Neuropsychiatric Disorder Associated with Streptococcal Infection (PANDAS) remains controversial due to the absence of rigorous clinical characteristics and biomarkers to differentiate it from childhood-onset OCD and tic disorders. A recent review notes that although in such patients the measurement of several autoantibodies remains in the normal range, PANDAS is indeed a new medical entity (Prato *et al.*, 2021). It is also stated in the same paper that efforts should be made to define proper diagnostic criteria so that PANDAS can be differentiated from autoimmune acute encephalitis, Sydenham's chorea, and several other syndromes.

It should be emphasised that there exist syndromes where aberrations of the immune system lead to major depression *without* signs of inflammation. For example, in many patients suffering from the Mast Cell Activation Syndrome (MCAS), the value of CRP is within the normal range. MCAS, for which diagnostic criteria (initially established in 2010) continue to be modified, is caused by the inappropriate activation of mast cells. This leads to their degranulation and the release of several inflammatory mediators, including histamine, serotonin, chymase, and cytokines. Typical triggers are stress, infections, menses, various foods, and beta-lactam antibiotics (such as penicillin). This syndrome is associated with serious neurologic and psychiatric manifestations, including major depression, generalised anxiety disorder, fatigue, 'brain fog', and headache. Although MCAS is apparently quite common, it is rarely recognised, and thus, patients can suffer for decades. A group of eight patients with significant neuropsychiatric disorders is presented in Weinstock *et al.* (2023); these patients remained undiagnosed for up to 40 years. Some studies suggest that SSRIs may induce mast cell degranulation, exacerbating instead of ameliorating the associated depression. On the other hand, benzodiazepines, apparently, stabilise the mast cell's membrane. Among the medications suggested for the treatment of depression in the MCAS is the tricyclic antidepressant doxepin, which exhibits anti-histaminergic activity (Theoharides *et al.*, 2019).

The above discussion suggests that, perhaps, efforts should be made to decipher the relations between depression and *dysfunctions of the immune system*, as opposed to simply the possible correlation with inflammation. This is consistent with the observation that among the prevalent autoimmune diseases, lupus erythematosus is the disease that often presents with symptoms of depression. Importantly, in this disease, the inflammation markers are within the normal range.

Immune mechanisms may also play a role in schizophrenia. Indeed, among the approximately 320 genes that increase the risk of schizophrenia, the strongest association is observed with the gene *complement component 4* (C4), which produces an inflammatory protein (Psychiatric Genetics Consortium, 2014).

11.2 Protein Misfolding and Messenger RNA Translation Defects

During their *de novo* synthesis, proteins are 'folded', acquiring a three-dimensional structure known as a super-secondary or tertiary structure. If the process of folding is abnormal, then the function of the resulting protein is compromised. For example, misfolded proteins clump inside neurons, and this eventually leads to apoptosis. *The concept of misfolded proteins provides a unified aetiological approach to a variety of neurodegenerative diseases.* The most important of these disorders are summarised below.

It was noted in Chapter 16 of Fokas (2024) that in a familiar form of Parkinson' disease, the gene associated with the protein *alpha-synuclein* is mutated, causing the misfolding of this protein. Also, as stated in Chapter 14 of Fokas (2024), protein misfolding is involved in some cases of both Alzheimer's and frontotemporal dementias.

In 1993, an international collaborative group, called the Gene Hunters, isolated and sequenced the mutant *huntingtin*, which is the gene causing the fatal form of Huntington's disease (MacDonald *et al.*, 1993). This mutated gene causes the insertion of additional glutamines into a normal protein, and this causes the production of an abnormally long version of this protein, leading to misfolding and neuronal death. Huntington's disease affects many areas of the brain, including the basal ganglia, cerebral cortex, hypothalamus, thalamus, and often cerebellum. This extensive involvement, in addition to resulting in chorea-type movements, causes dementia.

Protein misfolding is also implicated in 'chronic traumatic encephalopathy', which is a progressive degenerative disorder affecting people who have suffered repeated concussions. This is typically associated with contact sports, particularly boxing and American football.

In 1982, a novel form of protein misfolding was discovered by Stanley Prusiner (Nobel Prize in Medicine, 1997), who understood that the rare but fatal neurodegenerative encephalopathy, Creutzfeldt-Jakob disease, is caused by the misfolding of specific proteins called *prions*. These substances are highly infectious and have the unique feature that they can reproduce *without* containing any genetic material, i.e. they contain neither DNA nor RNA. Prions can be released by affected neurons and taken up by neighbouring cells, where they attach themselves to normal precursor proteins, causing them to fold abnormally. These misfolded proteins become prions and ultimately kill the cells that host them.[6]

Misfolding is also the cause of many non-neurological diseases, such as the common genetic disease alpha-antitrypsin deficiency. Alpha-antitrypsin (AAT) is a protein synthetised in the liver. It has a variety of functions, including protecting the lung tissue against the effects of the enzyme elastase, which was mentioned in the section on atherosclerosis. Misfolded AAT proteins accumulate in the 'hepatocytes', which is the medical term for liver cells, causing a specific liver disease. Furthermore, deficiency of AAT results in an excess production and secretion of elastase, ultimately causing a disease indistinguishable from emphysema, which is a chronic lung disease caused mainly by smoking.

---

[6] Interestingly, prion protein antagonists may be therapeutic targets for Alzheimer's disease (Cox *et al.*, 2019).

Another unified approach to neurodegenerative diseases is based on the observation that in many such diseases, including Alzheimer's disease, frontotemporal dementia, Parkinson's disease, Huntington's disease, amyotrophic lateral sclerosis, spinal muscular atrophy, and Charcot–Marie–Tooth disease, there occur aberrations in the messenger RNA (mRNA) translation. In some cases, the *integrated stress response* (ISR) causes an inhibition in a subset of mRNA. ISR is an important homeostatic mechanism, which is triggered in response to cellular stress and which attempts to restore homeostasis. The normal function of ISR is based on the interaction of four specific proteins (PERK, PKR, GCN2, and HRI) with the protein *translation initiation factor 2* (eIF2). Apparently, prolonged activation of the ISR can be harmful. In postmortem studies of patients suffering from various neurodegenerative diseases, increased brain levels have been reported of eIF2 and of the four proteins involved in ISR. Interestingly, in some diseases, such as the Charcot–Marie–Tooth disease, the activation of the stress response may promote neurodegeneration. Another mechanism responsible for aberrations of mRNA is the occurrence of mutations in genes encoding proteins that participate in translation. Importantly, several studies suggest that translation defects are not a consequence of neurodegeneration, but the cause of neurodegeneration. It is expected that specific inhibitors of the eIF2 kinase will soon be evaluated in clinical trials for the treatment of various neurodegenerative diseases (Storkebaum *et al.*, 2023).

What is the role of neuroinflammation in major neurodegenerative diseases, including Alzheimer's disease, frontotemporal dementia, Parkinson's disease, amyotrophic lateral sclerosis, and progressive multiple sclerosis (which may develop at the later stages of multiple sclerosis)? In such cases, in contrast to the typical neuroinflammation characterising multiple sclerosis, there occur only changes in the morphology of glial cells, especially in astrocytes and in microglia. These changes are accompanied by the presence of low-to-moderate levels of inflammatory mediators in the brain parenchyma. Such reactions are similar to those seen in stroke and traumatic injury (Ransohoff, 2016). The detection of inflammatory mediators in the autopsy of patients with Alzheimer's and Parkinson's' diseases led to the proposal of treating these patients with anti-inflammatory medications. However, as mentioned earlier, the treatment of patients with Alzheimer's disease did not provide any benefits (Ali *et al.*, 2019).

References

Agus, D. B. (2011). *The End of Illness.* New York: Simon and Schuster.

Ali, M. M., *et al.* (2019). Recommendations for anti-inflammatory treatments in Alzheimer's disease: A comprehensive review of the literature. *Cureus, 11,* e4620. doi:10.7759/cureus.4620.

Amarenco, P., *et al.* (2020). Ticagrelor added to aspirin in acute ischemic stroke or transient ischemic attack in prevention of disabling stroke: A randomized clinical trial. *JAMA Neurology, 78,* E1–E9.

Ashina, M. (2020). Migraine. In Ropper, A. H. (ed.). *New England Journal of Medicine, 383,* 1866–1876.

Baird, A. E. (2018). Paving the way for improved treatment of acute stroke with tenecteplase. *New England Journal of Medicine, 378,* 1635–1636.

Berkhemer, O. A., *et al.* (2015). A randomized trial of intraarterial treatment for acute ischemic stroke. *New England Journal of Medicine, 372*(1), 11–20.

Bhatt, D. L., *et al.* (2019). Cardiovascular risk reduction with icosapent ethyl for hypertriglyceridemia. *New England Journal of Medicine, 380*(1), 11–22.

Bullmore, E. (2018). *The Inflamed Mind: A Radical New Approach to Depression.* New York: Picador.

Cinoku, I. I., Mavragani, C. P., and Moutsopoulos, H. M. (2019). Atherosclerosis: Beyond the lipid storage. The role of autoimmunity. *European Journal of Clinical Investigation, 50,* e13195.

Cox, T. O., *et al.* (2019). Anti-PrPC antibody rescues cognition and synapses in transgenic alzheimer mice. *Annals of Clinical and Translational Neurology, 6*(3), 554–574.

Dantzer, R. and Kelley, K. W. (1989). Stress and immunity: An integrated view of relationships between the brain and the immune system. *Life Sciences, 44,* 1995–2008.

Dowlati, Y., *et al.* (2010). A meta-analysis of cytokines in major depression. *Biological Psychiatry, 67,* 446–457.

Dustin, M. L. and Colman, D. R. (2002). Neural and immunological synaptic relations. *Science, 298*(5594), 785–789.

Edwards, P., *et al.* (2005). Final results of MRC CRASH, a randomised placebo-controlled trial of intravenous corticosteroid in adults with head injury—Outcomes at 6 months. *Lancet, 365,* 1957–1959.

Evans, J. M., Donnelly, L. A., Emslie-Smith, A. M., Alessi, D. R., and Morris, A. D. (2005). Metformin and reduced risk of cancer in diabetic patients. *BMJ, 330*(7503), 1304–1305.

Fayad, P. (2023). Improving prospects. *New England Journal of Medicine, 388,* 1326–1328.

Fokas, A. S. (2024). *Ways of Comprehending: The Grand Illusion and the Essence of Being Human.* Singapore: World Scientific.

Germenis, A. (2019). *Anosia* (in Greek). Vouton: Crete University Press.

Hacke, W. (2018). A new DAWN for imaging-based selection in the treatment of acute stroke. *New England Journal of Medicine, 378*(1), 81–83.

Hanoun, M., *et al.* (2015). Neural regulation of haematopoiesis inflammation, and cancer. *Neuron, 86,* 360–373.

Hansen, M. K., *et al.* (2001). The contribution of the vagus nerve to interleukin -1 beta -induced fever is dependent on dose. *American Journal of Physiology-Regulatory, Integrative and Comparative Physiology, 280,* 929–943.

Kanashiro, A., *et al.* (2018). From neuroimmunomodulation to bioelectronic treatment of rheumatoid arthritis. *Bioelectronic Medicine (London), 1,* 151–165.

Katayama, Y., *et al.* (2006). Signals from the sympathetic nervous system regulate hematopoietic stem cells egress from bone morrow. *Cell, 124,* 407–421.

Khandaker, G. M., *et al.* (2014). Association of serum interleukin 6 and C-reactive protein in childhood with depression and psychosis in young adult life: A population-based longitudinal study. *JAMA Psychiatry, 71,* 1121–1128.

Køber, L. and Engstrøm, T. (2019). A more COMPLETE picture of revascularization in STEMI. *New England Journal of Medicine, 381,* 1472–1474.

Kotas, M. E. and Medzhitov, R. (2015). Homeostasis, inflammation, and disease susceptibility. *Cell, 160,* 816–827.

Li, S., *et al.* (2024). Reteplacse versus alteplase for acute ischemic stroke. *New England Journal of Medicine, 390,* 2264–2273.

Louveau, A., *et al.* (2015). Structural and functional features of central nervous system lymphatic vessels. *Nature, 523,* 337–341.

Luppino, F. S., *et al.* (2010). Overweight, obesity, and depression: A systematic review and meta-analysis of longitudinal studies. *Archives of General Psychiatry, 67,* 220–229.

MacDonald, M. E., *et al.* (1993). A novel gene containing a trinucleotide repeat that is expanded and unstable on Huntington's disease chromosomes. *Cell, 72*(6), 971–983.

Maes, M. (1995). Evidence for an immune response in major depression: A review and hypothesis. *Progress in Neuro-Psychopharmacology and Biological Psychiatry, 19,* 11–38.

Maes, M., *et al.* (1997). Increased serum IL-6 and IL-1 receptor antagonist concentrations in major depression and treatment resistant depression. *Cytokine, 9,* 853–858.

Moutsopoulos, H., Zampeli, E., and Vlachoyiannopoulos, P. G. (2018). *Rheumatology in Questions.* Heidelberg, Germany: Springer Verlag.

Nidorf, S. M., *et al.* (2020). Colchicine in patients with chronic coronary disease. *New England Journal of Medicine, 383,* 1838–1847.

Ordovas-Montanes, J., *et al.* (2015). The regulation of immunological processes by peripheral neurons in homeostasis and disease. *Trends in Immunology, 36,* 578–603.

Prato, A., *et al.* (2021). Diagnostic approach to pediatric autoimmune neuropsy-chiatric disorders associated with streptococcal infections (PANDAS): A narrative review of literature data. *Frontiers in Pediatrics, 9*, 746639.

Psychiatric Genetics Consortium. (2014). Biological insights from 108 schizo-phrenia-associated genetic loci. *Nature, 511*, 421–427.

Qiu, Z., Li, F., Sang, H., *et al.* (2025). Intravenous tenecteplase before thrombec-tomy in stroke. *New England Journal of Medicine, 393*, 139–150.

Ransohoff, E. (2016). How neuroinflammation contributes to neurodegeneration. *Science, 353*, 777–782.

Ridker, R. M., *et al.* (2017). Antiinflammatory therapy with canakinumab for atherosclerotic disease. *New England Journal of Medicine, 377*, 1119–1131.

Rosas-Balina, M., *et al.* (2011). Cholinergic-synthetizing T cells relay neural signals in a vagus nerve circuitry. *Science, 334*, 98–101.

Russo, M. and McGavern, D. B. (2016). Inflammatory neuroprotection following traumatic brain injury. *Science, 353*, 783–785.

Smith, R. S. (1991). The macrophage theory of depression. *Medical Hypotheses, 35*, 298–306.

Stone, G. W., *et al.* (2019). Five-year outcomes after PCI or CABG for left main coronary disease. *New England Journal of Medicine, 381*, 1820–1830.

Storkebaum, E., Rosenblum, K., and Sonenberg, N. (2023). Messenger RNA translation defects in neurodegenerative diseases. *New England Journal of Medicine, 388*, 1015–1030.

Sullivan, P., Wray, N., and PGC MDD Consortium. (2019). Genome-wide asso-ciation analyses identify 44 risk variants and refine the genetic architecture of major depressive disorder. *European Neuropsychopharmacology, 29*, S805.

Swedo, S. E., *et al.* (2022). Pediatric autoimmune neuropsychiatric disorders associated with streptococcal infections: Clinical description of the first 50 cases. In *Obsessive-Compulsive Disorder and Tourette's Syndrome*, pp. 184–191. Routledge.

Theoharides, T. C., Tsilioni, I., and Ren, H. (2019). Recent advances in our under-standing of mast cell activation–or should it be mast cell mediator disorders? *Expert Review of Clinical Immunology*, 639–656.

Toth, C. A. (2023). Optical coherence tomography and eye care. *New England Journal of Medicine, 389*, 1526–1529.

Vergallo, R. and Crea, P. (2020). Atherosclerotic plaque healing. *New England Journal of Medicine, 383*, 846–857.

Weinstock, L. B., Nelson, R. M., and Blitshteyn, S. (2023). Neuropsychiatric manifestations of mast cell activation syndrome and response to mast-cell-directed treatment: A case series. *Journal of Personalized Medicine, 13*, 1562.

Xiong, Y., *et al.* (2024). Tenecteplase for ischemic stroke at 4.5 to 24 hours with-out thrombectomy. *New England Journal of Medicine, 391*, 203–212.

Yan *et al.* (2025). Alteplase for posterior circulation ischemic stroke at 4.5 to 24 hours. *New England Journal of Medicine, 392*, 1288–1296.

Chapter 12

Gene Editing

The 2018 announcement by a Chinese scientist of the birth of twins whose genomes were edited during *in vitro fertilisation* was broadly condemned.[1] The unscrupulous act of this scientist provides a strong reminder of the challenges posed to humanity by gene editing. Transplantation of a variety of organs, such as the bone marrow, which is used extensively in medicine, does introduce into the patient foreign cells and therefore foreign DNA. However, gene therapy differs because its aim is to introduce modified DNA within the host's cells in a precise manner to achieve a desirable effect.

A gene is the basic unit of heredity. As summarised in Chapter 18 of Fokas (2024), each gene consists of a sequence of nucleotides that encode a protein. A change (*mutation*) of even a single letter of the gene may yield a defective protein, giving rise to a hereditary disease, such as sickle cell anaemia, haemophilia, thalassaemia, muscular dystrophy, and cystic fibrosis. As stated in Chapter 10, there are different types of mutations, which include the following: 'substitution mutations', namely, a nucleotide is substituted by another; 'insertions', where the same three letters in DNA that characterise a given amino acid are repeated many times (as it occurs in Huntington's disease); 'deletions', where three nucleotides are deleted (as it happens in the common form of cystic fibrosis). Following the complete characterisation of the human genome in 2001–2003,

---

[1]In this work, the technique CRISPR, which will be discussed later in this chapter, was used to edit the *CCR5* gene in the germlines of two embryos with the hope of conferring lifetime protection against the human immunodeficiency virus, which causes AIDS.

more than 4,000 mutations have been identified that can cause genetic diseases.

Incidentally, the function of only about 20% of the genome is fully understood. The question of the precise role of the remaining part of the genome in normal and pathological conditions remains open; apparently, this part is of vital importance for the implementation of the crucial decision of when and how a specific gene will be activated and expressed. It is speculated that the elucidation of the underlying mechanisms will have a major impact on the understanding and treatment of neurodegenerative and mental illness.

Until recently, it was not possible to substitute a defective gene with a normal one. Hence, the main strategy to cope with genetic diseases caused by a single defective gene was the use of certain techniques categorised under the name of *gene therapy*. There are two basic strategies: first, a gene is delivered to a long-lived postmitotic or slowly dividing cell, ensuring the expression of the gene for the lifespan of the cell. In this case, integration of the therapeutic DNA into chromosomes of the patient's cells is not required; instead, the transferred DNA is stabilised extra-chromosomally. Second, a gene is introduced into a precursor or stem cell so that it is passed to every daughter cell. Transduction of stem cells is generally an *ex vivo* process. If the therapeutic gene is transferred to somatic cells and not to an undifferentiated stem cell, a germ cell, or a gametocyte, an *in vivo* approach is used to deliver the gene. This is achieved via viral (retro-, adeno-, or adeno-associated virus) or non-viral (DNA plasmid or liposomes) vector systems. The transfer strategy makes use of the observation that viruses have the ability to incorporate (splice) their genetic material into the genome of infected cells. A viral vector is employed to deliver foreign DNA in the cells of a *specific defective organ* in order to express (or disrupt) the specific protein associated with the defective gene. In most circumstances, the gene-delivering vehicle is an adeno-associated virus (AAV) vector, which is a specifically engineered, nonpathological, and nonenveloped parvovirus. The approval in 2012 of gene therapy for the treatment of recurrent or severe pancreatitis in patients with the rare genetic disorder 'lipoprotein lipase deficiency' was the first gene therapy treatment approved in either the US or Europe.

In *ex vivo transduction*, cells are extracted from the patient, are transduced with the virus that contains the gene of interest, and are then returned to the patient. In 2016, *ex vivo* transduction, using haematopoietic and progenitor cells was approved by the European Medicine Agency for the treatment of a rare immunodeficiency fatal disorder (known as ADA-deficiency severe combined immunodeficiency). More recently, stem-cell gene therapy was used for patients with β-thalassaemia, where transfusion independence was achieved for up to 56 months (resulting in the conditional approval by the European Medicine Agency) (Thompson *et al.*, 2018). There are currently several ongoing trials for sickle cell anaemia. Overall, there are close to 1,000 ongoing clinical trials for somatic gene therapy in the US for a variety of genetic disorders.

Is it possible to replace part of a defective gene instead of supplanting it? This involves, first, causing a local break in the double-stranded DNA and, second, inserting the corrected gene. Due to a natural procedure, called *homologous recombination,* employed by organisms during mating and repair, the second step can occur spontaneously. Regarding the first step, i.e. causing a long break, after some earlier important developments,[2] a breakthrough called Clustered Regularly Interspaced Short Palindromic Repeats (CRISPR) was announced in 2012. A virus has the ability to incorporate its DNA in the invaded cell; hence, it is natural to expect that some microbes have developed specific defence mechanisms against this process. Indeed, in the early 1990s, several scientists identified specific repeated sequences in the genomes of several microbes. These particular sequences were later called CRISPR. Francisco Mojica, after analysing 19 different microbes, was able to match part of the above sequences with the DNA found in some viruses (Mojica *et al.*, 2005). Meanwhile, four specific genes, called *cas*, were found to occur adjacent to CRISPR. Following further developments, it became clear that the *cas* genes, guided by CRISPR, orchestrate the dismantling of invasive viral DNA.

In more detail, CRISPR is part of the DNA of bacteria or of archaea that consists of several short (of the order of 30) palindromic (nearly the same when read in either direction) letters of DNA and specific sequences, sandwiched between the short palindromic parts, which match perfectly with the DNA of known viruses infecting bacteria. CRISPR allows bacteria and archaea to protect themselves against any of the viruses corresponding to the sequences contained in CRISPR. Indeed, these microbes construct from the given CRISPR the corresponding RNA, which is then cut (via appropriate enzymes) into a collection of sequences of RNA corresponding to the individual viruses contained in CRISPR. These sequences can be used to destroy any of these viruses attacking the

---

[2] Such developments included the use of the enzyme I-SceI and the technique TALEN.

microorganism possessing CRISPR. This is achieved by *matching* the viruses' DNA with the associated part of the RNA generated via CRISPR. This matching procedure has motivated a simple and effective technique for identifying the precise point in the double-stranded human DNA that needs to be cut (Doudna and Sternberg, 2017). Jennifer Doudna and Emmanuelle Charpentier shared the 2020 Nobel Prize in Chemistry for the discovery of this remarkable technique.

This discovery provides a clear illustration of the importance of horizontal gene transfer, discussed in Fokas (2024). In this regard it is noted that in the early 1960s, Tsutomu Watanabe discovered an important implication of horizontal gene transfer that he called 'infective heredity' (Watanabe, 1963). Namely, bacteria, in addition to their standard DNA carried in a single circular chromosome, also contain an autonomous genetic element called a plasmid which codes for traits that might be useful in emergencies. Watanabe was able to prove that DNA carried in plasmid allowed *E. coli* to develop resistance to multiple antibiotics. Furthermore, shigella dysentery of patients with such resistant *E. coli*, also developed the same antibiotic resistance by receiving from *E. coli* the defending DNA sequence via horizontal gene transfer. In another example, after chickens were fed with tetracycline-laced food, it was found that within a week, bacteria developed in their intestines tetracycline resistance; furthermore, intestinal bacteria of farm workers also developed the same resistance over a period of one month (Levy, 2001).

Genome editing of early human embryos raises great concerns since it can lead to heritable changes transmitted through the germline. In 2015, following the first report that genome editing had been applied to human embryos *in vitro* (Liang *et al.*, 2015), various national academies organised the 'First International Summit on Human Gene Editing' to address related social, ethical, and regulatory issues. In that summit, the imminent clinical applications of somatic tissue editing to treat diseases were saluted, but at the same time it was stated that 'the clinical use of genome editing of human embryos was premature and irresponsible'. The second such summit, organised in 2018, reiterated strong opposition to clinical trials involving germline editing. In addition to these ethical considerations, there also exist recent scientific findings which raise serious concerns regarding the implications of CRISPR-Cas9: studies of different groups show extensive chromosomal damage in human embryos that have undergone CRISPR-Cas9 editing (Ledford, 2020). If these findings are further validated, it will mean that this technique may have unwanted effects (called 'off-target effects') by introducing deleterious mutations.

Despite the above serious concerns, there is widespread support for the prospect of using genome editing to prevent genetic diseases. For example, in the rare cases that couples carry certain defective genes, genome editing offers the only prospect for them bearing a healthy child. This occurs when both members of a couple carry two identical genes for a recessive disease or one of them carries two identical genes for a dominant disease (as is Huntington's disease). Gene editing could also be useful for the more common cases of couples *at risk* of transmitting an autosomal recessive or dominant genetic disease (autosomal DNA is carried by any of the chromosomes except the sex chromosomes, namely except the X chromosome for a female and the Y chromosome for a male). In this case, it has been argued that genome editing is unnecessary because *in vitro fertilisation* with preimplantation genetic diagnosis *can* identify disease-free embryos. However, the yield of such approaches is low and thus, even in these cases, genome editing could be beneficial.

On the other hand, it should be reemphasised that gene editing carries general risks. In particular, an attempt to use gene editing to reduce the risk of common, polygenic diseases, such as heart diseases, breast cancer, and Alzheimer's disease, could have unpredictable negative effects. For example, as mentioned in Fokas (2024), Alzheimer's disease is associated with the gene APOE4. However, it should be remembered that genes have a variety of effects, frequently of a contradictory nature; for example, APOE4 increases the susceptibility to Alzheimer's disease, but it also promotes better memory in young adults. In this context, it should be noted that various mutations arose as a protection against environmental threats. For example, the sickle cell trait arose as protection against malaria.

There is strong opposition to editing genomes in order to enhance the human traits of intelligence, memory, creativity, strength, etc. Clearly, such attempts would raise the spectre of eugenics. In addition, questions of cost and accessibility to a privileged few would lead to difficult ethical dilemmas of social justice and the further exacerbation of inequalities. Fortunately, traits such as intelligence emerge from the complex interaction of multiple genes with the environment; this discourages simplistic attempts of using gene editing. For example, the contribution of a single genetic locus to intelligence explainable by genetic traits is less than 0.05%. Thus, thousands of such genetic loci are needed to achieve 50% of inheritable intelligence (Plomin and von Stumm, 2018). In addition, MRI studies of children and adolescents suggest that intelligence is affected by the highly dynamic process of cortical thickening occurring during development.[3] The complex mechanisms mentioned above provide the best protection from the perils of such attempts, which would be a hubris of humans against the infinite wisdom of nature (Daley *et al.*, 2019).

12.1 Haemophilia Types A and B

Haemostasis requires a balance between procoagulants, which help the formation of clots, and anticoagulants, which inhibit clot formation. In the most common types of haemophilia, homeostasis is disrupted as a result of defective genes encoding the procoagulant factor VIII in the case of haemophilia A and factor IX in haemophilia B. The final outcome is impaired generation of thrombin, which is the key enzyme in the coagulation cascade, converting soluble fibrinogen into insoluble fibrin and allowing blood-clot formation. In haemophilia A, which has a prevalence of 17.1 cases per 100,000 males, clinical severity is measured by the level of factor VIII and the frequency of spontaneous bleeding episodes; patients with severe deficiency have less than 1% of normal values of factor VIII. Similarly is the case with haemophilia B.

For many years, the treatment of these disorders involved prophylactic therapy, with frequent intravenous infusions of clotting factor concentrations. The aim is to maintain a level of at least 3–5% of the active factor VIII (FVIII). Newly developed factor products with longer half-lives can reduce the injection interval to as little as once a week in haemophilia A and as little as once every two weeks in haemophilia B.

In haemophilia B, the development of such factors with a half-life four or five times longer than the earlier ones led to a substantial reduction in the number of required infusions by prolonging the interval of prophylactic infusions (Mannucci, 2020). For haemophilia A, the availability of such a factor was finally achieved with the development of a recombinant factor VIII fusion protein, called efanesoctocog-α. This protein consists of a recombinant factor VIII fused with part of the von Willebrand factor

---

[3] A correlation has been established between the growth of the thickness of the cortex and IQ: in children with scores above 120, the cortex started relatively thin but then matured rapidly, peaking in thickness around the age of 11; by the age of 18, the groups with high, average, and low IQs had the same thickness (Shaw *et al.*, 2006).

(VWF) (Chowdary, 2024). In healthy people, most of factor VIII forms a tight complex with VWF, and the presence of VWF stabilises the structure of factor VIII. In efanesoctocog-α, the exogenous VWF fragment has a similar effect on factor VIII, reducing substantially its plasma clearance. This treatment is also effective for children younger than 12 years of age; a once-weekly administration resulted in normal values of factor VIII for three days after administration and adequate levels for seven days. This led to effective bleeding prevention and, importantly, was not associated with serious adverse effects (Malec *et al.*, 2024).

About 30% of patients with haemophilia A and 10% with haemophilia B, when treated with exogenous factors VIII and IX, develop neutralising antibodies, called inhibitors, against these exogenous factors. The subcutaneous injection of emicizumab, which is a bispecific antibody that mimics the function of FVIII, because of its long half-life, needs to be administered only once every 1–4 weeks; it is now approved for both severe haemophilia A and haemophilia A with inhibitors.

An alternative approach is to bypass the exogenous administration of factor VIII by using specific monoclonal antibodies, such as concizumab, which increase VIII indirectly by inhibiting the tissue factor (van den Berg and Strivastava, 2023).

The development of gene therapies raises the possibility of curing haemophilia instead of lifelong treatment. Regarding the use of somatic gene therapy, a particular limitation of the AAV vector is its inability to incorporate a long gene. This had inhibited the use of this technique for haemophilia A since the gene *F8* (producing the factor VIII) is quite long. Remarkable progress has been recently achieved by encoding in AAV a *truncated* form of factor VIII. Durable efficiency and safety were reported in 15 patients who received a single infusion and who were monitored for three years (Pasi *et al.*, 2020). As of 2025, the only gene therapy licensed for haemophilia A is the one involving valoctocogene roxaparvovec. The approval followed a trial published in 2023, involving 132 participants who were monitored for two years (Mahlangu *et al.*, 2023).

In haemophilia B, the second most prevalent form of haemophilia, the coding sequence of *only part* of the gene FIX (which produces the factor XI) is manipulated (Ponder, 2011).[4] The first gene therapy for haemophilia B to be approved both in the US and Europe was a single injection

---

[4] Haemophilia B is also called Christmas disease, after a patient with this name was identified in 1952 with factor IX deficiency.

of an AAV vector expressing the Padua factor XI (etranacogene dezapar-vovec) (Pipe *et al.*, 2023). An alternative variant is the FIX factor IX (fidanacogene elaparvovec) (Cuker *et al.*, 2024). By monitoring 10 recipients for 13 years, it is now established that a single administration of gene therapy results in durable factor IX expression with no late-onset safety concerns (Reiss *et al.*, 2025). A very recent follow-up study showed that over a period of 3–6 years, this treatment was associated with either no or low-grade adverse effects (Rasko *et al.*, 2025).

Challenges of AAV-mediated gene therapies include the ineligibility of patients with preexisting anti-AAV antibodies, the development of liver dysfunction necessitating immunosuppressive treatment, and uncertainties regarding the durability of the expression of factors VIII and IX. Some of these challenges, apparently, can be overcome using the *ex vivo* technique. Specifically, in a recent breakthrough, haematopoietic stem cells from a patient were transduced with a lentiviral vector carrying a novel gene for factor VIII (*ET3*); transduced haematopoietic stem cells were transplanted into recipients. The use of lentiviral vectors overcomes the barrier of preexisting anti-AAV antibodies (Srivastava *et al.*, 2025).

12.2 Haemophilia C and Atrial Fibrillation

In specific disorders, it is advantageous to modify the levels of one or more of the coagulation factors. A key role in homeostasis is played by the 'coagulation cascade', which consists of the following: the 'extrinsic pathway', comprising factor VII and the tissue factor; the 'intrinsic pathway', comprising factors XII, XI, and IX; and the 'common pathway', involving factor X, prothrombin, and fibrinogen, whose activated forms are, respectively, factor Xa, thrombin, and fibrin.

The commonly occurring arrhythmia atrial fibrillation, which has already been mentioned a couple of times in the earlier chapters, affects 50 million people worldwide. It is associated with an increased risk of clot formation, which may result in an embolic stroke. Several studies have shown that anticoagulation therapy can be beneficial. Medications affecting factor XI are used for the prevention of this devastating possible side effect of atrial fibrillation. Targeting factor XI was motivated by the understanding that in haemophilia C, where there exists a congenital factor XI deficiency, there is a much lower incidence of bleeding than in other types of haemophilia. This suggests that factor XI is more important in clot formation than in bleeding. The current treatment of choice is riva-roxaban, but bleeding remains a serious concern (Angiolillo and Capodanno, 2025). In a recent study, the human monoclonal antibody abelacimab reduced, when compared to rivaroxaban, the occurrence of major or clinically relevant nonmajor bleeding by 69% and 62%, respectively. This impressive result was achieved despite the fact that doses of 90 mg and 150 mg reduced, respectively, the free factor XI levels from baseline by a median of 97% and 99% (Ruff *et al.*, 2025).

Incidentally, in recent years, there have been steady improvements in surgical interventions for eliminating or at least reducing substantially the occurrence of paroxysmal atrial fibrillation. It was already speculated in 1917 that this common arrhythmia was triggered by repetitive electrical discharges originating from the pulmonary veins (Mahida *et al.*, 2015). This is indeed the case, as firmly established by Haissaguerre and his group in the late 1990s. Since then, various surgical techniques have been employed which are based on percutaneous catheter ablation. The new technique of *pulse field* ablation uses intermittent high-intensity electrical discharges to induce pore formation in cell membranes and hence apoptosis. This development represents one of the most important innovations in the field of cardiac electrophysiology: in contrast to the earlier used techniques of thermal ablation by means of radiofrequency or cryoenergy, the new approach minimises the risk for collateral damage, including phrenic-nerve paralysis and pulmonary-vein stenosis (as well as the formation of atrioesophageal fistula) (Andrade, 2025).

12.3 Sickle Cell Anaemia and Thalassaemia

Sickle cell anaemia is recessive; namely, both copies of the relevant gene *HBB*, which encodes β-globin, must carry the mutation for sickle cell anaemia to occur. This disease is caused by a 'substitution mutation', in which an A is substituted by a T, resulting in the substitution of the amino acid glutamate by valine. This leads to abnormalities in the protein haemoglobin, which is the major oxygen-transporting component of red blood cells. In 1949, the double Nobel Laureate Linus Pauling correctly attributed sickle cell anaemia to a molecule defect; the precise mutation was identified in 1956. The defective gene gives rise to the defective haemoglobin S. Clinically, sickle cell disease is characterised by recurrent painful vaso-occlusive crises, progressive vasculopathy, and chronic haemolytic anaemia. Defective peripheral red cells express prothrombotic markers on their surface, which contributes to promoting coagulation abnormalities. Clinically, this leads to thrombotic episodes, pulmonary hypertension, and silent brain infarcts.

Severe β-thalassaemia is a recessive disease resulting from other pathogenic variants of the *HBB* gene, causing severe reduction or complete absence of the β-chains in adult haemoglobin. More than 200 mutations can cause transfusion-dependent thalassaemia. Ineffective erythropoiesis leads to bone marrow expansion and associated bone deformation, which explains characteristic features of β-thalassaemia patients, such as craniofacial protrusions. The evolutionary association between the thalassaemia carrier state and resistance to malaria explains the high prevalence in sub-Saharan Africa, the Middle East, and the Mediterranean basin (Taher *et al.*, 2021). In several countries, including Cyprus, Greece, and Italy, successful premarital and prenatal screening have reduced the number of affected persons.

Current treatments consist primarily of red cell transfusion accompanied with iron-chelation therapy to control iron overload. Haematopoietic stem-cell transplantation from an HLA-matched donor is a potentially curative option but is limited because less than 20% of patients have an HLA-matched donor; additional challenges include the risks of 'graft-versus-host' disease and the complications associated with immunosuppression.

Luspatercept, given subcutaneously every three weeks, is the most recently approved agent in the US and Europe for the treatment of adults.

Shortly after birth, foetal haemoglobin is replaced by adult haemoglobin. This is achieved by a developmental switch favouring the expression of β-globin instead of the γ-globin, which begins around week 12 of gestation and is completed by six months of age. In 1964, two Greek haematologists, Fessas and Stamatoyannopoulos, observed that patients with hereditary persistence of foetal haemoglobin have a milder course of β-thalassaemia (Fessas and Stamatoyannopoulos, 1964). Later, these and other authors speculated that stimulating the production of foetal haemoglobin might be beneficial to patients suffering from sickle cell anaemia or β-thalassaemia. Sixty years later, the discovery of CRISPR made it possible to apply a specific gene treatment, called exagamglogene autotemcel, to the gene *BCL11A*, which led to the reactivation of foetal haemoglobin. This resulted in transfusion independence in 91% of patients suffering from β-thalassaemia and in the elimination of vaso-occlusive crises in 97% of patients suffering from sickle cell anaemia (Locateli *et al.*, 2024; Frangoul *et al.*, 2024). As of 2025, two gene therapies are authorised by the FDA for sickle cell anaemia: exagamglogene autotemcel for a cost of US$2.2 million and lovotibeglogene autotemcel for a cost of US$3.1 million.

12.4 Additional Developments and Cost

In 2017, AAV therapy was approved by the FDA for the treatment of a rare form of autosomal recessive blindness caused by mutations of the gene which produces a retinal protein that is essential for recycling of visual pigment to photoreceptor cells (Chung *et al.*, 2019). The most common form of a genetic retinal disease in children is due to pathogenic variants of the gene *CEP290*, which produces an enzyme that is essential for the structural integrity and functioning of the sensory cilium of photoreceptor cells. The effect of gene editing in children and adults with CEP290 inherited retinal degeneration is discussed in Pierce *et al.* (2024); a sustained improvement in cone photoreceptor function was reported, which, for currently unexplained reasons, was not consistently associated with visual acuity.

In hereditary angioedema, a defective gene leads to an increased production of kallikrein, which, in turn, enhances the production of bradykinin. The latter increases vascular permeability, which gives rise to tissue swelling. Prophylactic treatment involves inhibitors of plasma kallikrein. The *in vivo* gene editing of the responsible gene (*KLKB1*) via CRISPR-Cas9 led to a durable reduction in total plasma kallikrein levels (Cohn *et al.*, 2025).

Alpha-1 antitrypsin deficiency is a genetic disorder associated with serious injury to the lungs and liver, often leading to emphysema and significant liver dysfunction. There is no current effective treatment for this disease, which incidentally remains underdiagnosed; it is estimated that in the US, approximately 100,000 individuals suffer from this condition. The recent correction of the underlying mutation (called *PiZ*) via the CRISPR technique led to increased levels in the blood of normal alpha-1 antitrypsin protein and to a substantial reduction in the mutant protein.

Leucocyte Adhesion Deficiency Type I (LAD-I) is caused by pathogenic variants of the gene *ITGB2*, which encodes the molecule CG 18. This surface molecule is crucial for the functioning of white cells, including their adhesion in the vascular endothelium. LAD-I is characterised by recurrent bacterial and fungal infections; 60% of children with the most severe form of the disease die, unless they undergo allogenic haematopoietic stem-cell transplantation. Nine children were pretreated with a monoclonal antibody targeting the interleukin-23 pathway to reduce hyperinflammation. Then, they received a single infusion of autologous stem cells that had been genetically corrected by a lentiviral vector encoding a copy of *ITGB2*. All children were alive after one year, and on average half their neutrophils expressed normal CD18 (Aiuti, 2025).

A variety of other diseases have been treated with gene therapies, including a type of spinal muscular atrophy caused by the mutation of the gene *SMN1*, which encodes the *protein survival motor neuron 1*. Systemic intravascular administration of AAV vectors targeting the liver or other organs has led to positive clinical outcomes in several serious, inherited diseases (High and Roncarolo, 2019).

The FDA classifies cellular immunotherapies, cancer vaccines, and other types of cells used for treatment with the name 'cellular therapies'. Products that modify or manipulate the expression of genes or alter the properties of living cells are considered 'gene therapies.' As of 2025, the FDA has approved 41 cell or gene therapies, with prices ranging from US$0.5 million to US$4 million. Global efforts have been intensified to find ways to cut the cost of gene therapies, which is currently prohibitive for the vast majority of patients.

References

Aiuti, A. (2025). The view from a milestone in gene therapy. *New England Journal of Medicine, 392*, 1745–1746.

Andrade, J. G. (2025). Heart rhythm interventions – devil in the details. *New England Journal of Medicine, 392*, 1548–1550.

Angiolillo, D. J. and Capodanno, D. (2025). Uncoupling thrombosis and hemostasis by inhibiting factor XI. *New England Journal of Medicine, 392*, 400–403.

Chowdary, P. (2024). Bioengineered factor VIII – more innovation for hemophilia A. *New England Journal of Medicine, 391*, 277–282.

Chung, D. C., *et al.* (2019). The natural history of inherited retinal dystrophy due to mutations of the RPE65 gene. *American Journal of Ophthalmology, 199*, 58–70.

Cohn, D. M., *et al.* (2025). CRISPR-based therapy for hereditary angioedema. *New England Journal of Medicine, 392*, 458–467.

Cuker, A., *et al.* (2024). Gene therapy with fidanacogene elaparvovec in adults with hemophilia B. *New England Journal of Medicine, 391,* 1108–1118.

Daley, G. Q., Lovell-Badge, R., and Steffann, J. (2019). After the storm-A responsible path for genome editing. *New England Journal of Medicine, 380,* 897–899.

Doudna, J. A. and Sternberg, S. H. (2017). *A Crack in Creation: Gene Editing and the Unthinkable Power to Control Evolution.* Massachusetts: Houghton Mifflin Harcourt.

Fessas, P. and Stamatoyannopoulos, G. (1964). Hereditary persistence of fetal hemoglobin in Greece: A study and a comparison. *Blood, 24,* 223–240.

Fokas, A. S. (2024). *Ways of Comprehending: The Grand Illusion and the Essence of Being Human.* Singapore: World Scientific.

Frangoul, H., *et al.* (2024). Exagamglogene autotemcel for severe sickle cell disease. *New England Journal of Medicine, 390,* 1649–1662.

High, K. A. and Roncarolo, M. G. (2019). Gene therapy. *New England Journal of Medicine, 381*(5), 455–464.

Ledford, H. (25 June 2020). CRISPR gene editing in human embryos wreaks chromosomal mayhem. *Nature News.* https://www.nature.com/articles/d41586-020-01906-4.

Levy, S. (2001). *The Antibiotic Paradox.* New York: Perseus.

Liang, P. *et al.* (2015). CRISPR/Cas9-mediated gene editing in human tri-pronuclear zygotes. *Protein Cell, 6,* 363–372.

Locateli, F., *et al.* (2024). Exagamglogene autotemcel for transfusion-dependent β-thalassemia. *New England Journal of Medicine, 390,* 1663–1676.

Mahida, S., *et al.* (2015). Science linking pulmonary veins and atrial fibrillation. *Arrhythmia & Electrophysiology Review, 4,* 40–43.

Mahlangu, J., *et al.* (2023). Two-year outcomes of valoctocogene roxaparvovec therapy for hemophilia A. *New England Journal of Medicine, 388,* 2023.

Malec, L., *et al.* (2024). Efanesoctocog-alpha prophylaxis for children with severe hemophilia A. *New England Journal of Medicine, 391,* 235–246.

Mannucci, P. M. (2020). Hemophilia therapy: The future has begun. *Hematologica, 105,* 545–553.

Mojica, F. J., García-Martínez, J., and Soria, E. (2005). Intervening sequences of regularly spaced prokaryotic repeats derive from foreign genetic elements. *Journal of Molecular Evolution, 60*(2), 174–182.

Pasi, K. J., *et al.* (2020). Multiyear follow-up of AAV5-hFVIII-SQ gene therapy for hemophilia A. *New England Journal of Medicine, 382*(1), 29–40.

Pierce, E. A., *et al.* (2024). Gene editing for CEP290-associated retinal degeneration. *New England Journal of Medicine, 390,* 1972–1984.

Pipe, S. W., *et al.* (2023). Gene therapy with etranacogene dezaparvovec for hemophilia B. *New England Journal of Medicine, 388,* 706–718.

Plomin, R. and von Stumm, S. (2018). The new genetics of intelligence. *Nature Reviews Genetics, 19*(3), 148–159.

Ponder, K. P. (2011). Merry christmas for patients with hemophilia B. *New England Journal of Medicine, 365,* 2424–2425.

Rasko, J. E. J., *et al.* (2025). Fidanacogene elaparvovec for hemophilia B-A multiyear follow-up study. *New England Journal of Medicine, 392,* 1508–1517.

Reiss, U. M., Davidoff, A. M. *et al.* (2025). *New England Journal of Medicine, 392,* 2226–2234.

Ruff, C. T., *et al.* (2025). Abelaciman versus rivoxaban for patients with atrial fibrillation. *New England Journal of Medicine, 392,* 361–371.

Shaw, P., *et al.* (2006). Intellectual ability and cortical development in children and adolescents. *Nature, 440*(7084), 676–679.

Srivastava, A., *et al.* (2025). Lentiviral gene therapy with CD34+Hhematopoetic cells for hemophilia A. *New England Journal of Medicine, 392,* 450–457.

Taher, A. T., Musallam, K. M., and Cappellini, M. D. (2021). β-thalassemias. *New England Journal of Medicine, 384,* 727–743.

Thompson, A. A., *et al.* (2018). Gene therapy in patients with transfusion-dependent β-thalassemia. *New England Journal of Medicine, 378*(16), 1479–1493.

van den Berg, H. M. and Strivastava, A. (2023). Hemostasis-a balancing act. *New England Journal of Medicine, 389,* 853–856.

Watanabe, T. (1963). Infective heredity of multiple drug resistance in bacteria. *Bacteriological Reviews, 27*(1), 87.

Part III

Breakthroughs in Physics

In a single year, in 1905, Einstein proved the existence of atoms, introduced the concept of photons, and established that, contrary to our erroneous intuition, time is not absolute. The last of these understandings elucidated the physical meaning of special relativity, which provides the proper refinement of Newton's classical laws of motion. In addition, Einstein derived the famous equation $E = mc^2$, relating energy, E, mass, m, and the speed of light, c. Later, Einstein developed the theory of general relativity, which provides a far-reaching generalisation of Newton's revered gravitational law. These astonishing contributions are discussed in Chapter 13.

Remarkably, Einstein also had a huge impact on the development of quantum mechanics, despite his scepticism regarding the probabilistic nature of the mathematical equations describing this mysterious world. The atomic world, starting with the classical prerequisites of Faraday and Maxwell, continuing with the transformative impact of Bohr and the brilliant insights of Schrödinger and Dirac, and concluding with the formulation of quantum electrodynamics, is discussed in Chapter 14. The dismantling of the atom, i.e. the understanding that each atom consists of electrons, protons, and neutrons, occurred in a single location, the

Cavendish Laboratory of the University of Cambridge. Remarkably, this is also the place where the first atomic fission was achieved. This fascinating story is reviewed in Chapter 15. In the 1930s–1950s, a plethora of new particles were discovered. Furthermore, it was understood at that time that most of these new particles, as well as protons and neutrons, are not elementary but contain an internal structure. Namely, they are all made up of a few basic particles. For example, protons and neutrons are composed of three *quarks* (whereas the electron is an elementary particle). The theory describing subatomic interactions and the associated ingenious experiments supporting this theory are discussed in Chapters 15 and 16.

In addition to offering the reader the opportunity to achieve a broad understanding of physics, Part III is also written for several additional reasons, including the following: it rebukes the position of the distinguished Cambridge scholar George Steiner that using merely our ordinary language, it is impossible to understand the utterly complicated language of advanced physics. I hope that after studying this part, the reader will appreciate that, even without using any mathematical formulas, it is still possible to appreciate the most important achievements that occurred both in theoretical and experimental physics. Another reason for reviewing the beautiful story of how our insight into physical processes continues to deepen is to reemphasise two concepts which are of crucial importance in our quest for knowledge: *continuity* and *interdisciplinarity*. Regarding continuity, it is noted that even the exceptional discoveries of Einstein were based on earlier breakthroughs. Interdisciplinarity is exemplified by discussing in Chapter 17 some of the far-reaching implications of nuclear physics in many different areas; we should not forget that the internet and 'cluster computers' were born as a result of technological questions raised at CERN. Incidentally, familiarisation with the mysterious world of quantum mechanics is a necessary prerequisite for understanding key elements of quantum computing.

Chapter 13

Associations and the Striking
Inventiveness of Einstein

Gerald Edelman has emphasised that neuronal circuits are not only interconnected but are also highly redundant: 'different circuits of neuronal groups have the same effect in perception and memory' (Edelman, 2006, p. 33). As noted in Chapter 2 of Fokas (2024), the connectivity of different local neuronal circuits allows the brain to bind circuits *associated* with different aspects of the same percept, giving rise to consciousness. The redundancy of neuronal circuits provides the basic mechanism for the phenomenal ability of the brain to make associations between distinct but broadly related notions. Actually, according to Edelman, 'redundancy almost inevitably leads to associations' (Edelman, 2006, p. 358). Also, he states, 'Metaphor is a reflection of the range and associativity of enormously complex and degenerate brain networks' (Edelman, 2006, p. 83).

In my opinion, there is a relationship between the brain's ability to make associations and creativity: *a brilliant brain can make far-flung associations*. This is best exemplified by the astounding achievements of Einstein. In what follows, some of these achievements are reviewed with the hope of illustrating the importance of this fundamental capacity of our brains.

13.1 Ernst Mach, Ludwig Boltzmann, and the Brownian Motion

One of the important contributions of Einstein was to settle the following question: 'do there exist atoms?'. This question, which had polarised physicists for a long time, was the basis of a historic debate in Vienna, at the turn of the 20th century, between Ernst Mach and Ludwig Boltzmann. Remarkably, both these great physicists were also professors of philosophy at the University of Vienna, appointed by the president of the university, Theodor Gomperz.[1] This illustrates the close relation between physics and philosophy that existed at that time. Before concentrating on the stunning contributions of Einstein, I briefly discuss the impact of Mach and Boltzmann.

Theodor Gomperz admired the radical philosophers Auguste Comte and John Stuart Mill. These were prominent exponents of *positivism*; namely, in contrast to traditional philosophers, such as Saint Thomas Aquinas and Georg Wilhelm Hegel, they rejected theological and idealistic positions and advocated knowledge based on the senses and on scientific facts. In 1880, Franz Brentano was forced to resign from his position as a chair in philosophy at the University of Vienna because of the controversy that arose after Brentano married even though he was a priest earlier in life. Following the recommendation of his son, Heinrich, Theodor Gomperz appointed in 1895 Ernst Mach to the chair vacated by Brentano.

I first encountered the contributions of Mach as an undergraduate in the Department of Aeronautics of Imperial College. The statement that an aeroplane flies, say, with a speed of 2 Mach, means that it flies with a speed twice that of the speed of sound. Supersonic aeroplanes, like the Concord, fly with speeds greater than the speed of sound, i.e. with Mach speeds of greater than 1. Subsonic jets, like the usual commercial aeroplanes, fly with speeds less than 1. The mathematical equations describing supersonic air flow differ drastically from those describing subsonic flow (the former equations are of the so-called hyperbolic type, whereas the latter are of the elliptic type). This reflects the fact that the underlying physics of the two situations is quite different. Supersonic airflow gives rise to the formation of shock waves, which generate high-pitched noise. Mach studied many important physical phenomena extensively, including

---

[1] The Gomperz family was one of the richest families of Vienna, along with the Wittgensteins, the Rothschilds, the Gutmanns, and the Ephrussis.

supersonic shock waves. In addition, Mach posed and tried to answer fundamental philosophical questions, including the following: technological progress is based on science. What is science based on? What is the true meaning of such fundamental physical concepts as force, acceleration, heat, and entropy? Mach discussed these issues in his three books: *The Science of Mechanics* (1883) (Mach, 2013), *Principles of the Theory of Heat* (1896) (Mach, 1986), and *Principles of Physical Optics* (published posthumously in 1921) (Mach, 2003). These books established a new philosophical discipline, namely, the philosophy of science. Mach's analysis is based on penetrating observations of the given phenomena, on the understanding of the history of the relevant science, and on placing primary importance on sensory impressions. The last aspect led him to physiology, where he made an unexpected discovery: he understood that the sense of balance resides in the inner ear. The related sense of proprioception was, as mentioned in Chapter 2 of Fokas (2024), the first new sense to be discovered since the discovery by Aristotle of the five classical senses.[2]

In his philosophical work, *Analysis of Sensations* (1886) (Mach, 1959), Mach attacked the 'metaphysical positions of Immanuel Kant' and laid the foundation of philosophical *empiricism*: all knowledge must be based on experience and all experience is based on perception. So, all knowledge is based on data obtained via the senses. According to Mach, 'the objects that we perceive consist merely of bundles of sense-data linked together in a regular way'. Also, he wrote, 'Among the relatively long-lasting patterns of memories, moods, feelings, etc., there is one pattern that is attached to a special body, and that pattern is called Ego'. Finally, he noted:

"I experience green, means that the element of greenness occurs within a certain pattern of other elements (sensations, memories) [...] When I die, these elements can no longer occur in their usual familiar groupings [...]. No Ego can remain. The Ego cannot be saved" (Sigmund, 2017).

---

[2]This discovery was made independently, and at approximately the same time, by Josef Breuer, a Viennese physician and co-founder with Sigmund Freud of psychoanalysis. These findings were later extended by Robert Barany, who was awarded the Nobel Prize in Medicine in 1913 for his work on the vestibular apparatus.

The last sentence had a profound influence on the writers of Vienna. In particular, the Viennese poet Hugo von Hofmannsthal, who wrote the librettos of most of the operas of Richard Strauss, attended Mach's lectures, and the Viennese writer Arthur Schnitzler adopted Mach's position in his *inner monologues*.

The Austro-Marxists hailed Mach's work as the rigorous foundation of materialism. However, Vladimir Lenin, in his book *Materialism and Empiriocriticism* (1908) (Lenin, 1962), attacked Mach as an idealist: by reducing the effect of matter into mere sensations, Mach was perceived as a threat to materialism!

In 1901, Mach had to resign from his position after suffering a stroke. Interestingly, he was replaced by another renowned physicist, Ludwig Boltzmann. Although Boltzmann was born in Vienna, he grew up in the town of Linz, which is the place where, in the early 17th century, Johannes Kepler wrote his magnum opus, *The Harmony of the World* (Kepler, 1997).[3] By the time he was 25, Boltzmann was considered one of the very few physicists to have a full understanding of Maxwell's work on electromagnetism and thermodynamics. Boltzmann's most important contribution is the so-called *kinetic theory of gases*. According to this theory, gases consist of *minute particles* that are constantly moving and colliding like billiard balls. The higher the *temperature* of the gas, the faster these particles move. The physical quantity of *pressure* is the result of the collision of these particles with the wall. Boltzmann's famous equations provide the mathematical description of this process. Furthermore, the statistical analysis of these equations yields a qualitative description of physical observables, including pressure. This description also provides a definition of the concept of *entropy*, which is a statistical measure of the disorder of a system at a microscopic level.[4]

Although Boltzmann used to joke that 'I disdain experiments the way a banker disdains coins' (Sigmund, 2017), he was appointed to a chair in experimental physics in Graz in 1878. His colleague, Walther Nernst, was awarded the Nobel Prize in Chemistry in 1920 for his work on electrochemistry and thermodynamics, but Boltzmann never received a Nobel Prize, despite the fact that in the period 1902–1906, he was nominated

[3]Linz is also the birthplace of the composer Anton Bruckner, who actually taught young Boltzmann piano.
[4]The equation defining entropy, namely $S = k \log W$, where S denotes entropy, k is a constant, log is an abbreviation for 'logarithm', and W denotes the number of the microstates of the gas, is engraved on Boltzmann' tomb in Vienna's First Cemetery.

every year. Boltzmann's great achievement was that, starting from first principles, he was able to provide an explanation of the laws of thermodynamics. His analysis was based on postulating the existence of some 'minute particles' which, following the terminology of ancient Greeks, were called *atoms*.

The introduction of atoms was highly controversial. Do atoms truly exist, or are they just mental objects playing the same role in physics that the idealised concept of points plays in mathematics? According to Mach, since atoms cannot be perceived by our senses, they cannot exist! Indeed, any time atoms were mentioned, Mach would ask: 'did you ever see one?' (Sigmund, 2017). This major controversy was the focus of the debate between Mach and Boltzmann that took place in 1896 in a lecture hall of the Vienna Academy, which, however, failed to resolve this basic question.

Boltzmann and Mach were interested in philosophy during their entire lives. At the same time, both emphasised the importance of mathematics. Mach wrote, '[…]without at least an elementary instruction in mathematics and science, man will remain a stranger in the world, a stranger in the culture that supports him' (Sigmund, 2017). Boltzmann wrote, 'I tried to become better informed about the basics of philosophy […] I first turned to Hegel; but what obscure, vacuous balderdash did I find there! My unlucky star ushered me to Schopenhauer […]'. At that time, Schopenhauer was highly reputable. Thus, the Viennese public, intrigued by such ferocious attacks on these well-known philosophers, was eager to attend Boltzmann's lectures. However, Boltzmann's philosophical lectures were highly mathematical and hence incomprehensible to a large section of the audience. Nevertheless, Boltzmann remained unapologetic: 'Mathematics is to science what the brain is to man'. Despite this high praise, he was aware of the limitations of mathematics: 'No equation ever represents any phenomenon with absolute precision. Each equation is an idealization […] and hence going beyond experience'. He was tormented by his inability to fully elucidate the role of mathematics in science and to resolve other major philosophical questions. Unable to come to terms with the nature of perception, he wrote, 'If none of these questions makes sense, then why can't we discard them, or what can we do to squelch them once and for all?" He attempted to get advice from the earlier occupant of his philosophy chair, Brentano, to whom he wrote, 'The irrepressible urge to philosophize is like nausea caused by a migraine […]. The sublime, majestic task of philosophy is to make things clear, to finally cure humankind of this migraine' (Sigmund, 2017). Severely disturbed by these contradictions, Boltzmann ended his life by hanging himself in 1906 at the age of 62. This took place in the castle, Duino, near Trieste, the same place that the Austrian poet and novelist Rainer Rilke began in 1912 writing his poet cycle, *Elegies.*

Mach died in 1916 at the age of 75. In his obituary, Albert Einstein wrote, 'Even those who view themselves as Mach's opponents, are hardly aware of how much of his viewpoint they have soaked up, so to speak, with their mother's milk'. After meeting Mach, the great psychologist and neuroscientist William James wrote, 'I do not think that anyone has ever left me with such a strong impression of unadulterated genius'[5] (Sigmund, 2017).

The controversy regarding the existence of atoms not only polarised the world of physics but also that of chemistry. The camps of Mach and Boltzmann boosted Nobel Prize laureates, including, respectively, Wilhelm Ostwald, who was awarded the Nobel Prize in Chemistry in 1909 for his work in catalysis, and Max Planck, a Nobel Laureate for his work in quantum mechanics.

The definitive answer to this question was provided by Albert Einstein in his seminal paper on Brownian motion, in his *Annus Mirabilis*, 1905. In 1827, the Scottish botanist Robert Brown observed, via a microscope, that tiny particles suspended in a liquid were constantly moving back and forth in an apparently random fashion. Einstein speculated that this motion was due to the collisions of these tiny particles with the *invisible atoms* and *molecules* making up the surrounding liquid. By making careful calculations based on the observed random Brownian motion, Einstein was finally able to calculate the size and speed of the invisible dancing partners of the visible (under the microscope) tiny particles. It is worth noting that Einstein had a difficult time deriving the correct result; his first attempts gave the wrong coefficient, and he had to struggle to obtain the correct value. The ingenious analysis of Einstein was soon afterwards fully confirmed experimentally by the French physicist Jean-Baptiste Perrin (Nobel Prize in Physics, 1926). The controversy was over—*atoms exist.*

Perrin's Nobel Prize-winning experiments gave rise to another, less known, controversy: Einstein observed a difference between the values of Avogadro's constant derived from these experiments and the value derived from Planck's calculation from black-body radiation.[6] Einstein

---

[5]In addition to his numerous scientific contributions, James was also a founder of philosophical *pragmatism.*

[6]The book *Einstein's mistakes* (Ohanian, 2008, p. 118) describes Einstein's first attempt to get the value of Avogadro's number in his dissertation: '[...] his execution of the

speculated that this was due to the incorrectly evaluated size of the spherules in Perrin's experiments and asked the Nobel Laureate to perform additional experiments and to use the viscosity formula introduced in his own dissertation. This discrepancy remained open until very recently, when it was shown that Einstein's speculation was correct (Škvarla, 2024).

Returning to Boltzmann's contributions, it is worth noting that Boltzmann's equations have become central pillars of physics, underpinning not only thermodynamics but also important technological applications, including the theory of semiconductors. Boltzmann's formulation has implications beyond physics. For example, in his influential book *What is Life* (1944) (Schrödinger, 1944), Erwin Schrödinger asks, 'Why should an organ like our brain […] of necessity consist of an enormous number of atoms […]?'. Schrödinger claimed that 'in order that its physically changing state should be in close and intimate correspondence with a […] thought', it is necessary that this physical system be highly ordered; this is achieved only if the number of atoms is extremely large. Indeed, according to the mathematical formulation of Boltzmann, the existence of an extremely large number of atoms is a necessary condition for 'order to emerge from disorder', i.e. for regular behaviour to emerge from the chaotic behaviour of the individual atoms. Interestingly, one of the examples elaborated by Schrödinger is Brownian motion: the movement of a tiny particle suspended in a liquid is completely irregular, but the movement of a very large number of tiny particles gives rise to the completely regular phenomenon of *diffusion*.

13.2 Associating Blackbody Radiation and the Motion of an Ideal Gas

As noted in Chapter 13 of Fokas (2024), light, X-rays, and radio waves are all examples of *electromagnetic waves*. These waves are distinguished by their different frequencies. The study of electromagnetic waves is facilitated by the introduction of the concept of a *blackbody surface*. This

---

mathematical program immediately turns into a comedy of errors. The editors of the modern edition of Einstein's Collected Papers attached more than thirty footnotes to the reprint of the dissertation, listing all the mathematical errors. These range from simple mistakes of sign, to typos in mathematical symbols, to wholesale omissions of terms in equations and errors in coefficients'.

is an idealised surface with the defining property that it *absorbs* every electromagnetic wave falling upon it. Namely, light incident on such a surface will be completely absorbed rather than reflected. So, the surface will appear black, and this justifies the terminology blackbody. In 1900, the leading German physicist of the time, Max Planck, introduced a truly revolutionary idea regarding the radiation emitted from a blackbody (under the assumption that this body is in thermal equilibrium with its environment). He claimed that a blackbody surface emits 'discrete packages' of energy at a characteristic frequency that depends on the speed of light and on the temperature of the surface. By quantifying his idea in terms of an appropriate mathematical formula, he succeeded in matching the available experimental data. Remarkably, these packages, which were later called *quanta*, take the form of integer multiples of a basic unit, later called *Planck's constant*. It turns out that the introduction of Planck's quanta was the first step towards the development of the transformative theory of *quantum mechanics*, which describes electromagnetism at very small scales, particularly at the atomic level.

Planck's formulation was in direct contradiction to a theory proposed a few years earlier by Hendrik Lorentz (Nobel Prize in Physics, 1902). Lorentz's theory was based on the understanding that radiation is electromagnetic waves described by Maxwell's equations. This implied that the energy is radiated from a body in a *continuous* manner.

Although Einstein could not explain Planck's result, he was very impressed with the remarkable agreement of Planck's formula with experimental data. So, Einstein went a step further than his contemporaries, who also could not explain the effectiveness of Planck's formula: Einstein accepted this result as given and investigated deeper its implications. For this purpose, he considered a *blackbody*. An approximate idealisation of such a body can be achieved by considering a cavity that has walls opaque to radiation and which has a small hole: electromagnetic waves incident on the hole will pass into the cavity, but if the cavity is sufficiently large, it is unlikely that these waves will be re-emitted. Remarkably, Einstein *associated* this physical situation with one that appears to be completely different: the case where a cavity contains an ideal gas held at a fixed temperature. In this situation, the (probability distribution of) the magnitude of the velocities of the molecules of this gas satisfies the so-called Maxwell–Boltzmann equations. By associating these two completely different phenomena, one involving electromagnetic waves and the other an ideal gas, and after performing several pages of mathematical calculations, Einstein reached the conclusion that light must consist of particles, later called *photons*!

In his analysis, Einstein speculated that an electromagnetic wave of frequency f, carries energy equal to hf, where h denotes Planck's constant. This formula was also of crucial importance for Einstein's brilliant explanation of the *photoelectric effect*, discussed in detail in Chapter 9 of Fokas (2024). In this regard, it is worth noting that in 1907, Robert Millikan began investigating the photoelectric effect with the goal of proving Einstein wrong. After 12 years, he proved that Einstein was actually correct! Interestingly, Millikan, who was a brilliant experimentalist, was introduced to the mysteries and beauty of physics by his Greek teacher at Oberlin College in Ohio. In 1903, Millikan added, in the usual mercury pump used to create a vacuum, a tube that contained charcoal immersed in liquid air. In this way, he managed to create a vacuum with pressure a billion times lower than the atmosphere. His experimental verification of the photoelectric effect led to the awarding of the Nobel Prize in Physics to Einstein in 1921 and to Millikan in 1923.

13.3 Associating Mass and Energy

The year 1905 is characterised as Albert Einstein's *annus mirabilis*. In my opinion, this characterisation is also valid for physics itself. Indeed, in addition to his complete analysis of the Brownian motion discussed earlier and his proof that light consists of photons, which was further discussed in Fokas (2024) in relation to the photoelectric phenomenon, Einstein also formulated the special theory of relativity. Finally, in the same year, as an afterthought on the latter theory, he derived one of the most famous equations ever written, namely $E = mc^2$. This equation relates energy E, mass m, and the speed of light c. Until then, energy and mass were regarded as two different concepts. In every physical situation, it was assumed that each of them is conserved. The *conservation of energy* and the *conservation of mass* were the two most important laws of physics. Thus, the *association* of these two seemingly different concepts was a stunning theoretical achievement. Furthermore, it provided the basis for the new, hugely important concept of atomic energy. Indeed, an atom consists of electrons, protons, and neutrons. The number of protons equals the number of electrons, and this number determines the specific *element*. For example, hydrogen atoms have one proton, whereas uranium atoms have

92 protons. The number of neutrons determines the specific *isotope*. For example, the specific uranium atom U-235 has 143 neutrons. When this particular atom absorbs a neutron, it breaks into two smaller atoms plus several neutrons. The mass of the latter combination is smaller than the mass of U-235. The difference in the masses multiplied by the very large number c^2 gives rise to a huge amount of atomic energy. This novel mechanism for generating energy can be used in electric plants but also, ominously, in atomic bombs. This type of bombs, called 'fission weapons' since they are based on the splitting of an atom, were used in Hiroshima and Nagasaki.

When two different concepts are associated, the more remote they are, the more profound is the potential importance of their association. In addition to the relation of energy and mass, this is also clearly illustrated by the association of gravity with acceleration that was established via the theory of general relativity. To explain this astonishing breakthrough, which occurred in the period 1905–1915, it is necessary to first briefly discuss the special theory of relativity.

13.4 Special Theory of Relativity

Until the introduction of this theory, the most important equations in physics were Newton's equation, written in 1687, describing the motion of a particle, and Maxwell's equations, written in 1886, which describe electromagnetic waves. Newton's equation of motion states that a force F acting on a particle of mass m causes this particle to move with acceleration a, where $F = ma$. Underlying this basic law was Newton's assumption of conservation of mass (Newton referred to mass as 'quantity of matter'). This assumption had been verified for a variety of circumstances. For example, Antoine Lavoisier checked for many chemical reactions that the total weight of the starting ingredients equals the weight of the material obtained after the reaction. Incidentally, Lavoisier was guillotined during the French Revolution, which prompted the mathematician Joseph Lagrange to state, 'It took them only a moment to cut off that head, but France may not produce another like it in a century'.

Several distinguished physicists realised that there was a contradiction between Newton's equation of motion and Maxwell's equations. This stemmed from the postulate underlying Maxwell's equations that the *speed of light is constant*, independent of the motion of an observer.

This led Einstein to revisit the concept of *relativity*. This fundamental concept, already understood by Galileo Galilei,[7] states that if the velocity of a particle is measured with respect to two different *frames of reference*, and if these frames move with different velocities, then the two velocities recorded by observers in these two frames will be different. In other words, there *does not exist an absolute velocity*; velocity can only be defined *relative* to a specific frame. For example, consider a fly buzzing inside a Concord aeroplane. The speed of the fly measured with respect to a frame of reference inside the aeroplane is rather low, so a passenger observes a fly moving in a usual way. On the other hand, the speed of the fly measured with respect to a frame of reference on the ground is affected by the motion of the aeroplane, so an observer on the ground possessing special binoculars will observe a fly moving with a supersonic speed.

It turns out that this concept of relativity implies the highly non-intuitive result that time also *cannot* be defined absolutely. This has unexpected implications for the notion of *simultaneity* between events recorded by observers travelling on different frames of reference.

It is interesting that the physics community thought at that time that the contradiction between Newton's and Maxwell's equations was due to an error in the latter. However, Einstein showed that this contradiction was due to the fact that Newton's equation needed to be corrected: the acceleration appearing in Newton's law must be defined with respect to the so-called *proper time*, which takes into consideration the fundamental concept of relativity.[8]

For the proper time to be well defined, Einstein reaffirmed that light has the unique property that indeed its *speed is absolute*, i.e. it is the same with respect to *all* frames of reference. For example, a beam of light inside a Concord has the same constant value, c, for both a passenger on this Concord and for an observer on the ground. This value is very large: it is

---

[7]Galileo was born in 1564, the same year that Michelangelo died. Galileo died in 1642, the same year that Newton was born.

[8]A necessary and sufficient condition for the laws of electromagnetism to look the same to two observers moving at a constant velocity with respect to each other is that Maxwell's equations remain the same (which in mathematical terms is called *invariant*) after performing a certain transformation involving both space and time coordinates. This transformation, which is called *Lorentz transformation*, was first derived by Lorentz and then perfected by Poincaré. In order for Newton's equation to remain invariant under this transformation, it is necessary that time be replaced with proper time.

approximately 3 times 100 million metres per second. If the speed of the particle is small in comparison with the speed of light, which is usually what happens in our macrocosm, then the proper time is approximately equal to the usual time, and the familiar laws of Newton provide a very good approximation of reality. More precisely, the proper time equals the usual time multiplied by $1/\gamma$, where the *Lorentz factor* γ is given by $\gamma = [1 - (v/c)^2]^{-1/2}$, with v and c denoting the speeds of the particle and of the light, respectively. If v/c tends to zero, then γ tends to one, and the proper time tends to the usual time.

It is worth noting that Hendrik Lorentz[9] and the mathematician Henri Poincaré had also developed, independently, a theory that is mathematically equivalent to the special theory of relativity. However, they failed to appreciate its deep physical meaning. The special theory describes how objects evolve in *local time* and also how different local times evolve *relative* to each other. If one thinks in terms of the dielectric pair of local versus global, then one naturally raises the following question: is there a concept of global time? This question cannot be discussed before the introduction of the general theory of relativity.

The development of the special theory of relativity illustrates several important concepts, including the following: first, the principle of *continuity* in the development of science, namely that knowledge progresses via steps involving several independent researchers—in this case, starting with Galileo; second, that the occurrence of an experimental or theoretical contradiction necessitates the replacement of an existing theory by a more general theory, providing a new Kuhnian paradigm; third, when, after sufficient accumulation of knowledge, the situation is mature for a 'breakthrough', this is typically achieved by more than one individual—in this case, by Lorentz, Poincaré, and Einstein.

In 1908, Hermann Minkowski, who was Einstein's professor in Zurich, building upon his former pupil's work, expressed the deep insight that although space and time are relative, the proper amalgamation of space-time is absolute. He famously wrote, 'Henceforth space by itself, and time by itself, are doomed to fade away into mere shadows, and only a kind of union of the two will preserve an independent reality'.

---

[9]Einstein wrote the following about the Dutch physicist Hendrik Lorentz: 'I admire this man like no other; I might say, I love him' (Einstein, 1995).

13.5 Associating Acceleration with Gravity

The different frames of reference used in special relativity move with constant velocities. What happens if the velocity of a frame is not constant? In other words, what is the form of the fundamental laws of physics for an observer on an accelerating frame of reference? In this regard, it is noted that according to classical mechanics, all objects falling under gravity move along identical paths independent of their masses. Indeed, Newton, in addition to his law $F = ma$ mentioned earlier, also postulated another fundamental law, namely the so-called *gravitational law*. This law states that the magnitude of the attractive force, F, experienced by two masses, m and M, at a distance r, is given by the formula $F = GmM/r^2$, where G is a constant. If M denotes the mass of the Earth, then using the two laws of Newton to equate the two expressions for the force F, namely the gravitational force and the force $F = ma$, it follows that m cancels, and $a = GM/r^2$. So, the acceleration of an object moving under the attraction of the Earth is independent of the mass of this object.

Newton's gravitational law has a feature which is conceptually puzzling: the force between two distant bodies is transmitted *instantaneously*. This strange effect worried greatly Newton himself, who wrote, 'That one body may act upon another at a distance through a vacuum without the mediation of anything else [...] is to me so great an absurdity that, I believe, no man who has in philosophic matters a competent faculty of thinking could ever fall into it'.

Maxwell's equations do not exhibit such paradoxical behaviour. This is the consequence of the fact that Maxwell's equations, in contrast to Newton's gravitational law, are equations which describe a *field*. In an attempt to introduce to the reader this difficult notion, I note that a field, instead of being localised at *discrete* points, is supported in a *region* of space. For example, the electric potential in a conductor is affected by every point in this conductor.

It is worth noting that the English scientist Michael Faraday had discovered, a few years before Maxwell's work, that when a magnetic field changes in time, it produces an electric field.[10] Maxwell

---

[10] Einstein kept a picture of Faraday on his study wall, alongside pictures of Newton and Maxwell. Incidentally, the American scientist Joseph Henry, while he was building electromagnets, he also discovered the above effect independently (and a bit later) from Faraday.

supplemented this fact with the understanding that, conversely, when an electric field changes in time, it produces a magnetic field. This led to his famous equations which support 'space filling and potentially self-regenerating fields'. This means that changing magnetic fields induce changing electric fields, which in turn produce changing magnetic fields, and so forth. These fields do not exert their effect instantaneously, but they *travel with the constant speed of light*.

Einstein knew that, according to the special theory of relativity, nothing can travel faster than the speed of light, so a force *cannot* act instantaneously between two masses. This implies that the *special relativity cannot be consistent with Newton's gravitational law*. Actually, as discussed earlier, *special relativity was born from the requirement of being consistent with Maxwell's equations, which express a field theory. On the other hand, Newton's gravitational law is not a field theory*. Hence, Einstein, in addition to understanding that special relativity demanded the replacement of Newton's gravitational law by a new theory, also knew the nature of the new law: *it had to be field theoretic*. So, he began searching for the *gravitational field*.

In order to implement this goal, Einstein conceived a thought experiment involving the motion of a particle, but without using Newton's gravitational law since he knew that this law cannot be valid. He considered a frame of reference fixed in a box falling in space towards the earth. Since the speed of the box increases as the box falls, the frame in the box is accelerating. An observer on this frame looking at objects contained in the box would think that the box is at rest. These objects behave as if there were no gravity. Today, this is the well-known phenomenon of weightlessness illustrated by videos of astronauts in space stations, but at that time it would have been difficult to imagine this phenomenon. This shows that the *mass* of a given object in a box falling under the influence of *gravity* does not affect its motion relative to an *accelerating* frame. Furthermore, this object does not 'feel' the effect of gravity. Therefore, the notion of gravity, just like the notion of velocity, is also *relative*, namely, it depends on the frame of reference used for its measurement. Scrutinising this situation deeper, Einstein concluded that the concepts of gravity and acceleration must be related. In fact, not only did he *associate* these two very different concepts, but he also postulated that they are *indistinguishable*. He called this identification the *principle of equivalence* and later, as mentioned in the introduction of Fokas (2024), referred to his eureka moment as 'the happiest thought of my life'.[11]

This association is indeed profound: on the one hand, the notion of acceleration is related to *motion*, which is described in terms of space and time, i.e. in terms of a four-dimensional *geometry* (the three coordinates needed for the characterisation of space and one for the specification of time). On the other hand, gravity is a *physical concept*. Hence, Einstein established a deep relation between a *four-dimensional geometry* and *physics*. In the words of the distinguished University of Princeton physicist John Wheeler, 'matter tells space how to curve, and space tells matter how to move'.

Returning to the question of whether there exists in our macrocosm a global time, the general theory of relativity implies that such a concept does not exist. The special theory of relativity implies that time is 'slowed down' by speed; the general theory of relativity implies that time is also slowed down by mass. Actually, the effect of the mass of a black hole on time is dramatic: at the boundary, called the 'horizon', time stands still! If one wants to think of a relevant *global* entity, one must think of the four-dimensional 'block space', coupling space and time under the influence of the gravitational field. Interestingly, as discussed in my forthcoming book, *From Ancient Greece to Viennese Modernism* (in preparation), Einstein's understanding of the special and general theories of relativity resolved a fundamental philosophical conflict between Aristotle and Newton.

Einstein's early attempts in 1907 to reformulate gravity so that it is consistent with the theory of special relativity failed. Starting in 1912, he began to realise that the only way forward is to employ certain highly non-intuitive geometrical notions (Einstein, 1916; Miller, 1998). Indeed, the geometry needed in the general theory of relativity is not the familiar Euclidean geometry but a complicated mathematical generalisation of this geometry known as *Riemannian geometry*. In this regard, Einstein was assisted by his friend Marcel Grossmann, who, in 1907, became a professor of mathematics at the Zurich Polytechnic (Einstein and Grossmann,

---

[11]Ernst Mach had claimed that a body's inertia depends not only on the body's mass but also on the distribution of mass in *all of space*. Einstein referred to this claim as 'Mach's principle' and noted that it pointed him towards the right direction. In 1910, when Einstein visited the ministry in Vienna to declare his religious status, needed for settling his appointment at the University of Prague, he made sure to pay his respects to Ernst Mach.

1913, 1914); Grossmann was a fellow student with Einstein at the Zurich Polytechnic (which, in 1911, was named ETH Zurich, Swiss Federal Institute of Technology in Zurich).

In 1916, the preeminent mathematician of that time, David Hilbert, who himself came close to discovering the general theory of relativity, declared that 'the physicist must become a geometer'. It turns out that the involvement of the Riemannian geometry makes the interaction of matter and space fiendishly complicated. As a matter of fact, it is not possible to write an analytical formula replacing the simple equation $F = GMm/r^2$ describing the attractive force between two particles of mass M and m at a distance r. In other words, it is not possible to write analytically the mathematical equation describing the motion of two particles (referred to as the two-body problem) in the framework of general relativity. Einstein himself co-authored a paper in 1938 where an approximation is presented in the case that v/c is sufficiently small, so that terms of order $(v/c)^2$ can be neglected (Einstein *et al.*, 1938). This approximation is called *post-Newtonian*. This result has been improved by many physicists, including the French physicist Luc Blanchet, who has derived an approximate formula, 4.5 pages long, in the case that terms of order $(v/c)^7$ are neglected (Blanchet, 2014). Blanchet's formula was cited in the announcement of the Nobel Prize in Physics in 2017 awarded for the recent experimental verification of the existence of gravitational waves (The Nobel Committee for Physics, 2017), since the mathematical formulation of these waves relies on the relativistic analysis of the two-body problem. Blanchet and the author have recently derived an approximation for the two-body case in another interesting case, which arises when the two masses are sufficiently small (Blanchet and Fokas, 2018). This approximation is called *post-Minkowskian*.

In forming his remarkable associations, Einstein was certainly aided by his unique ability for thought experiments and his broad education. Already at the age of 13, Einstein read Immanuel Kant's *Critique of Pure Reason*. This highly influential book, as well as several other books on philosophy, mathematics, and physics, was provided to Einstein by a young medical student, Max Talmud, who had lunch at Einstein's home once a week (at that time, organising such lunches was the custom in Jewish households in Munich). During 1902–1906, when Einstein was working at the Swiss federal patent office, he met regularly with three other young individuals, who were not research scientists. These friends spent many hours discussing books on diverse topics. This group, which

Einstein gave the grandiose name 'Olympia Academy', studied a variety of books, including Plato's *Dialogues*, several books by the great mathematician and philosopher Gottfried Leibniz, John Stuart Mill's *System of Logic*, David Hume's *Treatise on Human Nature*, and Henri Poincaré's *La Science et l'Hypothèse*. In 1904, Michele Besso became the fifth member of this group. This 'latecomer' had the honour to be explicitly acknowledged in Einstein's special relativity paper of 1905. This is because Einstein discussed with Besso extensively the concept of the proper time. One morning, 'without even saying hello', Einstein greeted Besso (who was also working in the same Swiss federal patent office) with the following phrase: 'thank you. I have completely solved the problem [...]. Time cannot be absolutely defined and there is an inseparable relation between time and signal velocity' (Einstein, 1970).

One of the members of the group wrote that Poincaré's book, *La Science et l'Hypothèse*, 'profoundly impressed us and held us spellbound for weeks on end' (Solovine, 1956). Indeed, the three fundamental problems solved by Einstein in 1905 were all clearly stated in *La Science et l'Hypothèse*, where Poincaré had formulated the three 'pressing physical problems' of the time:

- The resolution of certain contradictions associated with the concept of 'ether' (meaning 'upper air' in Greek), which refers to the mysterious space-filling substance needed for the propagation of electromagnetic waves. Einstein's *special theory of relativity* eliminated the need for the existence of ether.
- The understanding of the strange phenomenon that ultraviolet light can liberate electrons from metals. This phenomenon, known as the photoelectric effect, was explained by Einstein via his understanding that light consists of *photons* (as discussed in Chapter 13 of Fokas, 2024).
- The explanation of the erratic behaviour observed in Brownian motion. As discussed earlier, Einstein provided a quantitative analysis of this phenomenon via the introduction of *atoms*.

Remarkably, the same book also influenced Pablo Picasso: the treatment of a four-dimensional geometry presented by Poincaré helped Picasso to solve the problem of how to depict simultaneously different views of an object, a breakthrough first achieved in *Les Demoiselles d' Avignon* (1907). This painting and the revolutionary geometrisation of space achieved in Cubism will be discussed in my forthcoming book, *Cognition and Painting*.

Einstein was also influenced by the writings of David Hume. This Scottish philosopher, in contrast to Ernst Mach who emphasised the importance of sense perceptions, reached the conclusion that the exact laws of nature cannot be understood by relying only on sense perceptions or empirical data. Einstein wrote to Besso in 1948, recalling their discussions in the Olympia Academy, 'so far as I can recall, David Hume had a greater influence on me [than Mach]' (Miller, 1998). Indeed, as noted earlier, the special theory of relativity is based on the concept of proper time, and Lorentz as well as Poincaré were aware of the *mathematical* importance of this concept. Actually, the discussion about time in Einstein's paper of special relativity is very similar to the one presented in an article by Poincaré, *The Measurement of Time*, written in 1898 (Poincaré, 1902). However, Lorentz and Poincaré thought that the difference between proper time and real time was too small to be measured. Also, and more importantly, proper time depends on the relative motion of the associated clock used to measure time, and this dependence is counterintuitive. Lorentz and Poincaré accepted the mathematical significance of the proper time, but they rejected its *physical* significance. Only Einstein, apparently aided by Hume's philosophical positions on the importance of thinking versus senses, accepted both.

Despite Einstein's statement regarding Hume and Mach, the influence of Ernst Mach should not be underestimated: in addition to Mach's claim mentioned in footnote 10, Einstein, aware of Mach's position on the importance of measurability of physical quantities, insisted that all fundamental physical notions, such as space, time, and velocity, must be *measurable experimentally* or at least via *thought experiments*. So, he rejected theories based on ether since velocities relative to ether are unmeasurable.

Einstein's willingness to go against intuitive ideas generated by perceptions is consistent with his overall view of truth, which was essentially Platonic. In this connection, it is interesting to note that although he idolised the highly structured, 'mathematical' music of Bach[12] and eulogised the 'spontaneous' music of Mozart, he was somewhat less enthusiastic about Beethoven's music because, according to Einstein, it was 'created'.

---

[12] When a journalist asked Einstein's opinion about Bach, he replied: 'listen, play, love, revere – and keep your mouth shut'.

A Platonic viewpoint places both musical and physical truths in a world of ideals: *regarding music, beyond notes and instruments, there exists the ideal world of melodies. In physics, the Platonic world coincides with 'cosmos', where beyond observations, experiments, and theories, there exist universal laws written in beautiful mathematical language.* Both the above worlds cannot be 'created'. According to Plato, these worlds exist independently of us and can be reached through theorising, guided by aesthetics. Einstein's Platonic inclinations become evident from his remark regarding the music of Mozart that it is 'so pure that it seemed to have been ever-present in the universe, waiting to be discovered'.

Concluding this chapter, it should be emphasised that Einstein, as expected, was clearly aware of the importance of the unconscious in creating the 'building blocks' of conscious thinking. For example, in his response to a question from the journalist Carl Seelig regarding the definite birthdate of special relativity, he said:

> "Four or six weeks elapsed between the conception of the idea of special relativity and completion of the corresponding publication. But it would be hardly correct to designate this as the birth date, because earlier arguments and building blocks were prepared over a period of years [...]" (Seelig, 1954).

In 1946, Einstein defined unconscious thinking as 'free play with concepts' (Einstein, 1946). Also, according to him, creative thinking is nonverbal; for Einstein, creativity occurs through spontaneous visual imagery, whereas words 'were sought later only in a secondary stage '.

References

Blanchet, L. (2014). Gravitational radiation from post-Newtonian sources and inspiralling compact binaries. *Living Reviews in Relativity*, *17*(2), https://doi.org/10.12942/lrr-2014-2.

Blanchet, L. and Fokas, A. S. (2018). Equations of motion of self-gravitating N-body systems in the first post-Minkowskian approximation. *Physical Review D*, *98*, 084005.

Edelman, G. (2006). *Second Nature.* Connecticut: Yale University Press.

Einstein, A. (1916). Grundlagen der allgemeinen Relativitätstheorie. *Annalen der Physik*, *49*, 769–822.

Einstein, A. (1946). *Autobiographical Notes, in Albert Einstein: Philosopher-Scientist.* Chicago: Open Court.

Einstein, A. (1970). *The Meaning of Relativity.* Translated by E. P. Adams. Enlarged edition. Princeton, New Jersey: Princeton University Press.

Einstein, A. (1995). *The Collected Papers of Albert Einstein: Volume 5 (English). The Swiss Years: Correspondence, 1902–1914.* Princeton, New Jersey: Princeton University Press.

Einstein, A. and Grossmann, M. (1913). A generalized theory of relativity and a theory of gravitation. *Zeitschrift für Mathematik und Physik, 62,* 225–261.

Einstein, A. and Grossmann, M. (1914). Kovarianzeigenschaften der Feldgleichungen der auf die verallgemeinerte Relativitätstheorie gegründeten Gravitationstheorie. *Zeitschrift für Mathematik und Physik, 63,* 215–225.

Einstein, A., Infeld, L., and Hoffmann, B. (1938). The gravitational equations and the problem of motion. *Annals of Mathematics, 39*(1), 65–100.

Fokas, A. (2024). *Ways of Comprehending: The Grand Illusion and the Essence of Being Human.* Singapore: World Scientific.

Fokas, A. (in preparation). *The Golden Ages of Athens and Vienna.* Singapore: World Scientific.

Kepler, J. (1997). *The Harmony of the World.* Translated by E. J. Aiton, A. M. Duncan and J. V. Field. American Philosophical Society. (Original Publication: 1619 in Latin).

Lenin, V. (1962). *Materialism and Empirio-Criticism.* Translated by A. Fineberg. Edited by C. Dutt. Printed in the Union of Soviet Socialist Republics. (Original Publication: 1909 in Russian).

Mach, E. (1959). *The Analysis of Sensations.* Translated by C. M. Williams and S. Waterlow. New York: Dover. (Original Publication: 1886 in German).

Mach, E. (1986). *Principles of the Theory of Heat: Historically and Critically Elucidated.* Translated by T. J. McCormack. Translation revised and completed by P. E. B. Jourdain and A. E. Heath. Edited by B. McGuinness. Introduction by M. J. Klein. Dordrecht, Netherlands: D. Reidel Publishing Company. (Original Publication: 1896 in German).

Mach, E. (2003). *The Principles of Physical Optics: An Historical and Philosophical Treatment.* Translated by J. S. Anderson and A. F. A. Young. New York: Dover Publications. (Original Publication: 1921 in German).

Mach, E. (2013). *The Science of Mechanics: A Critical and Historical Exposition of Its Principles.* Translated by T. J. McCormack. Cambridge: Cambridge University Press. (Original Publication: 1883 in German).

Miller, I. (1998). *Albert Einstein's Special Theory of Relativity: Emergence (1905) and Early Interpretation (1905–1911).* Heidelberg, Germany: Springer-Verlag.

Ohanian, H. C. (2008). *Einstein's Mistakes: The Human Failings of Genius.* New York: W. W. Norton & Company.

Poincaré, H. (1902). *La Science et l'Hypothèse*. Paris, France: Flammarion.

Schrödinger, E. (1944). *What Is Life? The Physical Aspect of the Living Cell.* Cambridge: Cambridge University Press.

Seelig, C. (1954). *Albert Einstein: Eine dokumentarische Biographie.* Munich, Germany: Europa Verlag.

Sigmund, K. (2017). *Exact Thinking in Demented Times: The Vienna Circle and the Epic Quest for the Foundations of Science.* With a Preface by D. Hofstadter. New York: Basic Books. (Original Publication: 2015 in German).

Solovine, M. (1956). *Albert Einstein: Lettres à Maurice Solovine.* Paris, France: Gauthier-Villars.

Škvarla, J. (2024). Einstein–Perrin dilemma on the Brownian motion (Avogadro's number) resolved? *Archive for History of Exact Sciences, 78*, 833–881.

The Nobel Committee for Physics. (2017). *Scientific Background on the Nobel Prize in Physics 2017.* Stockholm, Sweden: The Royal Swedish Academy of Sciences.

Chapter 14

Generalisation, Mathematisation, Unification, and Quantum Electrodynamics

Physics provides the most profound examples of the relentless pursuit of the brain for *unification*. In order to discuss concrete cases of unification, it is necessary to review some of the basic laws of physics. Fortunately, the physical laws of our macrocosm take a very simple form. In addition to the Newtonian laws discussed in the earlier chapter, another example illustrating this marvellous but mysterious fact is Coulomb's law. This law expresses the force between two particles of electric charges, q_1 and q_2, at a distance r. The expression for the electrostatic force between these particles is very similar to the analogous expression for the Newtonian gravitational force: in appropriate units (called *electrostatic units*), it is given by the product of the two charges divided by the square of their distance. As noted earlier, Maxwell, using the insight of Faraday about the relation of electricity and magnetism, generalised this simple law and derived a set of equations, known as Maxwell's equations, which describe the laws of electromagnetism. In summary, *in our macrocosm, there exist two fundamental forces: gravity and the electromagnetic force. The associated phenomena are described by Newton's law of motion, Newton's gravitational law, and Maxwell's equations.*

These basic equations express fundamentally different types of physics: Newton viewed the world as consisting of discrete particles, whereas Maxwell introduced the concept of a *field*, already introduced in the earlier chapter. The fundamental difference between Newton's description, which is based on discrete points, and Maxwell's physics, which involves a region of space, is reflected in the corresponding mathematical formulations: Newton's equations are *ordinary differential equations*, whose solution depends only on one variable, namely time. In contrast, Maxwell's equations are *partial differential equations*, whose solution, in addition to time, also depends on the three spatial variables.

What happens in more exotic situations? If objects move very fast, then Newton's law of motion needs to be modified according to the special theory of relativity. In this case, as discussed in the previous chapter, this modification is very simple: the usual time must be replaced by the proper time.

Among all fast-moving particles, there exists a distinguished one, called *photon*, which moves with the highest possible speed, namely the speed of light. The equations of special relativity imply that a particle moving with the speed of light must be massless.

Maxwell's equations, as noted in the previous chapter, are consistent with the theory of special relativity. Hence, at the level of our macrocosm, even for objects moving with very high speeds, everything is well understood. Furthermore, Einstein's brilliance allowed him to go beyond our macrocosm. By replacing Newton's gravitational law with the general theory of relativity, he introduced a remarkable field theory that covers physical reality even at cosmological scales.

14.1 Quantum Mechanics

In addition to the physical reality described by the above physical laws, there also exists the highly complicated world of subatomic physics. In this exotic reality, which is known as the *quantum world*, the physical laws are very different than the laws valid in classical physics. A fundamental part of this reality was discovered at the Cavendish Laboratory of the University of Cambridge. As will be discussed in detail in the following chapter, in 1897, Joseph John Thomson discovered the electron; in 1917, Ernest Rutherford discovered the nucleus, and in 1932, James Chadwick discovered the neutron. The same year, Rutherford and two of his brilliant collaborators, by implementing the first ever atomic disintegration, proved that the nucleus is made up of protons and neutrons. All these remarkable achievements took place at Cavendish!

Electrons have a negative electric charge, protons have a positive charge, and neutrons have no charge. The energy of a given atom is determined by the dynamic interaction among its constituent particles. So, in order to describe atomic energy, it is necessary to derive a mathematical equation describing the underlying dynamics. The first successful attempt in this direction was made in 1913 by Niels Bohr (Nobel Prize in Physics, 1922), who studied the hydrogen atom. This is the lightest of all atoms, consisting of an electron and a proton. Bohr derived the famous *Bohr's model* by postulating that the electron moves with a constant speed in a circular orbit around the stationary proton. His approach illustrates the brain's process of generalising: *using current knowledge, the brain attempts to reach new horizons by making new associations.* Indeed, Bohr, guided by the prevailing knowledge of classical mechanics and Coulomb's law, employed the familiar Newton's law (force equals mass times acceleration) together with the expression for the force given by Coulomb's law mentioned earlier. For the formulation of Coulomb's law, Bohr used the fact that the proton and the electron have equal but opposite charges.

For a full appreciation of the efforts of Bohr, it must be noted that, following the seminal works of Planck and Einstein mentioned earlier, it was already understood that the physical laws at the microcosm are different from the laws of classical mechanics. How did Bohr capture the basic aspects of this *new physics*? There is a basic physical quantity called *angular momentum*, which for a particle moving on a circle is given by its mass times the speed times the radius of the circular orbit. Bohr attempted to capture the new physics, which is valid in the subatomic world, by *quantising the angular momentum*, i.e. following the works of Planck and Einstein, he assumed that the angular momentum equals an integer times Planck's constant.

The mathematical formulation derived by Bohr was based on the unrealistic assumption that the electron is rotating on a circular orbit with a constant speed. Later, Arnold Sommerfeld extended this model, deriving the so-called Bohr–Sommerfeld model by assuming an elliptic orbit.[1]

---

[1]Although Sommerfeld was nominated a record 84 times, he was never awarded the Nobel Prize. Remarkably, his first three graduate students, Werner Heisenberg, Wolfgang Pauli, and Peter Debye, became, independently, Nobel Laureates. Furthermore, his graduate student Hans Bethe and several of his postdoctoral associates, including Linus Pauling, also received the Nobel Prize.

Although Bohr's model gave spectacular agreement with certain experimental data, it did not capture the full richness of the subatomic physical reality. An important step forward was taken in 1923 by Louis de Broglie, the younger son of the French aristocrat Duc de Broglie. In his thesis, de Broglie, motivated by the understanding of Einstein that electromagnetic waves were composed of photons, which are particles, asked if the reverse was also true. Namely, could particles behave like waves? He argued that the answer is yes, establishing in this way the *dual nature, namely, the particle–wave nature of quantum-mechanical phenomena.* This duality is expressed by the simple equation $E = hf$, where E denotes the energy carried by a wave of frequency f.

A great leap towards the development of the ultimate mathematical formalism describing the quantum world was taken by the Viennese physicist Erwin Schrödinger, who derived a famous equation named after him. This equation provided, in my opinion, a paradigm shift in the formulation of physical theories. Indeed, all earlier fundamental equations of physics, namely the equations of Newton, Maxwell, and Einstein, *expressed mathematical relations among physical quantities that are intuitively well understood.* In contrast to this situation, the remarkable fact about Schrödinger's equation is that, although the wave function itself does not have a direct physical meaning, all physical properties of the associated quantum-mechanical motion can be determined via this entity. This is a new concept, which does not have an analogue in classical physics. In particular, the *probability* of finding a particle at a given point in space and at a given time equals the absolute value squared of the wave function. This fact, known as the 'Born rule', laid the foundation for the probabilistic interpretation of quantum mechanics. Namely, it implies that the position of a particle cannot be determined exactly but only probabilistically. This lack of the classical determinism characterising phenomena in the macrocosm was not acceptable to Einstein, who expressed his doubt about the validity of quantum mechanics with the famous statement, 'God does not play dice'.

It turns out that the better the position can be predicted, the worse the prediction for its speed. This relationship is known as Heisenberg' *uncertainty principle*, referring to Werner Heisenberg (Nobel Prize in Physics, 1932), who provided a mathematical expression quantifying this uncertainty. For example, if we know that the probability of finding a particle in a given space and time is 90%, then we know its speed with a probability of 10%.

The *kinetic energy* of a particle can also be computed in terms of the wave function. In contrast to classical physics, where a particle can be 'at rest', in quantum mechanics, a particle can never be still. It turns out that the more 'wavy' the wave function, namely, the greater its undulations, the higher the kinetic energy. The precise kinetic energy of a quantum-mechanical particle can be computed by applying a specific mathematical operator (called the Laplace operator) to the wave function.

Despite the probabilistic nature of quantum-mechanical phenomena, it is still possible to obtain precise answers by using a classical apparatus and making concrete measurements. The results of these measurements are described as the 'wave function collapse'. This awkward expression attempts to capture the fact that the measuring process 'overcomes' the uncertainty characterising quantum processes and gives rise to meaningful, well-defined results. Quantum phenomena occur at time and length scales that we cannot perceive with our senses. Hence, we cannot develop intuition for such phenomena, which in turn implies that it is very difficult for us to grasp notions such as wave function collapse. Actually, in the early 1920s, quantum notions gave rise to serious philosophical arguments, such as whether Schrödinger's cat is dead or alive, which remain unresolved until today.

An example of the importance of the Schrödinger equation is the use of this basic equation for determining the quantum-physical process of the decay of a radioactive particle. This process characterises the *transition* of one state to another. *This transition is determined by an appropriate solution of the Schrödinger equation.* This transition cannot be *deterministically* obtained from initial conditions; after computing the wave function, one can only calculate the *probability* for a given transition to occur.

In my opinion, *the triumph of quantum mechanics provides at the same time a triumph of mathematics.* Indeed, the new physical reality of the quantum world is defined purely via *mathematisation*. Taking into consideration that the numerical implications of Bohr's model are consistent with experimental values, the model suggested that the *classical electrostatic force* is 'somehow' relevant to the quantum world. But what precisely is this relevance? The miraculous answer to this puzzling question is that *the classical description of Bohr's model provides the basis for implementing the proper operations of mathematisation.* Indeed, in order to obtain the physical laws describing the quantum world, one does *not* need to invent a new force that replaces the electrostatic force. One simply calculates the potential energy corresponding to the electrostatic force and adds to this energy the kinetic energy of the electron. In other words, one computes classically the energy of the moving electron interacting with the nucleus. Then, one performs on this energy a specific *mathematical operation*. This consists of the implementation of certain well-defined *quantisation rules*. It is truly amazing that the abstract and formal process of certain mathematical operations performed on the classical energy gives the correct physical quantum description!

Incidentally, the success of Bohr's model is *a posteriori* easy to explain. The Schrödinger equation contains Planck's constant. Using the fact that this constant has a very small value, it is possible to show that a good *approximation* of the Schrödinger equation is indeed the equation obtained by Bohr. This approximation is referred to as the 'semiclassical limit'.

14.2   The Unification of Quantum Mechanics and Special Relativity

In the same way that the apogee of the brain's ability for *generalising* is reached in mathematics (Fokas, 2024), an analogous pinnacle of *unification* is reached in physics. Of course, unification is intimately related to generalisation. A beautiful illustration of this synergy is the *unification of quantum mechanics and special relativity*, achieved by the idiosyncratic and brilliant physicist of the University of Cambridge, Paul Dirac (who shared the Nobel Prize with Schrödinger in 1933). Dirac noted that, because the speed of electron is sufficiently large, a more accurate description of the hydrogen atom required the employment of the special theory of relativity instead of Newton's law of motion. This gave rise to a very difficult problem: the relativistic energy, in contrast to the classical one, involves the mathematical expression of 'square root', which made it impossible to apply the quantisation rules mentioned earlier. The only way forward was for this square root, somehow, to be eliminated. Dirac overcame this formidable difficulty by employing the remarkable ability of the brain to *generalise*. Indeed, he replaced the relevant *algebraic* expression with one involving certain abstract mathematical entities, now called *Dirac matrices*. In this way, in 1927, he 'factorised' the relevant square root, and he derived the famous Dirac equation.

This generalisation, in addition to overcoming the concrete problem at hand, had far-reaching *physical* implications. Namely, although Dirac's original goal was simply to find the correct relativistic formulation of the motion of the electron, his matrix equation did much more: in addition to describing an electron, it simultaneously described another particle, unknown at the time, and later called *positron*!

The first experimental confirmation of the positron was achieved in 1932 at the California Institute of Technology (Caltech) by Carl Anderson (Nobel Prize in Physics, 1936), who 'captured' this particle from *cosmic rays*. It is worth noting that when Anderson's results were published, Patrick Blackett (Nobel Prize in Physics, 1948) was contemplating publishing his own experimental evidence for the existence of positrons. His approach was not to use cosmic rays but to analyse nuclear reactions implemented at Cavendish. Blackett delayed publishing his results because he wanted to be absolutely sure about his findings since, at that time, the physics community was very sceptical about Dirac's theory. Wolfgang Pauli (Nobel Prize in Physics, 1945) called Dirac's equation 'nonsense', Niels Bohr (Nobel Prize in Physics, 1922) called it 'completely incredulous', and in 1928, Werner Heisenberg (Nobel Prize in Physics, 1932) stated that 'the saddest chapter in modern physics is and remains the Dirac theory'.

In this regard, it should be noted that *the more unexpected a specific result, the higher its potential value*. This is certainly the case with Dirac's discovery. In addition to predicting the existence of a new particle, more importantly, it led to the discovery of completely new physics. Namely, it established that to every subatomic particle, there corresponds a 'twin particle' (with opposite charge), called an *anti-particle*. In particular, the positron, which has a positive charge, is the anti-electron.

Anderson's discovery of the positron was based on the *cloud chamber*, which is a particle detector invented at Cavendish by Charles T. R. Wilson (Nobel Prize in Physics, 1927). This device consists of a sealed chamber containing supersaturated vapour of water. A highly energetic particle, by knocking off electrons from the gas (as a result of electrostatic interactions), creates *ionising radiation*, which reveals itself in the form of a trail of ions. These ions become condensation centres, around which a mist-like trail is formed, appearing as a 'cloud'. Electrons or photons give rise to thin, spider-web-like tracks, whereas alpha particles (to be discussed in the following chapter) cause short tracks.

Wilson constructed the first cloud chamber in 1896. By 1910, after the nature of the alpha and beta radiation was determined (to be discussed in the following chapter), Wilson formulated an ambitious goal: to use his invention to observe charged particles. Indeed, in 1911, using the cloud chamber, which Rutherford called 'the most original and wonderful instrument in scientific history', Wilson became the first person to observe and photograph the motion of individual alpha and beta particles. By the 1920s, various researchers placed cloud chambers between the poles of large magnets. The direction of the bending of the particle determined whether it was positively or negatively charged, and the magnitude of bending determined their energy since more energetic particles bend less.

In 1929, a Russian scientist found some tracks in a cloud chamber which exhibited very little bending, suggesting the presence of a particle with enormous energy. Robert Millikan, who had moved in 1921 from the University of Chicago to Caltech, asked his graduate student, Carl Anderson, to investigate these particles. During his graduate studies, Anderson had already improved the cloud chamber. By using alcohol instead of water vapour, he obtained brighter tracks, which were easier to photograph. For the new experiments, he made further modifications to the usual apparatus, finally constructing a chamber of the size of a car, which weighed about two tonnes. The operation of this machine required 425 kW, which was a substantial proportion of the power consumption of the entire Caltech campus, so he operated it at nighttime. Anderson, after analysing 1,300 photographs, found that 15 of them revealed a particle which had the same mass as the electron but an opposite charge. Although Anderson was unaware of Dirac's theory, which claims that this particle is the antimatter of the electron, he called the new particle positron.

Quantum-mechanical particles have novel properties not exhibited by the classical mechanical particles. One such property, called *spin*, expresses the intrinsic angular momentum of a particle spinning about its own axis. Spin takes values of only integers or half-integers. Dirac's equation characterises particles of spin ½ (like the electron). Particles of spin 0 and 1 are described by two different mathematical equations, called, respectively, the Klein–Gordon and Proca equations.

14.3   Quantum Electrodynamics and Beyond

If a quantum-mechanical particle moves with a speed which is comparable to that of light, then the Schrödinger equation must be replaced by Dirac's equation. The infinite complexity of nature is exemplified by the fact that there exist certain subtle physical effects which are *not* captured by Dirac's formulation. For example, the effect of the *magnetic moment* of the electron is not taken into consideration by Dirac's equation. Following the insight of Hans Bethe (Nobel Prize in Physics, 1967), the complete unification of quantum mechanics and special relativity was achieved by Shinichiro Tomonaga, Julian Schwinger, and Richard Feynman, who shared the Nobel Prize in Physics in 1965. The eventual theory that fully describes the dual, particle-like and wave-like, nature of light is called *quantum electrodynamics.*

The discovery of an ingenious mathematical technique developed by Feynman, known as *Feynman diagrams*, made it possible to compute with great accuracy various quantum quantities. Remarkably, the theory of quantum electrodynamics can be used to calculate physical quantities with an error of the order of one over a billion!

What are the implications of quantum electrodynamics on the concept of time? It turns out that the value of Planck's constant, h, which is of crucial importance in quantum theory, together with the fundamental assumption that nothing can travel faster than light, implies the following astonishing (but irrefutable) conclusions regarding the limiting values for which time and space make sense: there does not exist a unit of time smaller than 10^{-44} s , i.e. a hundred millionth of a trillion of a trillion of a trillion of a second. Similarly, the notion of space becomes meaningless below a length of space approximately smaller than 10^{-33} cm , i.e. a millionth of a billionth of a billionth of a billionth of a centimetre. It must be emphasised that a plethora of experimental results over the past 100 years are consistent with the validity of quantum theory and with the upper limit of the speed of light. Consequently, the above two numbers represent the absolute limits of our current knowledge about space and time. Alternatively, they could mean that both time and space are 'granular'. Discreteness is prevalent in the quantum world. For example, energy takes the form of quanta, light consists of photons, and the energy of electrons in atoms can take only certain discrete values. Hence, perhaps both time and space are 'granular', requiring a completely new treatment beyond the current approach of the 'continuous' approximation.

Summarising the discussion presented in the previous and current chapters, it can be stated that *quantum electrodynamics provides the transcription of Maxwell's theory of electromagnetism into relativistic quantum mechanics. This theory together with the theory of general relativity are the two fundamental theories governing phenomena all the way from the atomic level to the limits of the observable Universe.*

But the story is far from over: it turns out that although the electron is indeed an elementary particle, i.e. it does not have an internal structure, protons and neutrons *do* possess such structure. Namely, each of them consists of three elementary particles, called *quarks*. The theory describing the interaction of the elementary particles is called the Standard Model. This theory introduces two new forces (in addition to gravity and the electromagnetic force described earlier): the *strong force* that holds the quarks together and the *weak force* responsible for the radioactive decay of atoms. The Standard Model is not nearly as complete as the theories describing general relativity and quantum electrodynamics (Griffiths, 2014). In particular, gravity has not, so far, been unified with the other three forces.

The quest for unification raises the ultimate question of whether it is possible to develop a theory encompassing the *infinitely large* and the *infinitesimally small*. From purely epistemological considerations, it is expected that this question has a positive answer. This expectation is further strengthened by the following developments. On the one hand, physicists have begun to understand the quantum fluctuations of the primordial radiation at an epoch close to the birth of the Universe. Crucial steps towards this new insight were the discovery in 1965 of the cosmic microwave radiation, along with significant improvements in the experimental techniques for observing such radiation, and the recent remarkable theoretical advances in cosmology (I hope to return to these far-reaching developments in a future volume). On the other hand, huge experimental advancements at CERN allow physicists to approach energies corresponding to the high temperatures of the primordial structure of the Universe. Therefore, data from the infinitely large and the infinitesimally small are beginning to overlap. This overlapping regime should be characterised by a grand unified theory based on the quantum extension of general relativity.

Perhaps the most important equations in physics are the equations of Newton, Maxwell, Einstein, Schrödinger, and Dirac. It is interesting that these equations do not assign any physical meaning to the *direction* of time. Indeed, these equations are agnostic regarding the distinction between the past and the future. The only physical law that provides a possible explanation for why time marches forward is the second law of thermodynamics (with the first law being the conservation of energy): entropy increases. Thus, according to this fundamental law, by following the entropy of a system, we can distinguish the future from the past.

Boltzmann appreciated that our understanding of this associated *global* time, called thermal time, is a reflection of the enormous complexity at the *local* level. After all, entropy is the measure of the huge number of the microscopic states characterising the macroscopic state of a given system. One may expect that at microscopic scales, quantum effects should be important. This is indeed the case, as rigorously established by Alain Connes, who proved that the thermal time (i.e. the time determined by a macroscopic state) and a time implied by *non-commutativity*, which is a fundamental property of quantum phenomena, are aspects of the same physical mechanisms.

Interestingly, quantum mechanics provides another possible approach to imposing an arrow in the direction of time: according to some physicists, the description of quantum phenomena via the Schrodinger equation is incomplete because this equation does not take into consideration the critical role of actual measurements. There is a view that, during the measurement of a quantum-mechanical entity, for example an experiment to determine the position of an electron, which corresponds to the so-called 'collapse' of the wave function (characterised by the Schrodinger equation only probabilistically), a time direction is imposed (Carroll, 2010).

It is worth noting that a radical approach to dealing with the inability of the fundamental equations of physics to distinguish between the past and the future is to derive equations which do not contain a time variable. The relevant theory, known as loop quantum gravity, was pioneered by John Wheeler and Bryce DeWitt (Rovelli, 2004).

It is interesting that among the five great physicists mentioned earlier, three of them, namely Newton, Maxwell, and Dirac, were at the University of Cambridge. Maxwell, who was a fellow at Trinity College, Cambridge, derived his equations at King's College London, but returned to Cambridge in 1871 as the first Cavendish Professor of Physics. The transformative impact of Cavendish on modern physics will be discussed in the following chapter .

References

Carroll, S. (2010). *From Eternity to Here.* New York: Dutton.
Fokas, A. S. (2024). *Ways of Comprehending: The Grand Illusion and the Essence of Being Human.* Singapore: World Scientific.
Griffiths, D. (2014). *Introduction to Elementary Particles.* New Jersey: Wiley.
Rovelli, C. (2004). *Quantum Gravity.* Cambridge: Cambridge University Press.

Chapter 15

Dismantling the Atom and the Astonishing Contributions of Cavendish Laboratory

Next to the entrance of the Eagle Pub in Cambridge, UK, there is a blue plaque documenting the following event: on 28 February 1953, the Nobel Laureates Francis Crick and James Watson interrupted the patron's lunch to announce that 'We have just discovered the secret of life'. Indeed, they had deciphered the double-helix structure of DNA. A couple of blocks away from this pub is the location of the old Cavendish Laboratory of the University of Cambridge. This 'Mecca of experimental science', which opened in 1874, had a tremendous impact on modern science. Until 1974, when it was relocated to a new building in West Cambridge, 26 Nobel Prizes were awarded to researchers in this world-famous place.

The laboratory's moniker honours the 18th-century physicist and chemist Henry Cavendish. This remarkable scientist, who was born in 1731 to an aristocratic family (his father was Lord Charles Cavendish), began his studies at St. Peter's College, Cambridge, but he left after three years without earning a formal degree. He joined his father in London, where he set up his own home laboratory. He was notoriously shy, speaking only to one person at a time and avoiding women.

Cavendish is best known for the discovery of hydrogen (which he called 'inflammable air'). He produced it by adding certain acids to specific metals. Robert Boyle had produced hydrogen earlier, but it was Cavendish who realised that hydrogen was an element. Also, in his laboratory, he produced carbon dioxide, which he correctly identified as exhaled air. By using a modification of the 'torsion balance', which was built by the geologist John Michell, Cavendish determined the density of the Earth to within 1% of the current value (the result was published in 1798). For his 1778 paper titled 'General Considerations on Acids', he won the most prestigious prize of the Royal Society, the Copley Medal. Cavendish died in 1810.

For centuries, home laboratories, such as the one used by Cavendish, were the norm. At that time, students were not formally exposed to laboratory work. Attitudes began to shift in the mid-19th century. By 1869, Oxford University had begun to build its Clarendon Laboratory, and Lord Kelvin set up a physical laboratory in an old wine cellar, which he also used for teaching his students. Soon after, a Cambridge committee issued a report calling for the establishment of a physical laboratory similar to the one in Oxford. However, there were insufficient funds to realise the committee's vision until a wealthy university chancellor, the Seventh Duke of Devonshire, William Cavendish, who was a member of the same family as Henry Cavendish, donated the funds. A young physicist, James Clerk Maxwell, was selected as the first professor of experimental physics at the University of Cambridge and was given the task of organising a laboratory.

Maxwell realised that the bulk of Cavendish's scientific findings remained unpublished. In 1879, Maxwell published the results of the electrical experiments that Cavendish conducted 100 years earlier. He was astonished to realize that Cavendish had anticipated the fundamental notions of electrical potential, Ohm's law, Coulomb's law, and the concept of the dielectric constant. There was even a draft of a manuscript describing his mechanical theory of heat, anticipating modern thermodynamics. Maxwell was so impressed with these results that he renamed the laboratory after Cavendish.

Maxwell died at the young age of 48, before the impact of his revolutionary results, namely, his mathematical unification of electricity and magnetism, as well as his work on thermodynamics and statistical physics, was fully appreciated. He was succeeded by Lord Rayleigh (Nobel Prize in Physics, 1904), who discovered the chemical element argon and also made fundamental contributions in fluid mechanics. In 1884, Rayleigh was succeeded by J. J. Thomson (Nobel Prize in Physics, 1906), whose discovery of the electron marked the starting point of the modern revolution of physics. Thomson was succeeded by Ernest Rutherford (Nobel Prize in Chemistry, 1908), who discovered the atomic nucleus and also made crucial contributions regarding the nature of radiation. In 1932, under the leadership of Rutherford, James Chadwick (Nobel Prize in Physics, 1953) discovered the *neutron,* and in the same year, Rutherford's students, John Cockcroft and Ernest Walton, using the first *particle accelerator*, experimentally produced the first controlled nuclear disintegration (for which they were awarded the Nobel Prize in Physics in 1951). In addition, Cockcroft and Walton provided the first experimental verification of the famous Einstein's equation, $E = mc^2$. A couple of years earlier, the mathematical physicist of St. John's College, Cambridge, Paul Dirac, derived his famous equation predicting the positron and establishing, as mentioned in the previous chapter, that with every elementary particle, there also exists an anti-particle. As noted earlier, Patrick Blackett (Nobel Prize in Physics, 1948) was contemplating publishing experimental evidence for this particle when the Caltech physicist Carl Anderson published his results, where the positron was identified via the analysis of cosmic rays.

After Lawrence Bragg (joint recipient with his father, William Bragg, of the Nobel Prize in Physics in 1915) became director of the Cavendish in 1938, he continued his work on X-ray crystallography. The use of this technique by Rosalind Franklin contributed to the discovery of the structure of DNA by Crick and Watson. In 1954, Bragg was succeeded by Nevill Mott (Nobel Prize in Physics, 1977), whose pioneering research led to the development of what is now called *condensed matter physics*. The Cavendish Laboratory was relocated to its new location in West Cambridge in 1974, under the directorship of Brian Pippard.

15.1 X-Rays and Electrons

In 1895, in Wurzburg, Germany, Wilhelm Röntgen (Nobel Prize in Physics, 1901) placed on a wooden table an apparatus called the *cathode-ray tube*. This is a glass tube of the size of a small wine bottle, from which most of its air has been removed via the use of a vacuum pump. There is a metal electrode at the end of the tube (the negative cathode) and another electrode halfway inside the tube (the positive electrode). When the electrodes are supplied with high-voltage electricity via two wires, a glow appears inside the tube, revealing the so-called cathode rays. Although Röntgen had used this apparatus many times before, this particular day he observed something completely unexpected: a small phosphor-coated screen on the other side of the lab was emitting a green-coloured light!

To his great surprise, the green light continued to be present even after he placed paper or a hard rubber between the tube and the screen. These mysterious rays could even go through a wooden door. As soon as he turned the cathode off, the light was eliminated. When he placed his hands in front of the tube, he noted that the "darker shadow of the bones is seen". Indeed, the rays travelled through the skin and flesh but not so easily through bone or metal. He also established that an aluminium foil prevented the propagation of the rays. Since, in science, a typical symbol for an unknown quantity is '*x*', he called these rays 'X-rays'. After arranging in January 1896 a live demonstration of his discovery to the Wurzburg Physical Society, hundreds of scientists possessing a ray tube reproduced his findings.

What was the origin of the mysterious X-rays? Röntgen established that these rays did not have the same reflection and refraction properties as visible light or ultraviolet or infrared light. The basic nature of the X-rays remained a mystery until the transformative discovery by J. J. Thomson of the electron. Thomson, after studying mathematics at Trinity College, Cambridge, was appointed, at the age of 27, professor and director of the Cavendish Laboratory. In contrast to Röntgen and the German school, who believed that the cathode rays were a type of light, he believed that these rays were composed of *particles*.

After using a magnet to force the rays to bend and hit an electroscope, which is an apparatus capable of recording electric charge, Thomson saw that this device was recording a negative charge. However, consistently with the earlier results of Heinrich Hertz, who was the leading expert in experiments involving electromagnetic waves, Thomson found that, while—as he expected—a large electric voltage deflected the rays, to his surprise, a smaller voltage had no effect. In other words, it appeared that the rays were behaving like negatively charged particles for large voltages but as light for small voltages. He finally understood that the small amount of air left in the tube was becoming electrically charged and was cancelling the expected effect of the small voltage. Thomson, by using meticulous experiments utilising different electric and magnetic fields, was able to determine the charge and mass of these particles. He concluded that these particles were not any of the known atoms or molecules at the time. In April 1897, he demonstrated some of his experiments at the Royal Institution of London and announced the discovery of the *electron*, the first subatomic particle. The mass of the electron is about 2,000 times smaller than that of hydrogen, which is the lightest atom.

The discovery of the electron led to an explanation for the origin of X-rays: the high voltage across the cathode gives rise to electrons, which are attracted by the positive cathode. Some of them hit the bottle glass, emitting energy, which reveals itself as the glow inside the tube. Furthermore, if the electrons lose a sufficient amount of energy, they can create a high-energy form of light- electromagnetic radiation, the X-rays.

15.2 Radioactivity and the Nucleus

Now we know that the atom is made up of a tiny nucleus surrounded, at a great distance (in comparison to the volume occupied by the nucleus), by a number of electrons. The nucleus, which accounts for more than 99.9% of the mass of the atom, consists of a number of protons equal to the number of electrons as well as of several neutrons. At the time of the discovery of the electron, chemists and the majority of physicists thought that electrons had nothing to do with atoms. The idea of Democritus that the atom is the smallest form of matter was still prevalent. Actually, in 1900, a debate took place on this topic between the Oxford-trained chemist Frederick Soddy (Nobel Prize in Chemistry, 1921) and Ernest Rutherford, a student of Thomson. Rutherford claimed that, since electrons originate from matter, they must be part of the atom.

In 1896, Henri Becquerel (Nobel Prize in Physics, 1903), motivated by the discovery by Röntgen, while studying the light-emitting effect of uranium crystals, discovered the phenomenon of *radioactivity*. A couple of years later, Marie Curie discovered that thorium emitted radiation, and later, with her husband, Pierre, after discovering the radioactive substances polonium and radium, introduced the terminology 'radioactivity' (they shared the Nobel Prize in 1903).

During his graduate studies at the Cavendish, Rutherford became interested in radiation and demonstrated that there were at least two different types of radiation: *alpha* and *beta radiation*. In 1899, Becquerel showed that the latter consists of electrons, whereas in 1907 Rutherford showed that alpha radiation consists of helium atoms that have lost two electrons and hence are positively charged. Actually, there also exists a third type of radiation, the gamma rays, consisting of high-energy light (like the X-rays).

After Rutherford took a professorship at McGill University in Canada, he continued his investigation of radiation. Together with his first

graduate student, Harriet Brooks,[1] while investigating the element thorium, they discovered a fascinating phenomenon. First, thorium was turning into a gas, and second, when this gas was brought in contact with other
objects, it gave them a 'magical property': the objects would spontaneously emit alpha, beta, or gamma radiation. It was later understood that
atoms of thorium *spontaneously* changed their form, transforming from
solid thorium into gaseous thorium. Actually, Rutherford, together with
Soddy, discovered something even stranger: the radioactive elements
decay spontaneously, emitting alpha and beta particles, and in this way
transform themselves into different elements. These Nobel Laureates also
discovered that radioactive decay follows an *exponential law*. In particular, there is a specific time, known as a 'half-life', during which half the
mass of a given radioactive material changes into another type. For
example, the half-life of oxygen-15 is two minutes. This means that if one
starts with 100 atoms of this substance, after two minutes, only 50 atoms
will remain; the other 50 will be transformed into nitrogen-15.

Incidentally, there exist substances with the same chemical structure
but with different atomic number, i.e. with different masses. This is a
consequence of the fact that they possess a different number of neutrons.
These different substances, which, as noted earlier, are called *isotopes*
(a terminology due to Soddy), may have different radioactivity properties.
For example, almost all of potassium exists in the stable form of K-39
(where the number 39 indicates that its mass is 39 times the mass of the
hydrogen atom). However, 0.0012% of potassium exists in the radioactive
form of the isotope K-40 (which has one neutron more than K-39). Since
bananas contain potassium, they contain a minuscule amount of radioactivity. Actually, our bodies contain several radioactive isotopes, so, techni-
cally, we are all radioactive, but the relevant amount of radioactivity is too
small to cause ill effects.

While at McGill, Rutherford observed that, whereas alpha particles
hitting a photographic plate formed a clear image, this image became
fuzzy after the placement of a thin piece of metal in front of the

---

[1] Brooks won a fellowship for her doctoral work with Rutherford and, in 1902, went to
Cavendish to work with Thomson, becoming the first woman researcher at Cavendish.
When Rutherford moved to the University of Manchester, he attempted to hire her, writing
a reference letter claiming that Brooks was the most prominent woman physicist in radioactivity next to Curie. However, Brooks decided to get married and move back to Canada,
abandoning physics.

photographic plate. He concluded that this could be the result of the scattering of the alpha particles, perhaps caused by collisions with the atoms of the metal. In 1907, he returned to the UK as a professor at the Physics Department of the University of Manchester and decided to further investigate this phenomenon. For this purpose, he asked Ernest Marsden and Hans Geiger to conduct a number of meticulous experiments, which finally led to the discovery of the nucleus.

At that time, the prevailing view of the structure of the atom was that of Thomson, namely, that the atom was a sphere with a positive charge, with negatively charged electrons embedded *within* it (which was referred to as the 'plum pudding model'). However, the reality is very different: atoms are not solid entities, which had been the prevailing view for millennia, but *they consist mostly of empty space*! The core of the atom is made up of a tiny positively charged nucleus; electrons, which are indeed part of the atom, circulate at a very large distance (in comparison to the size of the nucleus) from the nucleus. In the outstanding book by Suzie Sheehy, *The Matter of Everything* [Bloomsbury Publishing, 2023], the following illuminating analogy is made: if the atom is the size of a cathedral, then the nucleus is the size of a fly and the electrons circulate at the walls of the cathedral.

In order to establish the existence of the nucleus, Marsden and Geiger created a powerful source of alpha particles by using a particular radioactive substance. Furthermore, they invented a very accurate way of observing the reflection of a single alpha particle. Incidentally, an alpha particle is about 7,000 times heavier than an electron and is positively charged. So, it takes a substantial force, such as the repulsive force due to the interaction with another positively charged entity, to make it change its path. As the thickness of the gold foil placed in front of the photographic plate was reduced, it was observed that, as expected, the particles would travel through it more easily. However, no matter how thin the foil was, approximately one in every 8,000 alpha particles appeared to bounce off the foil. The only explanation for this fact was that *the atom consists of empty space except for a tiny nucleus*. In the extremely rare case that an alpha particle encountered this nucleus, it was reflected back. The relevant experiments were conducted in 1907–1908 and published in 1909. The associated theoretical explanation was published in 1911. Interestingly, during this period, Rutherford enrolled in a course in mathematics in order to make sure he could provide an explanation which was mathematically sound.

Rutherford appreciated that the revelation of the unexpected structure of the atom, as well as the understanding of how radioactivity works, marked the beginning of a new epoch in physics.

His enthusiasm about this new revolutionary understanding was expressed in the statement, 'We are living in the great age of science', as recorded by one of his collaborators in Cambridge, the chemist C. P. Shaw, who later became a distinguished writer.

15.3 The Splitting of the Atom

By 1917, Rutherford had reached the conclusion that the nucleus contained positively charged particles, later called *protons*. He also speculated that the nucleus, in addition to protons, also contained particles which were neutral, later called *neutrons*. After being appointed as the director of Cavendish, his main goal became the experimental proof that the nucleus was indeed composed of protons and neutrons. However, the tools under his disposal were rather limited. As a source of particles, he was using alpha particles generated from the radioactive elements of radium or polonium. In addition, he was using a specific target and a scintillation screen.

A fundamental problem in the experimental verification of protons and neutrons was identified by Cockcroft: alpha particles, as well as protons, are positively charged; therefore, in order to reach the positively charged nucleus, they had to be accelerated at a very high speed so that they would overcome the repulsive nuclear force. In other words, these particles had to overcome the so-called 'Coulomb barrier'. Cockcroft calculated that the needed acceleration required a voltage of the prohibitively high value of 10 million volts! How could this extremely high voltage be generated? Cockcroft knew that other physicists in the US, such as Merle Tuve, were trying to achieve high voltages via the use of a Tesla coil. But still, it appears that this extremely high voltage could not be reached in the near future. Fortunately, it turns out that the required voltage was much smaller than originally estimated. This is a consequence of *quantum tunnelling*, which is another example of the unexpected behaviour of particles in the mysterious world of quantum mechanics. The tunnelling effect was understood by the Russian theoretical physicist George Gamow. He argued that, during the process of radioactive decay, before the alpha particles escape an atom, they are trapped in the nucleus. Hence, during this process, before the alpha particles can harness the Coulomb force, which allows them to be expelled from the atom, they must first escape the trapping force holding them inside the nucleus. In the same way that during radioactivity, somehow, some alpha particles manage to *leak away*, escaping the nuclear force holding them inside the nucleus, perhaps, during the reverse process of the bombardment of the nucleus, some particles could *sneak inside* the nucleus. Gamow formulated a model describing the probability of alpha particles leaking out of uranium, and in this way, he was able to predict the half-life of this radioactive element. Gamow, after visiting Niels Bohr, presented arguments that this idea would also work in reverse, concluding that it would only take 1/16 of the energy computed originally for the alpha particles to overcome the Coulomb barrier. Cockcroft, after studying Gamow's paper, concluded that not only alpha particles but also protons could use quantum tunnelling to go through the Coulomb barrier. His computations revealed that this would require energy as low as 300 kiloelectron volts. It is not clear whether Gamow or Cockcroft was the first to communicate to Rutherford this exciting idea. However, there is no doubt that Rutherford made his decision based entirely on these theoretical calculations. In 1929, he told Cockcroft, 'Build me a one million-electron-volt accelerator'.

This was easier said than done. Luckily, during the First World War, Cockcroft had interrupted his studies in mathematics and had worked in the electrical engineering company 'Metrovick' (Metropolitan Vickers). As a result of his knowledge of the technical expertise of this company, he assigned to Metrovick the task of providing him with the needed transformer to step up the voltage to 300,000 volts. However, the existing transformers were too large to go through the doorway of the Cavendish. A device, called a rectifier, was also needed to transform the high-voltage AC into a DC source (the negative part of the AC wave would decelerate instead of accelerating the particles). Also, it was crucial that the entire apparatus avoided sharp edges since electric fields reach very high values on such edges. As a source of particles, Cockcroft and his close collaborator Walton decided to use protons, which they generated by using an *anode ray tube*. This is an apparatus similar to the cathode ray tube, but it is filled with hydrogen gas, with a large voltage applied between the cathode and the anode. The large electric field rips apart the hydrogen, creating protons, which travel towards the negative anode. Then, the protons travel towards the main accelerating section, a 1.5-metre-long glass vacuum, where the protons accelerate further via a high voltage applied to two metal electrodes. By 1930, Metrovick had delivered a compact transformer, which was transported down to the basement of Cavendish.

Also, the bulb-shaped tubes for the rectifier and the accelerating section of the anode ray tube, which were difficult to construct, were replaced with the more reliable glass cylinders. This modification was invented by the Caltech physicist Charles Lauritsen, who together with Merle Tuve and Ernest O. Lawrence were considered the three great American pioneers of nuclear physics. In April 1932, a historical event took place: by bombarding the nucleus of lithium-7 with accelerated protons, Cockcroft and Walton were able to split the nucleus of the lithium into two alpha particles. In this way, the use of the first *particle accelerator* had caused *nuclear disintegration*.

As predicted by Gamow, this amazing result was achieved with the relatively low voltage of 250 kiloelectron volts. Importantly, the sum of the masses of the proton and the lithium was slightly higher than that of the masses of the two alpha particles. Computing the difference in these masses and multiplying the resulting number by the square of the speed of light, Cockcroft and Walton arrived at an energy of 8 MeV, which was indeed the energy that the alpha particles were carrying. In this way, Einstein's famous equation, $E = mc^2$, was confirmed!

In early 1932, another major discovery was made at Cavendish: for 12 years, James Chadwick was quietly performing a series of experiments trying to capture the neutron. Finally, by scrutinising the results of Irene Curie and Frederic Joliot (this couple shared the Nobel Prize in Chemistry in 1935), who had bombarded beryllium using alpha particles generated from polonium, he demonstrated that the outcome of this process was not the production of gamma rays, as wrongly claimed by the French investigators. Actually, this process generated a particle with mass approximately equal to the mass of the proton. Finally, he had discovered the neutron!

Starting in the mid-1930s, with regard to important developments in physics, a shift occurred from Europe to the US. The experimental discovery of the positron at Caltech was the prodrome of this shift. These developments are discussed in the following chapter.

Chapter 16

Co-operation and the Primacy of Experimental Physics

16.1 The Discovery of a Plethora of New Particles

One of the major developments in physics made during the early 1930s was the discovery of many new particles, beyond protons, electrons, and neutrons. These particles were detected either in ingenious laboratory experiments after they were created through the acceleration and collision with existing particles or were 'captured' via sensitive apparatus when they arrived on Earth as *cosmic rays*. As discussed in Chapter 14, the first cosmic ray particle to be detected was the positron, using a modification of the cloud chamber.

This particle detector was instrumental in discovering *cosmic radiation*, whose existence had remained a mystery for a rather long time. Indeed, in the early 1900s, scientists in Berlin, Cambridge, and Vienna observed that their instruments appeared to pick up some extra radiation. What was the origin of this radiation? Marie Curie discovered that many minerals are radioactive; in particular, she spent years working in a shack mine analysing a mineral called pitchblende. Thus, it was natural to speculate that the radiation was coming from the earth. Early attempts to verify this speculation by showing that the radiation up in a hot-air balloon was less than that on the ground failed because the movement of the balloon interfered with the operation of the electroscope, which at that time was the main apparatus used to measure radiation. The German Jesuit priest and physicist Theodor Wulf improved the electroscope by making it more robust. In 1909, using his new electroscope, he measured the radiation levels at the base of the Eiffel Tower, as well as at its top, i.e. at a height of 300 metres. To his surprise, the radiation persisted, which suggested that the radiation was not coming from the earth.

Some scientists began to think that the radiation may be coming from space. This was indeed proven by the Austrian physicist Victor Hess (Nobel Prize in Physics, 1936). Using the newer electroscope, he established that there was far more radiation at the altitude of 5,300 metres, which was reached via a hot-air balloon.

What is the cause of this *cosmic-ray radiation*? Apparently, somewhere in space, forceful interactions create radiation which is sufficiently powerful to travel across large distances and reach the Earth. It is thought that this radiation is formed in supernovae and perhaps is accelerated by the strong gravitational field around black holes. It is now known that cosmic rays consist mostly of highly energetic protons. When these protons collide with atoms in our atmosphere, they produce a plethora of particles, including positrons. Incidentally, although the Sun is not the source of cosmic rays, it does influence cosmic-ray radiation: the Sun constantly produces material—the 'solar wind'—creating the *heliosphere*, which is a huge bubble surrounding the planets of the solar system. When the Sun is less active, the heliosphere is weaker, allowing more cosmic rays to reach the Earth.

Carl Anderson, after his great success of capturing the positron, decided to look for additional cosmic-ray particles. In 1935, together with his graduate student Seth Neddermeyer, he made measurements for six weeks in Colorado at an elevation of 4,300 metres. They discovered the *muon* and the *anti-muon*, which, like the electron and the positron, have a negative and a positive charge, respectively. Their mass is about 400 times larger than that of the electron, and they have a very short lifetime, decaying in 2.2 millionths of a second, transforming into electrons.

In 1947, Cecil Powell and Giuseppe Occhialini published their results regarding the discovery of another cosmic-ray particle, the *pion*, which appears in three forms, namely, with a positive charge, with a negative charge, and with no charge. Instead of using a cloud chamber, these physicists used an improved version of a new technique, called *nuclear emulsions*, which is based on the use of a special type of photographic plate. This technique was developed by Marietta Blau at the Radium Institute in Vienna. She showed that her method could be used to detect cosmic rays, and together with her graduate student, Hertha Wambacher, she spent four months in the Austrian Alps, running experiments. Blau was Jewish, and in 1938, on the eve of Anschluss, she fled Germany, finally moving to the

US with the help of Einstein. Sadly, Wambacher, a Nazi party member, continued to publish their joint results derived with Blau without any mention of her former thesis advisor.

It is interesting that the pion was first discovered by Bibha Chowdhuri, who in the period of 1942–1945, in collaboration with the famous physicist D. M. Bose, performed a series of experiments in Sandakhpu, India, using the nuclear emulsions technique. Surprisingly, although Powel gave explicit reference in one of his key papers to the work of Chowdhuri, the Nobel committee awarded the Nobel Prize in Physics in 1950 to Powel, ignoring the work of the two female physicists, Blau and Chowdhuri.

In 1939, as discussed in the following chapter, Ernest O. Lawrence designed a *circular* particle accelerator called the *cyclotron*. Also, he significantly influenced the invention of the *synchrotron*, which is an accelerator with the distinguishing feature that it emits the so-called *Schwinger radiation*. The synchrotron was originally constructed for accelerating electrons, but it was later realised that, since a proton is 2,000 times heavier than an electron, protons would emit considerably less radiation than electrons. So, it was decided to use proton accelerators. However, taking into consideration that it is much more difficult to bend protons than electrons, it was necessary to either use much stronger magnets or to construct much larger accelerators.

After the war, with financial support from the US government, Lawrence, together with Luis Alvarez (Nobel Prize in Physics, 1968) and other collaborators, by incorporating a key idea (about phase stability) from Edwin McMillan (Nobel Prize in Chemistry, 1951), built the *synchrocyclotron*. This reached an unprecedented beam energy of 350 megaelectron volts (MeV). In 1949, this achievement allowed them to discover the electrically neutral version of the pion. In this way, this particle was found using an accelerator instead of cosmic rays. This motivated Alvarez to formulate a clear and ambitious goal: to build a proton accelerator sufficiently powerful that all known particles found in cosmic rays, and perhaps additional ones, could be produced. This would involve smashing a beam of protons on a specific target, causing the disappearance of the original protons and facilitating the conversion of some of the released energy into the creation of new particles.[1]

---

[1] In this process, most of the released energy takes the form of kinetic energy carried away by the debris. The associated mathematical analysis involves the use of various conservation laws, including the conservation of energy, angular momentum, and electric charge, as well as the invariance of several quantum numbers.

A similar goal was formulated at the Brookhaven Laboratory, which in 1953 switched on the *Cosmotron*, a 23-metre ring involving 288 magnets, each weighing six tonnes; it reached an energy of 3.3 gigaelectron volts (GeV). A year later, a team at Berkeley led by Alvarez and colleagues turned on the *Bevatron*, a 41-metre accelerator that reached an energy of 6.2 GeV. By 1954, using these new accelerators, in addition to pions, muons, and positrons, a variety of other particles were discovered.[2]

More particles were soon discovered using the *bubble chamber* developed by Donald Glaser (Nobel Prize in Physics, 1960). This is similar to the cloud chamber, but it is filled with a superheated transparent liquid, such as hydrogen (Glaser first used diethyl ether, a liquid used as an anaesthetic). Since the liquid is a thousand times denser than air, the chance of capturing a particle travelling in this chamber is a thousand times higher than in a cloud chamber. In 1958, the team in Berkeley discovered the particle, Y*(1385), which has a half-life of the order of 10^{-23} seconds.[3] Utilising these remarkable experimental innovations, the number of new particles soon surpassed 200! It was clear that a deeper theoretical understanding was needed.

16.2 The Strong Force and the Experimental Verification of Quarks

What is the nature of the nuclear force that binds the protons and neutrons together in the atomic nucleus? In 1934, Hideki Yukawa (Nobel Prize in Physics, 1949) claimed that this force was different from the two known forces, namely gravity and the electromagnetic force, and proposed that it be named the *strong force*. A year earlier, Enrico Fermi (Nobel Prize in Physics, 1938) proposed an additional force, the *weak force*, in order to explain the phenomenon of radioactive beta decay.

---

[2] Among these particles, there were several with a mass in the neighbourhood of 500 MeV, as well the following particles which are heavier than the proton: the neutral lamda, the two charged sigmas, and the negative xi.

[3] The star indicates that it belongs to the so-called class of 'resonance' particles, whereas its name reveals that it has a mass of approximately 1,385 MeV. Its extremely small half-life implies that, even if it travels with the speed of light, it only moves a distance less than the width of the proton before decaying.

I was lucky to be at the Department of Applied Mathematics at Caltech in 1975–1980, during the period when the two towering figures of modern physics Richard Feynman and Murray Gell-Mann were at the Physics Department. As noted in Chapter 14, Feynman, in addition to contributing to the completion of the theory of quantum electrodynamics, also introduced the most powerful analytical tool used in particle physics: the Feynman diagrams. Gell-Mann played a crucial role in the development of *quantum chromodynamics*, the theory describing the dynamics of strong interactions. In particular, in 1956, he introduced the abstract property of *strangeness*.[4] It was already known from classical physics that matter possesses fundamental properties, such as *mass* and *electric charge*. Quantum mechanics introduced new properties, such as the *spin*. The importance of these properties becomes clear through the fact that they are preserved during interactions. Gell-Mann suggested that the property of strangeness is preserved during strong particle interactions but not during the process of decaying; in other words, weak interactions do not preserve strangeness.

In 1961, Gell-Mann and, independently, Yuval Ne'eman, imposed some order to the confusion caused by the discovery of the plethora of new particles. They proposed a particle classification based on the properties of spin, electric charge, and strangeness. The pions (and the kaons) belong to the group of *mesons*, which consists of eight particles, all of spin 0. The protons and neutrons (along with the lambda particles) belong to the *baryons*, a group of eight particles, all of spin ½. In addition, there exists another group of baryons consisting of 10 particles of spin 3/2, which contained many of the newly discovered strange particles (such as the delta, sigma, and xi). Interestingly, this decuple contained a particle, called omega minus, which had not been discovered at that time. In 1964, at the Brookhaven Laboratory, a group of physicists led by Nicholas Samios discovered this missing particle. Incidentally, prior to this discovery, the Cosmotron had been upgraded and was supplemented with an enormous bubble chamber and a 400-tonne magnet.

The triumph of this theoretical prediction motivated Gell-Mann and, independently, George Zweig (a student of Feynman) to make the revolutionary proposal that protons, neutrons, pions, etc., are *not* elementary particles but consist of certain fundamental constituents. Gell-Mann called

---

[4]This property was also introduced, independently, by Abraham Pais, Tadao Nakano, and Kazuhiko Nishijima.

these basic constituents *quarks*, borrowing this term from James Joyce's novel *Finnegans Wake*. He suggested that there exist three types of quarks: the 'up', 'down', and 'strange' quarks. Actually, the current theory of quantum chromodynamics is based on six types of quarks, called *flavours*: the 'up', 'down', 'charm', 'top', 'strange', and 'bottom', as well as six anti-quarks. Each quark has spin ½, a specific mass, and a specific fractional electric charge. All quarks have a strangeness of zero, except for the strange quark. Interestingly, the quarks have a new abstract property called *colour* charge. This property plays a crucial role in the associated theory, and this justifies the terminology 'chromodynamics' ('colour' in Greek is *'chroma'*). Each meson consists of two quarks, and each baryon consists of three quarks. For example, the neutral pion consists of either an 'up' quark and an 'up' anti-quark or a 'down' quark and a 'down' anti-quark. A neutron consists of an 'up' and two 'down' quarks, whereas a proton consists of a 'down' and two 'up' quarks. The masses of the 'up' and 'down' quarks are almost identical, which explains the fact that the proton and the neutron have almost the same mass. The strangeness of a particle can be computed by adding the strangeness of its quarks. So, the nucleus has zero strangeness.

One of the triumphs of modern experimental physics was the verification of the existence of quarks, which was achieved at the Stanford Linear Accelerator Centre (SLAC). In 1967, after the establishment of a joint collaborating effort between SLAC and the Massachusetts Institute of Technology (MIT), Richard Taylor of the University of Stanford, together with Henry Kendall and Jerome Friedman of MIT, began a series of experiments, which finally led to the firm conclusion that protons and neutrons, indeed, contain three quarks.[5] In addition, in collaboration with CERN, which provided comparative data for the electrically neutral *neutrinos*, it was established that quarks have fractional electric charge.

---

[5] The final proof of the quark composition of protons and neutrons involved the synergy of theory and experiments, as well as the work of several physicists over several years. A crucial role was played by a formula (the cross-section) involving the angle of the inelastic scattering of electron–proton and two functions called 'structure functions'. Feynman contributed to early models via the introduction of the so-called 'partons'. Since neutrons decay within minutes, in order to prove that neutrons contain three quarks, high electron beams were passed through liquid deuterium, whose nucleus consists of a proton and a neutron. An excellent review of the remarkable efforts and ingenuity involved for the justification of the quark model is Riordan (1992).

This achievement earned Taylor, Kendall, and Friedman the Nobel Prize in Physics in 1990.

The experimental verification of quarks illustrates several important concepts; particularly, the importance of *analogical thinking, cooperation, continuity,* and *interdisciplinarity.* Indeed, the basic idea used by the three Nobel Laureates is reminiscent of the approach used by Rutherford, Geiger, and Marsden for the discovery of the nucleus: electrons were accelerated until they travelled with a speed very close to the speed of light, acquiring an energy of 20 GeV. This allowed these electrons to penetrate into protons and neutrons. When they encountered a quark, they bounced back. The scattered electrons were directed through a device called *magnetic spectrometer,* which, by measuring the bending of these electrons via the use of a magnetic field, provided accurate estimates for the energy and angles of the scattered electrons.

It is important to note that according to quantum chromodynamics, in the same way that the electromagnetic force is mediated by the massless and neutral *photon,* the strong force is mediated by a massless and neutral particle, called the *gluon* (the name alludes to the fact that these mediators 'glue' the quarks together). Further experiments at SLAC verified the existence of gluons. In addition, it was revealed that protons and neutrons actually possess a much more complicated structure. First, they consist of an approximately equal number of quarks and gluons. Second, in addition to the basic three quarks (called 'valence' quarks), there also exists a 'sea' of quarks–anti-quarks, which must be taken into consideration for the full description of the mass and interaction properties of protons and neutrons.

Regarding the importance of continuity and interdisciplinarity, it should be noted that the *linear* accelerator constructed at Stanford has its origin in the ideas of Rolf Widerøe (discussed in the following chapter), which also provided the starting point for the construction of the cyclotron. The Stanford accelerator, in contrast to the Cavendish accelerator where a single high voltage was used, is based on a series of cavities activated via *radiofrequency.*

Interestingly, the development of this technique did *not* arise in research into the area of particle physics, but rather in efforts to detect unknown objects via the use of radio signals. In 1939, the British physicist Sir Robert Watson-Watt invented a system which could locate moving objects via the use of short-wave radio signals and an antenna. This was achieved by emitting radio signals and then analysing the reflected waves after these signals had struck the moving object. By 1939, when the Second

World War broke out, several such systems, called 'Radio Detection and Ranging', or *radar*, were set up along the south and east coasts of Britain. These systems, which were used for the detection of enemy ships and air-craft, replaced the older giant concrete sound mirror installations built along the south coast of England between 1915 and 1930.

It was soon realised that the early radars needed three key improve-ments. First, for the detection of certain objects, such as the German U-boats, it was required that the signals have shorter wavelengths. Second, in order for the signals to reach objects located at longer dis-tances, it was necessary to have more powerful transmitters. Third, in order to mount a radar on a plane, it was necessary to invent lighter and more compact radars. These improvements were achieved via the devel-opment of two different inventions: the discovery of the *klystron* by the Stanford University researchers Robert Hansen, Russell Varian, and Russell's brother Sigurd; and the construction of the *cavity magnetron* by the University of Birmingham physicists John Randall and Harry Boot.[6]

The klystron was a can-sized device containing an electron beam travelling through a series of cavities. This flow of electrons, modulated by radio signals, created a specific dynamic state called *resonance*, which gave rise to the emission of waves of much shorter wavelength than the familiar radio waves. The mechanism of resonance was also used in the cavity magnetron, but now, as the electrons moved from the main cavity to the neighbouring ones (called 'petals'), a magnet caused the emission of electromagnetic waves of unprecedented power.

In September 1940, when the bombing of the UK had intensified, a top-secret British delegation, which included John Cockcroft, reached Washington to share with the US government the invention of the magne-tron. During these meetings, the US contingent revealed the invention of the klystron, which admittedly, had limited use since it was lacking sufficient power. This led to the US government funding a project of

---

[6]The Varian brothers were born to theosophist parents and early on participated in the run-ning, in California, of the utopian community of Halcyon. After establishing independent careers in electronics and aviation, they came together with Hansen, an ex-roommate of Russell from graduate school, who was already a professor of physics at Stanford University. In 1948, the two brothers founded 'Varian Associates' to market the klystron and some of their other inventions. This company was the first to move to the 'Stanford Industrial Park', providing the seed for the development of the 'Silicon Valley'.

recruiting several physicists at MIT and creating the Rad Laboratory, which at its peak employed 4,000 people.

After the war, when the klystron and the cavity magnetron were declassified, Hansen revisited his original idea, which was to use these techniques to build a new type of electron accelerator. Hansen formed a team of physicists at Stanford, which included Edward Ginzton, who had known Hansen and the Varian brothers when they all were graduate students at Stanford University. By 1947, a compact linear electron accelerator was built, known as *LINAC*. The first such accelerator, Mark I, reached an energy of 6 MeV. Before Mark II became operational, which reached an energy of 33 MeV, Hansen died from a chronic lung disease, and Ginzton became the director of the project. By 1953, Mark III was approaching the original goal of 1 GeV. In 1961, after many efforts and after overcoming a series of political hurdles, Ginzton's team secured the funding needed for building SLAC.

16.3 The Weak Force and the Experimental Discovery of Neutrinos

Since the early 1900s, experiments involving beta decay gave rise to a conundrum. During beta decay, a given atom gives rise to a different atom and an electron. Using one of the most basic conservation laws of physics—the law stating that the momentum is conserved—it is possible to compute the energy of the emerging electron. However, measurements revealed that the electron did not possess the expected energy! It should be emphasised that in the case of alpha and gamma radiation, the revered law of conservation of momentum did provide the correct answers. How could this mystery be resolved? Niels Bohr suggested that the conservation of momentum, one of the cornerstones of physics, was *not* applicable. Wolfgang Pauli (Nobel Prize in Physics, 1945), who was tormented by this paradox, ignoring the advice of Peter Debye (Nobel Prize in Chemistry, 1936) 'to simply not think about it', proposed, in 1930, the following explanation: a tiny, electrically neutral particle was also generated in this reaction, carrying away the missing energy. In 1933, Enrico Fermi (Nobel Prize in Physics, 1938) elevated this idea to a theory and called the missing particle *neutrino* ('a little neutral element'). This particle, in addition to having no electric charge, was also supposed to be massless, so it seemed that it would be impossible to be detected.

Actually, in 1934, Sir Rudolf Peierls and Hans Bethe (Nobel Prize in Physics, 1967) presented calculations showing that a neutrino created during beta decay could pass through the entire Earth *without* interacting with matter.

Despite this seemingly unsurpassable difficulty, Frederick Reines and Clyde Cowan, after their chance meeting at the Kansas City airport, decided to undertake the Herculean task of detecting neutrinos. Their idea was to use the following interaction: a proton captures a neutrino, turning it into a neutron and releasing a positron. Reines and Cowan, who decided to use a nuclear reactor as a source of protons, had the brilliant idea of verifying this reaction by using two 'signatures', namely by observing two gamma ray flashes, five microseconds apart. Indeed, first, the emerging positron would annihilate an electron, creating a gamma ray (which is precisely the mechanism used in positron emission tomography). Second, the absorption of the emerging neutron by a nucleus would also emit a gamma ray, five microseconds later.

A major step towards their final success in detecting neutrinos was the development of a purely electronic method of detecting the gamma rays, avoiding the analysis of millions of photographs required in the use of cloud or bubble chambers. A crucial additional step was the elimination of external sources of radiation producing gamma rays. In particular, an early setup near a nuclear reactor in Hanford, Washington, had to be abandoned after they discovered that the 'noise' affecting measurements was not coming from human-made radiation but from cosmic rays. So, it was decided to move underground! A new setup was built at the Savannah River inside a nuclear reactor in South Carolina, which consisted of three layers. The detector was located 12 metres beneath the reactor, and electronic cables carried the signals to a trailer at ground level. The proof of the existence of neutrinos, as does happen with the solution of many very difficult problems, did not involve a 'eureka moment' but involved a series of several important steps. Finally, their apparatus allowed Reines and Cowan to catch each hour a few of the 100 trillion neutrinos emitted by the nuclear reactor each second. Furthermore, they were able to measure the consequence of the generation of these neutrinos. Twenty-five years after Pauli's prediction, the team at the Savannah River sent to Pauli the following telegram: 'We are happy to inform you that we have definitely detected neutrinos'. Pauli, who interrupted a meeting he was attending at CERN to read out this telegraph, apparently wrote back, 'Everything comes to those who know how to wait', which however never reached

Reines and Cowan. In 1995, Reines was awarded the Nobel Prize in Physics for this discovery (sadly, Cowan had died 13 years earlier).

Another Nobel Prize in Physics for the discovery of neutrinos was awarded in 2002 to Ray Davis and Masatoshi Koshiba. Davis' attempts were based on the results of a theoretical paper by Bruno Pontecorvo, who established that the interaction of a neutrino with an atom of chloride yields a radioactive atom of argon.[7] The reason that in his first attempts Davis failed to detect neutrinos is that his apparatus was capable of detecting only neutrinos, as opposed to anti-neutrinos, whereas nuclear reactors and beta decay produce anti-neutrinos. Following this failure and motivated by the fact that the Sun produces a tremendous number of neutrinos, which, travelling with the speed of light, reach the Earth approximately eight minutes later, he decided to capture the neutrinos coming from the Sun. By 1965, an enormous cavern was excavated deep inside a mine in South Dakota and a 380,000-litre tank was built, which was filled with dry-cleaning liquid (which contains chlorine atoms). Using this facility, after collecting a few dozen of radioactive argon atoms, Davis was able to establish the presence of solar neutrinos. However, there was a serious problem that he was unable to solve for the next 20 years: he could find only a third of the number of neutrinos predicted by his collaborator, the theoretical astrophysicist John Bahcall.

Koshiba, after his failed efforts to verify experimentally proton decay, as predicted by one of the prevailing theories of that time, modified his elaborate Kamiokande facilities, which were built in 1983. This finally led to the construction of the Super-Kamiokande, a neutrino observatory located 1,000 metres beneath Mount Ikeno, Japan, in the Mazumi Mine. Using this facility, the team led by Koshiba, in 1988, verified the results of Davis and also provided evidence that the neutrino is *not* massless. Incidentally, the best current estimate for the mass of neutrino was obtained in 2022, following the KArlsruhe TRItium Neutrino (KATRIN) experiments conducted in Germany by a team of international collaborators. The mass of the neutrino is less than 0.8 eV/c^2, i.e. hundreds of thousands of times less than the mass of the electron, which until the discovery of the neutrino was the lightest known particle.

---

[7]Pontecorvo was an Italian-Jewish physicist who early on worked as an assistant to Fermi. A dedicated communist, in 1950 he defected to the Soviet Union, where he lived until the end of his life.

The resolution of the mystery of the 'missing neutrinos' in the experiments conducted by Davis was already alluded to in a theoretical paper by Pontecorvo, written in 1957. It is now known that there exist three types of neutrinos: the standard neutrino (now called the electron-neutrino), the muon-neutrino, and the tau-neutrino. On the way to Earth, some of the electron-neutrinos transform into the other two types, and since the apparatus built by Davis could only detect electron-neutrinos, he only detected a third of the total number. The experimental confirmation of this fact was provided in two different experimental setups and led to another Nobel Prize in Physics for the discovery of neutrinos, awarded in 2015 to Takaaki Kajita and Arthur McDonald. The first experimental evidence for the transformation of the solar neutrinos came from the team led by Kajita, in 1998, at the Super-Kamiokande detector. The second came in 2001 by a team led by McDonald, who, after graduating from Caltech and becoming a full professor at Princeton University, was appointed director of the Sudbury Neutrino Observatory, Canada, a facility a mile underground, inside a nickel mine in Ontario.

What is the nature of the weak force? Its defining property is the ability to change the flavour of a given quark. For example, during the beta-minus decay, a proton is converted to a neutron, emitting an electron and an anti-neutrino. In this particular case, the weak force transforms a proton into a neutron by changing a 'down' quark to an 'up' quark.

It turns out that during the early Universe, shortly after the Big Bang, the electromagnetic force and the weak interaction were unified into the *electroweak* force. Experimentally, this unification was confirmed at CERN using the *Large Electron–Positron* collider. The development of a theoretical framework for this unification was achieved in the early 1960s by Sheldon Glashow of Harvard University, Abdus Salam of Imperial College, and Steven Weinberg of Princeton University, who were awarded the Nobel Prize in Physics in 1979.

Incidentally, the Pakistani Salam is the only Muslim from an Islamic country to ever be awarded a Nobel Prize in science. After obtaining degrees in mathematics and physics at St. John's College, Cambridge, Salam obtained his PhD at the Cavendish, and after a short visit to Pakistan, he was appointed, in 1954, to a chair in mathematics at the University of Cambridge. In 1957, he moved to Imperial College, where he established the Department of Theoretical Physics. Salam, a tireless advocate of science, in 1964, founded in Trieste, Italy, the famous International Centre for Theoretical Physics, where he served as its director until 1993. When I graduated from the Aeronautics Department of Imperial College in 1975, it was already speculated that Salam would win the Nobel Prize.

In the same way that the electromagnetic and strong forces are mediated by the photon and the gluon, respectively, the weak force is mediated by three other elementary particles: the positively charged W^+boson, the negatively charged W^-boson, and the neutral Z boson. These particles were first detected in CERN in 1988.

16.4 The Standard Model

In 1969, Robert R. Wilson, a student and protégé of Ernest Lawrence, appeared before the U.S. Congress asking for US$250 million for the construction of Fermilab, which began operating in 1972. This facility, housing the most ambitious accelerator built in the US, involved the initial acceleration of protons via a linear accelerator, as well as the use of superconducting magnets.[8] After their initial boost, the accelerating protons formed a proton beam circulating around the 6.28-kilometre circumference of the main ring. Soon after it became operational, Fermilab broke the world record for a proton beam, reaching by 1975 an energy of 500 GeV.

An early experimental leader at Fermilab was the Columbia University physicist Leon Lederman, who in 1988, shared with his Columbia University colleagues, Jack Steinberger and Melvin Schwartz, yet another Nobel Prize in physics awarded for neutrinos: they established that the electron-neutrinos generated via a proton beam are distinct from the muon-neutrinos.

At that time, the following classes of particles were known: the electron, the muon, the tau (which was discovered in 1975 at SLAC), and the three neutrinos. In addition, the 'up', 'down', and 'strange' quarks had been discovered. Lederman speculated on the existence of additional, heavier quarks. In 1970, Glashow, John Iliopoulos, and Luciano Maiani predicted the existence of the 'charm' quark, which was discovered simultaneously and independently at the National Laboratory and at SLAC; this discovery was achieved in connection with the particle J/ψ, which consists of a 'charm' quark and a 'charm' anti-quark. In 1977, the discovery of

---

[8]This technology had a significant impact on the development of the magnetic resonance interference apparatus.

upsilon was announced, the heaviest particle ever discovered and the first new particle to be discovered at Fermilab. This particle consists of the 'bottom' quark and the 'bottom' anti-quark; in this way, the existence of the 'bottom' quark was also established.

By then, Fermilab was no longer the largest accelerator in the world. CERN had built a seven-kilometre ring, called the *Super Proton Synchrotron*, reaching an energy of 450 GeV. CERN already had a long history. In 1954, 12 European countries, namely, Belgium, Denmark, France, Germany, Greece, Italy, the Netherlands, Norway, Sweden, Switzerland, the United Kingdom, and Yugoslavia, sponsored the creation of a new laboratory, the 'Conseil European pour la Recherche Nucleaire', known as CERN. In contrast to the constitutions of several US laboratories, it is explicitly stated in the constitution of CERN that it 'shall have no concern with work for military requirements and the results of experimental and theoretical work shall be published or otherwise made generally available'. In other words, CERN is an 'open' institution dedicated to peace.

Meanwhile, Wilson had a new goal: to build a facility where particles could smash directly into each other instead of hitting a fixed target, i.e. to build a *particle collider* instead of a particle accelerator. The *Tevatron*, containing a new ring built underneath the main ring, became operational in 1986; in this way, Wilson's dream was realised: a proton and an anti-proton beam slammed into each other. In 1995, using this facility, the 'top' quark was discovered, with precisely the mass predicted in 1977 by Martinus Veltman, who shared the Nobel Prize in Physics in 1999 with his former student Gerardus "Gerard" 't Hooft.[9]

In 1973, two different papers were published, independently, in *Physical Review Letters*: one by David Gross of the University of Santa Barbara and by Frank Wilczek of MIT, and one by David Politzer of Caltech. These authors, who were awarded the Nobel Prize in Physics in

---

[9]Leon Lederman, who became the director of Fermilab in 1978 after Wilson, had become frustrated with running this facility and advocated the construction of what was to be called *Superconducting Super Collider*; it would have leapfrogged CERN's corresponding facility. This project was approved in 1983 by the U.S. Congress, and by 1992, 22.5 kilometres of the relevant tunnel were already built, at the cost of US$3 billion. This was the period when the US was in recession, and the relevant budget for the construction of this facility had ballooned to US$12 billion. In 1993, despite the efforts of President Bill Clinton, the U.S. Congress decided to end this project.

2004, eliminated a possible fundamental inconsistency of particle interactions. Indeed, it is well known that the Coulomb electrostatic force between two charged particles and the gravitational force between two masses become infinitely large as the distance between the two particles tends to zero. If the strong force behaved the same way, then strong interactions would become infinitely strong at short distances. However, the Nobel Laureates proved that the strong force exhibits *asymptotic freedom*, meaning that when quarks are close to each other, the strong force becomes so weak that the quarks behave like free particles. The opposite happens at large distances: as the distance between the quarks increases, the force also increases, just like it happens when stretching a spring. This property, which is known as *confinement*, provides an explanation for the fact that quarks are never found in isolation. When they are produced in particle accelerators, they always appear in the form of 'jets', which consist of many colour-neutral particles. Actually, almost always, these jets consist of mesons (two quarks) and baryons (three quarks) clustered together.[10]

Following these astonishing developments, a new comprehensive theory, the 'Standard Model' was firmly established (Veltman, 2003). It involves three 'particle generations' of increasingly larger mass: 'down' quark, 'up' quark, electron, and electron-neutrino; 'strange' quark, 'charm' quark, muon, and muon-neutrino; and 'bottom' quark, 'top' quark, tau, and tau-neutrino. The Standard Model also includes the mediators of the electromagnetic, strong, and weak interactions, which are the photon, the gluon, and the three bosons, W^+, W^-, and Z, respectively.

An important ingredient of the theory, the famous *Higgs boson*, was still missing! This particle was essential to the theory holding together the Standard Model: it provided the mechanism needed for the mediators of the weak force, the W^+, W^-, Z bosons to acquire mass. The existence of this particle was predicted in 1964 in three different theoretical papers, authored by Peter Higgs, by Robert Brout and Francois Englert, and by Gerald Guralnik, Carl Hagen, and Tom Kibble. Remarkably, these papers were published the same year, independently, and in the same journal, *Physical Review Letters*. The experimental discovery of the so-called

---

[10]In the early 2000s, there were claims of the experimental discovery of pentaquark, namely of particles consisting of four quarks and one anti-quark. In 2015, CERN reported that the analysis of the decay of the bottom lambda baryon was consistent with the existence of a pentaquark. In 2022, CERN, finally, announced the discovery of the $P_{\psi s}$ (4338) pentaquark.

'God's particle' required yet another major experimental development. The *Large Hadron Collider* (LHC), built by CERN, is located 100 metres underground, beneath the border of France and Switzerland. This enormous system accelerates two beams of hundreds of billions of protons to a speed almost equal to the speed of light, reaching an energy of 7 TeV, focuses these beams to less than the width of a hair, and then collides them together. In July 2012, CERN held a press conference, streamed live to the auditorium in Melbourne, where a major conference on theoretical physics was taking place. CERN's director general announced, 'We have a discovery'. The Higgs boson had been discovered! It has zero spin, zero electric charge, zero colour charge, and a very large mass of approximately 125 GeV/c^2. The following year, Englert and Higgs were awarded the Nobel Prize in Physics.

It should be emphasised that despite the above incredible successes, many questions remain open. These questions range from the elucidation of several puzzling properties of some of the basic particles involved in the Standard Model to certain fundamental open issues. For example, matter can be 'left-handed' or 'right-handed', depending on the direction that a particle spins in relation to the direction of its motion. Surprisingly, all neutrinos are left-handed and all anti-neutrinos right-handed. More importantly, the Standard Model makes the wrong assumption that the neutrinos are massless.

A fundamental weakness of the Standard Model is that it has failed to unify the gravitational force with the electromagnetic, strong, and weak interactions. A major—rather disheartening—problem is that the Standard Model accounts for only about 5% of the matter-energy content of the Universe. The remaining 95% of the Universe is composed of what is, at the moment, *not* known: dark matter and dark energy (NASA-Science, 2024, 2025). Furthermore, currently, there does not exist any experimental evidence providing a suggestion for the origin of the dark sector.

Methodologically, it is suspicious that the Standard Model contains 17 elementary particles, with a huge discrepancy between their masses, as, for example, becomes evident by comparing the masses of the almost massless electron-neutrino and the massive Higgs boson. Obviously, there is nothing special about the number 17. In an ideal world, one would expect a very small number of particles, perhaps the trinity of the photon, the electron, and the neutrino, which could introduce the fundamental properties of quantum electromagnetic radiation, electric charge, and mass, respectively.

It is worth noting that it is often stated that 'gravity is unbelievably weak compared to the other three forces' (Sheehy, 2023, 279). However, very recently, it was established theoretically that, under certain circumstances, the gravitational force can be quite large. In particular, this can occur when, first, the post-Minkowskian approximation of the general theory of relativity is valid (Blanchet and Fokas, 2018) and, second, the particles move very fast. For example, it is shown in Fokas and Floratos (in preparation) that the gravitational force between two particles with the mass of the 'up' quark moving with a speed analogous to the speed of the 'up' quark and 'up' anti-quark forming the neutral pion, is of the order of 10^9 of the classical Newtonian gravitational force.

References

Blanchet, L. and Fokas, A. S. (2018). Equations of motion of self-gravitating N-body systems in the first post-Minkowskian approximation. *Physical Review D*, *98*, 084005.

Fokas, A. S. and Floratos, E. (in preparation). The gravitational force in the first post-Minkowskian approximation and its ultra-relativistic limit.

NASA-Science. (2024). Dark Energy. Accessed 2025. https://science.nasa.gov/mission/roman-space-telescope/dark-energy/.

NASA-Science. (2025). Dark Matter. Accessed 2025. https://science.nasa.gov/dark-matter/.

Riordan, M. (1992). The discovery of quarks. *Sience* (*Science*), *256*, 1287–1293.

Sheehy, S. (2023). *The Matter of Everything: How Curiosity, Physics, and Improbable Experiments Changed the World.* New York: Knopf.

Veltman, M. J. G. (2003). *Facts and Mysteries in Elementary Particle Physics.* Singapore: World Scientific.

Chapter 17

Interdisciplinarity and the Huge Impact of Particle Physics

The discovery of electrons led to the revolution of *electronic* as opposed to *electrical* devices. In the latter, electrons flow through wires, whereas in electronic devices electrons flow through a vacuum. The first such device is the *Fleming valve*, whose construction was based on combining Thomas Edison's trial-and-error invention of the light bulb and Thomson's understanding of the nature of cathode rays. In 1899, Thomson showed that the carbon filaments that were burning inside Edison's bulb were emitting electrons via the so-called *thermionic* emission. Following this understanding, the electrical engineer Sir John Ambrose Fleming produced the thermionic (Fleming) valve. Soon after, a third filament was added, which allowed for the amplification and modulation of electrons. The electronic revolution had begun!

Rutherford understood that *radioactivity* could be used for keeping track of time. In this regard, he established that uranium-238 decays through several intermediate steps to lead-206. By comparing the quantities of uranium-238 and lead-206 in a sample of the mineral pitchblende, and by showing that it takes 4.5 billion years for half of a given quantity of uranium-238 to decay, he concluded that the Earth was much older than previously thought.[1] Actually, the current techniques for determining the

---

[1]An earlier estimate of the age of the Earth was derived by Lord Kelvin (William Thomson), who early on was a strong believer in the indestructibility of matter. However, he later accepted radioactivity and was forced to pay a bet made with Lord Rayleigh.

age of fossils, rocks, and a variety of artefacts originate from the ideas of Rutherford. For example, using such techniques, we know that dinosaurs were wiped out 65 million years ago, presumably by an asteroid hitting the Earth.

In addition to its use for determining the age of a given object, radiation has had a plethora of other applications. For example, it is used for sterilising mail. This became widely used after the poisoning attempts made using anthrax in the US in 2001.

The interaction of *cosmic rays* with the atmospheric nitrogen creates carbon-14, which is a radioactive isotope. As is well known, carbon combines with oxygen to form carbon dioxide, which is absorbed by plants through photosynthesis. So, when animals and humans digest plants, together with carbon-12, they also absorb a small amount of the radioactive carbon-14, which has a half-life of 5,730 years. In the 1940s, Willard Libby (Nobel Prize in Chemistry, 1960) realised that by comparing the amounts of carbon-14 and carbon-12 in a sample of wood, bone, or some other organic material, he could calculate the age of this specimen. This so-called *radiocarbon dating* has had a profound impact on archaeology.

Another effect of the cosmic rays is the creation of beryllium-7 and beryllium-10 via the interaction with atmospheric oxygen. These isotopes have a half-life of 1.4 million years and 53 days, decaying, respectively, to boron-10 and lithium-7. These stable isotopes build up in layers of ice in Antarctica and Greenland, so drilling into the ice provides an accurate method of dating.

The understanding of the physics of the *photoelectric phenomenon*, together with the exploration of certain remarkable properties of *semiconductors*, led to a plethora of applications. Many useful devices were constructed, which are based on the use of a *photodiode*. Examples include remote controls, a variety of sensors (such as those that turn lights on when someone enters a room or dispense hand sanitiser), and GPS sports watches that monitor heart rate. A photodiode is made of a semiconductor, which is used to form a junction that allows electricity to flow more easily in one direction than the other. As a consequence of the photoelectric effect, when light shines on a photodiode, it generates electricity. For example, a remote control contains a light-emitting diode (LED). When the button is pressed, light travels to the photodiode, which sends a signal, say, to an air conditioner. The impeding photons free up electrons, electricity is generated, and the air conditioner is turned on.

The invention of *particle accelerators* found remarkable applications beyond particle physics. Indeed, after the smashing of the atom at Cavendish in 1932, many companies, both in Europe and the US, began building high-voltage machines. In particular, by 1937, Westinghouse developed a 5 MeV machine using a method developed by the American physicist Robert Van de Graaff. Some examples of important applications follow.

In order to make a semiconductor useful, it is necessary to add to it tiny amounts of other elements, such as phosphorous or gallium in the case of silicon (the preeminent semiconductor used in electronic devices). This is achieved by using a particle accelerator, a process called 'ion implantation'.

A procedure called 'ion beam analysis' uses a particle accelerator to verify or refute the authenticity of artwork. The sample is placed in the line of the particle beam, and a detector picks the ions that are scattered backwards. Then, by employing a technique called 'Rutherford backscattering spectrometry', it is possible to determine the atomic composition of the artwork. Many museums possess such facilities. For example, a 37-metre-long installation is used at the Louvre museum in Paris, located 15 metres below the famous glass pyramid.

A variety of products, including computer chips, are routinely strengthened using particle beams. Harry Gove and Mayer Rubin realised that the use of a particle accelerator makes it possible to achieve accurate carbon dating using a much smaller quantity of a given specimen. In 1987, using this new technique, these investigators established with a confidence level of 95% that the Turin Shroud was medieval and not 2,000 years old.

Muons, as a result of their ability to travel through dense objects unhindered, provide a powerful method for scanning very dense objects. For example, in 2017, this method was used to discover a hidden room in the Khufu's Great Pyramid in Gaza, the first new structure to be discovered since the 19th century. Earlier, this technique, known as *muon tomography*, had allowed Hiroyuki Tanaka of the University of Tokyo to explore the internal structure of the Mount Asama volcano of Japan.

In order to support the production of *radars* at the Rad Laboratory, many companies, including General Electric and Westinghouse, built magnets and electronic equipment. In 1945, Percy Spencer, an engineer at one of these companies, named Raytheon, observed that a chocolate bar had melted in his pocket while he was standing in front of a magnetron. This led to Raytheon filing a patent for the construction of the first *microwave*. Incidentally, it was understood early on that a klystron beam could be used to inform the pilot of a plane of the distance from the ground. This led to the company Sperry Gyroscope licensing the klystron for commercial and military applications of radars. Several applications were further explored by the Varian Associates, the commercial company of the Varian brothers.

In addition to medical applications that will be discussed in the following section, *linear accelerators* have a variety of other uses, from security scanning at ports and borders (since they can penetrate larger and denser objects which cannot be adequately scanned by standard X-rays) to removing potential pathogens from certain foods. Recently, they have been used for treating wastewater from factories (avoiding chemicals) and even cleaning noxious fumes from power plants.

One of the greatest contributions of particle physics in general and of CERN in particular is the invention, by Sir Timothy Berners-Lee, of the World Wide Web.[2] This hugely important development arose from the requirement at CERN to manage and communicate increasingly large volumes of data. Storage capacity is measured in units called 'bytes.' A megabyte (MB), or 1 million bytes, is the memory needed to store a typical song; the storage needed for a high-resolution photo taken with a smartphone is about 5 MB. A gigabyte (GB), or 1,000 MB, is enough memory to store a standard-definition video of one hour. A common laptop computer has a disc that can typically store 256 or 512 GB. A terabyte (TB), or 1,000 GB, is a huge amount of memory; the entire collection of books in the U.S. Library of Congress, the largest library in the world, can be stored in about 10 TB of memory. Nowadays, large data centres and cloud storage providers offer storage capacity in the petabyte (PB) range, where 1 PB equals 1,000 TB.

In 1989, the calibration and experimental data associated with the running of the *Large Electron Proton collider* were of the order of GB and

---

[2]Actually, Berners-Lee invented all three components underpinning the web: the Hypertext Markup Language (HTML), which remains the formatting language for the web; the Uniform Resource Locators (URLs), the unique addresses used to access the various resources in the web; and the Hypertext Transfer Protocol (HTTP), which is the communications protocol used to connect servers and distribute information.

TB, respectively. How could this 'computing problem' be addressed? By employing new computational technologies, Sir Timothy provided a highly effective solution to this problem.

The operation of the *Large Hadron Collider* (LHC) generated new unprecedented computational needs. The associated annual data output was soon approaching 100 petabytes per year. CERN's datasets were becoming too large to be handled locally. The only solution to this challenging problem was to share the data with many other centres. This motivated CERN to establish an international network of computing centres, connected via fibre optics, which ensured extremely fast connections. This led to the Worldwide LHC Computing Grid, usually referred to as 'the Grid', consisting of more than 200,000 servers located in different parts of the world. The Grid has had a plethora of applications beyond particle physics. For example, early on, it was used for the analysis of 140 million chemical compounds, which finally led to the production of a new anti-malaria medication. Importantly, the shared-resources approach of CERN motivated many companies to adopt a similar technique. Indeed, 'cloud' services, like the one provided by Google, use a similar methodology, with the only difference being that, instead of distributing their data and computing power in many different computers worldwide, they store them in corporate cloud server warehouses.

17.1 The Impact on Medicine

Within a year of the 1896 discovery of *X-rays*, the Glasgow Royal Infirmary installed the first hospital-based X-ray imaging apparatus. In addition to their crucial use in radiology, X-rays also provide the basis of computer tomography (details can be found in Fokas, 2024).

In 1933, Irene Curie and Frederic Joliot, in their Nobel Prize-winning discovery, showed that it is possible to create *radioactive* elements by bombarding stable elements with alpha particles. In particular, after striking an aluminium foil with alpha particles, they produced a radioactive isotope of phosphorus. Until this discovery, the only way to obtain radioactive elements was to painstakingly extract them from their natural ores. The construction of new radioactive elements was greatly facilitated by the development of the *cyclotron*, for which Ernest O. Lawrence was awarded the Nobel Prize in Physics in 1939. Although Lawrence was aware that his childhood friend and neighbour Merle Tuve[3] was constructing an apparatus capable of reaching a voltage of 1 MV, he thought that there must be a better way to accelerate particles. In 1929, he read a paper by the Norwegian physicist Rolf Widerøe, which had a transformative impact on Lawrence's academic life. In this paper, it was claimed that, instead of using a single very high voltage to accelerate particles, it would be better to apply an oscillating voltage to a series of metal tubes lined next to each other with gaps between them. Lawrence understood that this idea was impractical because in order to reach a high voltage, an incredibly long series of tubes was needed. However, he immediately thought of how to bypass this difficulty: instead of the particles travelling in a line of tubes, they could be bent around a circle, so that the same accelerating gap could be used many times. Moreover, he was aware that the bending could be achieved via the use of a magnetic field. By the early 1930s, Rutherford, together with one of his graduate students, built a cyclotron of a 30 cm diameter, which, using a voltage of just 3,000 Volts, was able to produce protons reaching an energy of one million eV. The impact of the cyclotron was greatly enhanced by a new discovery: in 1932, the same year that Chadwick discovered the neutron and Cockcroft–Walton smashed the atom, Harold Urey (Nobel Prize in Chemistry, 1934) of Columbia University, discovered a new isotope of hydrogen, called deuterium, whose nucleus consists of a proton and a neutron. Within a few days of the discovery by Curie and Joliot of the artificial production of radioactive elements, Lawrence, by using his cyclotron to bombard table salt with the nucleus of deuterium, discovered radio-sodium. As he expected, the higher the energy of the cyclotron beam, the higher the yield of the radio-sodium. Soon thereafter, radio-phosphorus was also produced.

Phosphorus is the most abundant mineral in the body after calcium and plays a crucial role in the formation of bones and teeth. Could the discovery of the phosphorus radioisotope have medical implications? In the summer of 1935, John, the younger brother of Lawrence, of the University of Yale, visited Ernest's Radiation Laboratory at Berkeley with the goal of exploring whether the newly discovered radioisotopes could be used for the treatment of cancer. Since John was a haematologist, it was natural for him to investigate leukaemia. After inducing leukaemia in a group of mice, he injected them with radioactive phosphorous produced

---

[3] Rutherford and Tuve, as children, were building radio equipment that allowed them to communicate via Morse code.

by the cyclotron of his brother. Two weeks after the treatment, the group of mice that had received the radioactive phosphorous were alive and in good health, whereas all the mice in the 'control' group that were not injected with the radioisotope were dead. The era of the life-saving radioactive treatment had begun.

In 1937, the chemist Glenn Seaborg (Nobel Prize in Chemistry, 1951) discovered the new radioisotope iron-59, which was immediately used in treating a variety of blood diseases. A year later, Seaborg discovered cobalt-60, which is the most widely used source of radiation in medicine, with numerous uses, from sterilisation of medical equipment to cancer treatment.

It is worth noting that until the invention of the cyclotron, four elements were still missing in the periodic table of Dmitri Mendeleev, namely, those with atomic numbers 43, 61, 85, and 89. The first of those elements, technetium, was discovered in 1937 by the Italian physicist Emilio Segrè (Nobel Prize in Physics, 1959), after convincing Lawrence to mail him a thin foil of molybdenum that was part of Lawrence's cyclotron, so that Segrè could try to determine the type of radioactive elements that were present in this foil. The other three were discovered in the next few years. The predominant naturally occurring isotope of technetium, technetium-99, has a half-life of 211,000 years. So, it is very difficult to find it in nature since almost all of it has decayed away. However, it is straightforward to construct technetium via a cyclotron. In 1938, Segrè and Seaborg, using the cyclotron, produced a variant of technetium-99, called technetium-99m,[4] with a half-life of six hours. This is the most widely used radioisotope in medical imaging. Importantly, more than 50 different radioisotopes are currently created and used in nuclear medicine.

Incidentally, the cyclotron paved the way for going beyond the known periodic table. Indeed, during the period of 1944–2010, 24 new elements were synthesised with atomic numbers 95–118. These 'synthetic' elements were created by forcing additional protons into the nucleus of an element with an atomic number lower than 95. This was achieved using a particle accelerator, employing a nuclear reactor, or during the explosion of an atomic bomb. The first synthetic element to be made was curium, which

---

[4]The letter 'm' stands for 'metastable'. The nucleus of this type of radioisotope first appears in an excited state, which later becomes stable, usually via the emission of gamma rays.

was synthetised by Seaborg and collaborators via the process of bombarding plutonium with alpha particles.

In 1946, Robert Wilson, a student of Lawrence, published a paper suggesting the use of high-energy protons for the treatment of cancer. At the time, it was already known that X-rays, as well as a beam of electrons, can kill cancer cells. The advantage of using protons is that these heavier particles can penetrate deeper inside the body and can also deposit their energy in a well-focused domain. The first human was exposed to proton beams in 1954 at the Berkeley Radiation Laboratory. Today, more than 100 centres around the world use particle therapy consisting of protons or deuterons. They usually use cyclotrons, which are currently quite compact because they use superconducting magnets. Some centres use the synchrotron.

Despite the superiority of the proton beam, the use of electrons and X-rays remains more common than the use of protons or ions because the relevant apparatus is smaller and less expensive. This involves a metre-long linear accelerator located above the bed where the patient lies. The accelerator boosts electrons to approximately 25 MeV, and these electrons hit a metal target, where they slow down, emitting X-rays (as it happens in a cathode ray tube). Finally, these X-rays are directed towards the patient via a device called a *collimator*. In the US, the first linear accelerator was developed by Edward Ginzton and Henry Kaplan at the Stanford University Medical School.[5] In 1986, after completing my medical degree at the L.M. Miller School of Medicine of the University of Miami, I began my internship in medicine at Stanford University. During my interview, the above achievement of this distinguished medical school was explicitly mentioned. The first patient to be treated by Kaplan with the Stanford linear accelerator was a child with a retinoblastoma in the right eye, which was also threatening his left eye. The child survived into adulthood with normal vision in his left eye. Incidentally, the first use of a linear accelerator in medicine took place in 1953 in the Hammersmith Hospital, London, using an apparatus constructed by the Medical Research Council Radiotherapeutic Unit of the same hospital.

---

[5]Kaplan made extensive use of radiation therapy for the treatment of Hodgkin's lymphoma, a cancer that was fatal until the development of radiotherapy. Kaplan's motivation for using radiotherapy was the death of his father from lung cancer, which was also the cause of Kaplan's death himself, despite the fact that Kaplan was not a smoker.

Interestingly, Varian remains one of the two main companies that dominate the market in medical accelerators. The second is Elekta, formed in 1972 by the Swedish neurosurgeon Lars Leksell. This company was formed in connection with the *Gamma Knife*, which is an apparatus capable of delivering highly focused beams of gamma radiation to tumours and a variety of brain lesions.[6] More than a million people have benefited from this painless computer-guided treatment.

17.2 Structural Biology

Starting with the awarding of the Nobel Prize to William and Lawrence Bragg in 1915, so far, 27 more Nobel Prizes have been awarded in connection with the use of the powerful imaging technique of X-ray crystallography. In their original work, the Braggs fired an X-ray source onto a salt crystal. The Nobel Laureates understood that by analysing the resulting beautiful diffraction pattern, they could decipher the molecular structure of the crystal itself. Using this technique, Dorothy Hodgkin (Nobel Prize in Chemistry, 1964) revealed the structure of penicillin in 1949, of B-12 in 1955, and of insulin in 1969. These structures, as well as those of graphite, graphene, haemoglobin, myoglobin, and countless proteins, were determined using conventional X-rays. After the structure of a given protein is determined, it is deposited in the Protein Data Bank, where it can be freely used for the investigation and exploration of the properties of this protein, including for drug development.

In 1961, the U.S. National Bureau of Standards announced that the light emitted during the operation of the particle accelerator *synchrotron* had various remarkable properties. Specifically, it can be extremely intense, it has laser-like coherence, and it can cover the full electromagnetic spectrum, from X-rays to visible light to infrared light (depending on the strength of the associated magnetic field and the energy of the accelerated electrons). This light is also polarised, meaning that the oscillations of the light waves occur in a single plane with all waves vibrating in the same direction. The use of the synchrotron to produce this type of unique light has had a huge impact on crystallography. For example, Sir John Walker (Nobel Prize in Chemistry, 1997) determined the structure of adenosine

---

[6]Conditions treated by the Gamma Knife include acoustic neuroma, arteriovenous malformations, trigeminal neuralgia, and tremors.

triphosphate, which is the molecule that transports and stores energy in all plants and animals. Roger Kornberg (Nobel Prize in Chemistry, 2006) elucidated the fundamental role of messenger RNA, and Venki Ramakrishnan (Nobel Prize in Chemistry, 2009) unravelled the structure of ribosomes. Importantly, in less than six weeks after six sequences of the SARS-CoV-2 virus were published, a group of Chinese scientists deposited in the Protein Data Bank the molecular structure of the main *protease* associated with this virus. This proteolytic enzyme, in addition to its standard use of breaking down proteins, is also vital for the replication of this virus, providing an important target for drug development.

It is worth noting that the synchrotron has a long and illustrious history. Its development goes back to Ernest Walton, who, in the late 1920s, suggested to Rutherford the construction of a *circular* accelerator, in which accelerated particles would go around, indefinitely, inside a circular tube. Walton presented calculations showing the following: in order for the particles to be pushed back towards the centre of the tube so that they would not fly off, it is necessary that the applied magnetic field decrease as the radius of rotation increase. Walton failed to implement his construction, and this is why he joined the programme suggested by Cockcroft for building a *linear* accelerator. However, by the early 1940s, an apparatus based on this idea, called a *betatron*, was constructed. In contrast to the cyclotron that was used to accelerate alpha particles and positrons, the betatron was used to accelerate electrons, reaching a speed of 99.99% of the speed of light (the name 'betatron' reflects the fact that beta radiation consists of electrons). In late 1945, Lawrence, in his efforts to eliminate the need for the use of the giant magnets necessary for the operation of the cyclotron and the betatron, came up with yet another new idea: he proposed the construction of an accelerator in which the strength and frequency of the magnetic field vary according to the acceleration of the electrons. This required a delicate synchronisation between the imposed magnetic field and the induced acceleration, motivating the name synchrotron. The new machine was built by General Electric and began operating in late 1946. Soon after, General Electric physicists, together with the Caltech physicist Robert Langmuir, observed a very bright bluish spot coming from the synchrotron. Langmuir immediately realised the origin of this phenomenon: it was due to the *Schwinger radiation*. Indeed, just after the betatron had become operational, two physicists from the Soviet Union published a paper showing that when charged particles are bent around a circular arch, they emit radiation (this is a straightforward consequence of the application of the law of conservation of momentum). Julian Schwinger, who, as stated in Chapter 14, later shared with Feynman and Tomonaga the Nobel Prize in Physics, revisited this result and showed that this radiation formed a tight beam pointing along the path of the movement of the electron.

Incidentally, interdisciplinarity often leads to discoveries within physics itself. This is beautifully illustrated by the impact of the synchrotron on the understanding of the origin of *radio emissions* in astronomy. The cosmic radio waves were first identified in 1933 by the Bell Laboratory engineer Karl Jansky, who called them 'star noise'. In 1937, the radio engineer Grote Reber funded and built the first radio telescope, with which he identified sources of radio waves in the constellations of Cygnus and Cassiopeia. In 1945, it was shown that the Sun also emits radio signals, and by the 1960s, it had become clear that various celestial objects, including the Milky Way (our own galaxy), were emitting radio waves. In 1974, the British radio astronomer Antony Hewish was awarded the Nobel Prize in Physics for the discovery of *pulsars*, which are highly compact stars that emit radiation from their poles. Actually, pulsars were first discovered by Jocelyn Bell Burnell in 1967, who was apparently not considered by the Nobel committee due to her status as a PhD student of Hewish at that time (in 2018, Bell Burnell was the recipient of the prestigious Breakthrough Prize; she donated the US$3 million awarded to her to fund scholarships in physics with the goal of enhancing diversity). What is the origin of these emissions? It is now clear that many celestial objects, including the Milky Way and pulsars, possess strong magnetic fields. The bending of charged particles by these magnetic fields causes the emission of radiation, just as it happens in the synchrotron.

17.3 The Manhattan Project

Concluding this section, it is worth noting that several of the physicists mentioned in this part of the book played a crucial role in the Manhattan Project, which employed approximately 100,000 people, most of whom were unaware that the goal of this project was to build an atomic bomb. Among these scientists were Neddermeyer and von Neumann, as well as the Nobel Laureates Alvarez, Bethe, Bohr, Chadwick, Cockcroft,

Compton, Fermi, Feynman, McMillan, Lawrence, Libby, Segrè, Urey, Wilkins, and Robert Wilson.[7] For example, Lawrence's cyclotron was used to separate uranium isotopes, and Lawrence himself supervised an important project building a large number of *calutrons*, which were specific apparatuses used for the separation of radioactive isotopes. Alvarez met Robert Oppenheimer before the war when he was working at the Berkeley Radiation Laboratory, where Oppenheimer was the head of the theoretical group supporting this laboratory.[8] One of the many contributions of Alvarez to the Manhattan Project was the development, together with the physicist Lawrence Johnston, of a set of microphone transmitters to be parachuted from an aircraft to measure the strength of the blast wave caused by the atomic explosion. In this way, using a formula obtained by von Neumann, the energy released from the bomb could be calculated. For this reason, in August 1944, Alvarez and Johnston were in the B-29 aircraft accompanying the plane that dropped the 'Little Boy' atomic bomb in Hiroshima. Three days later, Johnston was in the aircraft accompanying the plane that dropped the 'Fat Man' atomic bomb in Nagasaki. Within the next few months, the effect of these bombs killed between 90,000 and 146,000 people in Hiroshima and between 60,000 and 80,000 people in Nagasaki, approximately half of them dying the first day.

One physicist who refused to participate in the Manhattan Project was Lise Meitner, nicknamed by Einstein 'the German Marie Curie'. After moving from Vienna to Berlin, she became an assistant to Max Planck. Although she was the first female professor of physics in Germany, Meitner had to flee Germany because of her Jewish heritage. She was the first to realise that, in addition to emitting alpha and beta particles, nuclei could be split completely, releasing enormous amounts of energy. Her nephew, Otto Frisch, elaborating on Meitner's idea, introduced the term 'nuclear fission'. Meitner's colleague, Otto Hahn, published the first

---

[7]Additional Nobel Laureates not mentioned earlier include Edward Teller, James Franck, and Joseph Rotblat. The last of these, who had strong reservations against the use of science for developing devasting weapons, left Los Alamos at the end of 1944, when it became clear to him that Germany had abandoned plans to develop a nuclear bomb.

[8]Alvarez's sister was a part-time secretary of Lawrence, and through her introduction, Alvarez, as soon as he completed his PhD at the University of Chicago, moved to the Berkeley Radiation Laboratory.

evidence of nuclear fission, without listing Meitner as a co-author, in order to avoid prosecution for collaborating with a Jew. For this work, Hahn was awarded the Nobel Prize in Physics in 1944, but surprisingly, the crucial contribution of Meitner was overlooked.

Reference

Fokas, A. (2024). *Ways of Comprehending: The Grand Illusion and the Essence of Being Human.* Singapore: World Scientific.

Part IV

The Importance of Unconscious Processes and Mapping the Brain

We overestimate the importance of the information we become directly aware of, as opposed to the information for which we have only indirect access. This means that *we are biased towards consciousness*. For example, Descartes's famous aphorism 'I think therefore I am' (*'cogito ergo sum'*), taking into consideration that – in Descartes' context – thinking is a conscious process, suggests that consciousness is the essence of humanity. In my opinion, the emphasis on awareness and the neglect of the crucial importance of unconscious processes undermine our appreciation of the essence of being human. As stated in Fokas (2024), *the occurrence of unconscious processes is a necessary but not a sufficient condition for awareness*. Indeed, there exist an uncountable number of unconscious processes that never give rise to consciousness; from the plethora of mechanisms that ensure that an organism is in a state of homeostasis to the mysterious creative processes of artists and scientists.

Since every conscious experience is preceded by an unconscious process, the principle of continuity implies that the full understanding of conscious mental functions, including learning, perception, and creativity, is impossible without studying the continuum of unconscious–consciousness.

The American psychologist and neuroscientist Stephen Kosslyn, in the foreword of Benjamin Libet's book, *Mind Time* (Libet, 2005), writes:

"[Libet] is famous for discovering that we unconsciously decide to act well before we think we have made the decision to act. This finding has major implications for one of the deepest problems in philosophy and psychology, namely the problem of *free will*".

The pioneering experiments of the late German neurophysiologist, which are reviewed in detail in Chapter 18, initially evoked scepticism and then amazement. Furthermore, they led several distinguished philosophers to conclude that 'free will does not exist'. It is shown in Chapter 18 that, by adopting the framework introduced in Fokas (2024), the conclusions of Libet's experiments become predictable rather than being unexpected. Furthermore, the question of the existence of free will becomes moot. Related to Libet's experiments is the question of the proper definition of *subjective* time, i.e. our conscious sense of time. The perception of time is deeply related to the concept of rhythm, so it will be discussed further in my forthcoming book, *Cognition and Music*, where music and especially opera is discussed.

Chapter 19 begins by noting that one of the many important open questions in medicine is the understanding of the precise neurobiological mechanisms underlying the powerful and pervasive effects of *placebo* and *nocebo*. The elucidation of these mechanisms will undoubtedly have a major impact on clinical practice. This position becomes clear by reviewing several 'mysterious syndromes' where, apparently, unconscious processes are of fundamental importance. In this chapter, it is emphasised that, since an illness is crucially affected by the given cultural-social-political environment, as well as by the distinguishing characteristics of the patient, it is imperative that a *global* approach be followed, wherein all these factors are analysed, rather than a *local* approach, focusing on specific symptoms.

Chapter 20 discusses in detail the crucial importance of sleep. Its vital role in our health was clearly appreciated by the ancient Greeks, who elevated it to the level of a god. According to Greek mythology, Hypnos, the god of sleep, is the son of Nyx (Night) and Erebus (Darkness) and the twin brother of Thanatos (Death).

Electroencephalography (EEG) is of vital importance for sleep studies. This brain imaging technique, along with magnetoencephalography

(MEG), has the important advantage, in comparison to other functional imaging techniques such as PET, SPECT, and fMRI, that they provide information about brain activity in, essentially, real time. EEG, MEG, and the recently developed imaging methodology of optogenetics are reviewed in Chapter 21. Additional approaches to mapping the brain are also discussed in this chapter.

The historian Yuval Noah Harari laments our limited and fragmented approach to life, which he claims is expressed via our reduced capacity to 'smell', 'pay attention', and – most importantly – 'dream'. These effects, he warns, could 'end up downgrading humans' (Harari, 2015, p. 421). As an eternal optimist,

> *I have always believed that society usually solves the problems that it itself creates.*

In particular, the astonishing advances in algorithms tend to place the spotlight on consciousness at the expense of scrutinising and appreciating unconscious processes. However, technology, which is the result of algorithms, may provide tools for penetrating the world of the unconscious, allowing us to explore its mysteries and enjoy its beauty.

I speculate that the next great human endeavour will be the discovery of ways to explore the enormous wealth of knowledge and capabilities hidden in our unconscious.

Some very early attempts in this direction are discussed in Chapter 21.

References

Fokas, A. (2024). *Ways of Comprehending: The Grand Illusion and the Essence of Being Human.* World Scientific.

Harari, Y. N. (2015). *Homo Deus: A Brief History of Tomorrow.* Harvill Secker.

Libet, B. (2005). *Mind Time: The Temporal Factor in Consciousness.* Harvard University Press.

Chapter 18

Time Lag in Conscious Sensory Perception and Free Will

In the 1970s, Benjamin Libet (1916–2007), in a series of ingenious experiments, derived estimates for the time required for certain unconscious processes to give rise to awareness. In particular, he observed a specific activation in the brain, called 'readiness potential', that occurs approximately a third of a second *before* we become aware of making the conscious decision to act. This finding generated a lot of interest beyond neuroscience, with some distinguished philosophers considering this work as evidence for the non-existence of free will.

If this finding is analysed within the framework developed in Fokas (2024), then it becomes clear that Libet's famous result becomes utterly predictable. Indeed, it is postulated in Fokas (2024) that:

Every conscious experience is preceded by an unconscious phase.

This implies the following:

The awareness of 'deciding to perform a given act' must be preceded by the unconscious decision to act. The readiness potential is the local brain excitation reflecting this unconscious process.

When we become aware of our decision to act, we are simply 'informed by our brain' of what the brain already 'has decided'. The importance of Libet's work is not establishing that the unconscious decision precedes awareness of the decision, but estimating the time needed for the transformation from an unconscious process to the conscious decision.

By viewing Libet's experiments in this framework, the question regarding the relationship between these experiments and the existence of free will becomes meaningless. Indeed:

> *Since unconscious and conscious processes form a continuum, it is epistemologically wrong to identify free will only with consciousness, ignoring the crucial role of the unconscious processes. Our thoughts, feelings, and behaviour are determined by our genes, as well as by prenatal and postnatal learning. The above factors not only shape our consciousness but also influence our unconscious processes.*

Incidentally, regarding prenatal processes, it is worth noting that newborn babies of women who listened to specific music to relax during pregnancy stop crying and open their eyes upon hearing the same music. The great sensitivity of the unborn children to melody may explain why French babies cry with a rising intonation, whereas German babies cry with a falling intonation, perhaps reflecting the different intonation contours of the two languages. Babies can also remember scent and taste stimuli present when they were in the womb (Swaab and Hedley-Prole, 2014, 34).

Libet was aware that many distinguished scientists and philosophers of the first three-quarters of the 20th century were sceptical of whether consciousness, which reflects *introspective and highly subjective* processes, could be studied in an *objective way*. His scepticism was influenced by the philosophy of logical positivism, which, as mentioned in Fokas (2024), accepted as admissible scientific data only *directly observable* events. In addition, behaviourists argued that the study of the mind alone is unimportant because they were convinced that behaviour can be predicted by the conditions that precede it and follow it without any need to analyse the mysterious processes taking place in the brain. Libet's scepticism was overcome by developments in cognitive psychology, where it was established that the analysis of brain mechanisms *can* elucidate behaviour. Libet, embracing this modern understanding and under the influence of Karl Popper, decided that introspective reports of conscious experience should be treated like other types of *objective data*. This view, in addition to being embraced by cognitive psychologists, was also shared by several leading experimental neurophysiologists, including Lord Adrian, Sir John Eccles, and Roger Sperry. Libet postulated that externally observable events, such as flexing the wrist, and inner observable neural events, such as the readiness potential, although interrelated, are phenomenologically *independent categories*. Hence, their precise relationship can be deciphered only via the simultaneous observation of these two separate phenomena.

18.1 Time Lag in Sensory Perception

Before investigating the interaction between conscious and unconscious processes occurring when making a decision, Libet first investigated the time delay involved in sensory awareness. In particular, he studied the time lag between the perception of certain sensory signals in the brain and the subsequent conscious awareness of these signals. For these studies, he had available patients with permanently implanted stimulating electrodes in certain locations along the sensory pathway of their brains. These electrodes were placed in the brains of patients suffering from intractable pain by the neurosurgeons Y. Hosobuchi and N. M. Barbaro of the University of California, San Francisco, following earlier work by Bertram Feinstein.

Libet began his investigations by concentrating on simple sensory experiences devoid of emotional aspects that often distort objective reports. Sensory experiences, including those of pain, colour, harmonies, and odours, are often referred to as *qualia*. However, Libet did not distinguish qualia from other types of experiences, emphasising that they all share the common defining feature of *awareness* (this defining feature of consciousness is discussed in Chapter 19 of Fokas, 2024).

The earliest studies conducted by Libet took place in 1958 at the Mt. Zion Hospital in San Francisco, in collaboration with the neurosurgeon Bertram Feinstein, who had introduced in the Western United States the technique of 'stereotaxic neurosurgery' (Feinstein *et al.*, 1960). This technique, pioneered by Feinstein's mentor, the distinguished Swedish neurosurgeon Lars Leksell, is based on fixing a three-dimensional coordinate frame on the patient's skull. After a target point is chosen in the brain, the coordinates of this point are mapped in advance on the above frame. Then, a therapeutic electrode or a probe reaches this point using one of several possible paths. At that time, this technique was mainly used for inactivating certain deep structures in the brain via a heating probe in order to relieve Parkinsonian tremor. Since the brain tissue does not have pain fibres, a local anaesthetic was applied to the scalp, and the patients were awake during the procedure.

Early on, Libet had access to these patients in the operation room for about 30 minutes. However, in the 1960s, Feinstein began leaving the inserts in the brain for days or a week to allow the therapeutic lesion to be made in stages. Also, later, Feinstein and his associates began treating patients suffering from intractable pain by placing stimulating electrodes permanently in their brains. This allowed Libet to study such patients at length.

It was known prior to the work of Libet that a very weak electric stimulation of the skin gives rise to awareness. Moreover, Harvey Cushing had observed that the electric stimulation of the brain yields the sensation of tingling that the subject experiences as coming from the skin or body structure mapped to the specific part of the brain that is stimulated (and not coming from the brain). Libet, by systematically investigating these two processes, identified the following paradox. Although awareness arises even if a *single weak electric pulse* is applied to the skin, it is necessary to apply to the brain many such pulses of a *duration of at least half a second* to elicit awareness. As explained in the following, the resolution of this paradox is related to the phenomenon of time lag associated with sensory awareness.

In more detail, a stimulus consisting of a single pulse of current of very weak strength and of duration 0.1–0.5 milliseconds (a millisecond is one thousandth of a second) creates a signal that propagates in the following manner. It first travels along a large bundle of neurons at the back of the spinal cord until it reaches the *medulla*, which, as noted in Chapter 3 of Fokas (2024), is the lowest portion of the brain stem. There, these fibres cross over to the other side, which is the reason that the representation of a sensation appears on the hemisphere opposite to the side of the peripheral origin of the stimulus. Then, the signal proceeds along a large ascending bundle of axons of the medulla called *medial lemniscus*, until it reaches specific *nuclei* (groups of cells) in the *thalamus*. Finally, after 14–50 milliseconds from the time of the skin stimulation, the message arrives in the sensory cortex. The precise time of arrival depends on the distance between the brain and the point of skin stimulation. For a shorter path, such as that starting from the head, the time needed is 14–20 milliseconds, whereas for a longer path, such as that starting from the foot, the time needed is 40–50 milliseconds.

In the cortex, the signal generates an activation, which is expressed as an electric potential, known as the 'primary evoked potential'.[1] This activation, which lasts for about 125 milliseconds, is the neural signature of the unconscious perception of the touch sensation by the brain. This does *not* yield sensory awareness. The evoked potential generates a sequence of additional evoked potentials, called 'event-related potentials', of duration of approximately 350 milliseconds. The primary evoked potential is exhibited only in a highly *localised* small area of the primary sensory cortex, whereas the later event-related potentials are *broadly distributed* in the cortex. Awareness takes place only *after* the appearance of these additional evoked potentials. Apparently, this extended activation reflects 'neuronal calculations' by several local neural circuits associated with additional features of the sensation of touch. The global integration of these local activations yields the *mental image* of the specific touch experience.

Under general anaesthesia, the primary evoked potential remains intact or it is enlarged, but the event-related potentials disappear. If the strength of the single pulse applied to the skin is lowered until the subject experiences nothing, the event-related potentials disappear, but the primary evoked potential remains.

In summary, a single weak electric pulse applied to the skin gives rise to a short, localised activation in the corresponding area of the sensory homunculus. This cortical activation, which is completed within 150–175 milliseconds after the initial skin stimulation, does not yield awareness. If the electric pulse is sufficiently strong so that the localised activation is accompanied by further activation which is broadly distributed in the cortex and which lasts approximately for an additional 350 milliseconds, then and only then does awareness take place.

The above discussion implies that a sensory experience requires cortical activation that lasts for at least half a second. This is consistent with the cortical stimulation experiments of Libet. In order to elicit awareness, it was necessary to apply to the sensory cortex a train of pulses of duration of at least half a second. After such a stimulation, the sensation of touch was experienced by the subjects at the part of the skin associated with the corresponding part of the sensory homunculus. A similar situation was observed when the electrical stimulation was applied in the medial lemniscus.

---

[1] A current of a typical pulse has a strength of approximately 0.3 milliamperes, and it yields an evoked potential with a strength of approximately 50 millivolts.

The time lag in sensory awareness provides an explanation for experiments carried out by Arthur Jensen, who was trying to determine the 'reaction time' of different groups of people (Jensen, 1979, 231). In these routine tests, Jensen was asking subjects to press a button as quickly as possible after the appearance of a specific signal. The reaction time was in the 200–300 millisecond range. However, when he asked the subjects to deliberately try to delay their reaction by 100 milliseconds, the reaction time jumped to the 600–800 range. A possible explanation of this large jump is the following. There is independent evidence that the reaction time is achieved with no awareness of the signal. The deliberate effort of increasing this time apparently requires for the subjects to *become aware* of the specific signal. Taking into consideration that awareness requires approximately an additional 350 milliseconds, this explains the inability of the subjects to achieve a small increase of the reaction time.

The time lag occurring in sensory awareness is counterintuitive. When, for example, we tap our fingers on the table, we do not perceive any time lag; we are sure we experience this event occurring in 'real time'. This implies that there is a difference between actual time and 'subjective' time. Libet has claimed that our *subjective time is retroactively referred to the moment of the occurrence of the primary evoked potential*. In other words, our brain tricks us into making us feel that our awareness takes place simultaneously with the appearance of the first brain activation. This is consistent with Leonard Mlodinow's statement that:

> "...the brain is a decent scientist but an *outstanding* lawyer [...]. In the struggle to fashion a coherent, convincing view of ourselves and the rest of the world, it is the impassionate advocate that usually wins over the truth seeker" (Mlodinow, 2014, 201).

Evidence that the subjective time coincides with the moment of the occurrence of the primary evoked potential is provided by the following experiment. After a train of electric pulses of half a second duration was applied to the sensory cortex of a normal subject, a single pulse was applied to the skin at different times after the start of the cortical stimulation. Even when the skin stimulation was delayed with respect to the start of the cortical stimulation, the subject reported that the sensation of the skin was experienced *before* the cortically induced sensation. This is consistent with the fact that the cortical stimulation does not generate a primary evoked potential. Hence, for the cortical stimulation, the real and subjective time coincide, and awareness of the cortical stimulation is felt approximately half a second later. Indeed, when the skin stimulation was delayed by approximately half a second, the subject reported feeling both sensations simultaneously.[2] Interestingly, electric stimulation applied to the medial lemniscus does produce a primary evoked potential. In this case, there is no delay between the time of experiencing this stimulation and the skin stimulation since, now, both events are referred to the moment of the occurrence of the primary evoked potential.

This discussion elucidates the difference between real and subjective (unconscious) time. Incidentally, the meaning of real time continues to be debated. In this regard, it is noted that Einstein worried a lot about the deeper meaning of this fundamental concept. According to the philosopher Rudolf Carnap, in their conversations, Einstein expressed the opinion that, whereas natural time can be described via physical processes, subjective time, particularly the importance of the present, should be analysed with the aid of psychology. When Michele Besso, who, as mentioned in Chapter 13, was a close friend of Einstein, passed away in 1955, only a month before Einstein's own death, the great physicist wrote to Besso's grieving family:

"Now he has departed from this strange world a little ahead of me. That signifies nothing. For those of us who believe in physics, the distinction between past, present and future is only a stubbornly persistent illusion."

The statement that '…the distinction between past, present and future is only a stubbornly persistent illusion' is consistent with the theory of general relativity discussed in Chapter 13, where our *block universe* is characterised as a four-dimensional space, with past and future having no particular meaning other than going up or down the time axis. This is in agreement with the recounts of Karl Popper that Einstein accepted the block universe view (Popper, 1992). But Einstein's statement, '*Now* he

---

[2] In further supporting evidence of Libet's position, when stimuli were applied simultaneously to both hands of a female who had suffered a right-hemisphere stroke, she felt the right-hand stimulus before the one on her left hand (which was affected by the stroke). Since there was no evoked potential in the right hemisphere, the patient could not refer the sensation of touch in her left hand to an earlier time. This means that for this hand, the actual and subjective times are the *same*, and hence, she felt the left-hand sensation about half a second later.

has departed ... *ahead* of me', clearly expresses the subjective importance of the present.

Time lag is not only a feature of sensory sensations. It occurs in any transformation from an unconscious process to awareness. For example, if a boy steps in front of a car, a competent driver breaks after about one-seventh of a second. At this moment, the *brain* of the driver perceives the boy, but the driver is not aware of him. Slamming the brake is not a simple reflex. It involves recognising the appearance of the boy and deciding to try to avoid hitting him. This complex mental function is carried out unconsciously. The driver becomes aware of the series of these events only about half a second after the initial appearance of the boy. However, this delayed awareness is subjectively antedated, referring to the earlier time when the brain perceived the boy. So, the driver erroneously believes to have become aware of the boy immediately.

18.2 Time Lag in Decision-Making and Free Will

Libet has established that a time lag in awareness occurs not only in events where the initiating stimulus is external but also when the stimulus is endogenous. An example of such a case is 'the decision to act', where Libet found experimentally the following: subjects became aware of 'deciding' to carry out a specific act approximately 150 milliseconds before the execution of this act. Also, and this is the result that created controversy, there was brain activation approximately 400 milliseconds *before* the subjects became aware of their decision to act.

These experiments were motivated by earlier works of the German neurophysiologists Kornhuber and Deecke (1965), who had observed an activation in the brain about 0.8 seconds before subjects performed a voluntary act. This activation, which occurred in the premotor cortex, was called 'readiness potential'. The long interval between brain activation and action made Libet suspicious that, maybe, extensive brain excitation took place before the conscious intention to carry out the act.

In Libet's experiments, subjects were asked to perform freely the following voluntary act: to suddenly flex the wrist 'at any time they felt like doing so;'. The first problem faced by Libet was how to achieve the accurate reporting by the subjects of the precise time *when they made the conscious decision to execute this act*. This problem was solved by placing the subjects in front of a specially designed large clock in which, instead of the usual hand measuring seconds, there was a fast-moving bright spot that could complete a full circle in 2.56 instead of 60 seconds. This faster clock allowed the subjects to report time more accurately than a usual clock. Supplementary experiments established that this method of reporting had an error of at most 50 milliseconds. The subjects were asked to report after the end of the experiment the precise time that they made the conscious decision to act. The time when the flexion took place was determined by using *electromyography*. This technique detects muscle movement by employing a sensor attached to the body. Brain activation was determined via electroencephalography, which, as will be discussed in Chapter 21, records on the scalp the values of the electric potential.

The subjects were instructed to avoid preplanning and to act spontaneously.[3] The average value of the onset of the readiness potential was 550 milliseconds before flexion. The average reported time of awareness was 200 milliseconds before flexion, which was corrected to a value of 150 milliseconds due to the error in reporting.

The pioneering studies of Libet have been repeated and extended by several investigators. It is now clear that the brain's activation in preparation for executing a motor task takes place even earlier than the time suggested by Libet. For example, John-Dylan Haynes and collaborators instructed participants to press a button with either the right or left index finger whenever they felt the urge. These researchers, instead of using electroencephalography, used fMRI to monitor the activation of the subject's brain (Soon *et al.*, 2008). Remarkably, there was activation several seconds before the action. From this activation, it was possible to predict whether the right or left finger would press the button but only with 60% accuracy (it should be recalled that fMRI's time resolution is rather poor, namely, of the order of a few seconds). More accurate results were obtained by studying patients with electrodes surgically implanted in their brains for the treatment of epilepsy. This allowed recording from single neurons in real time. In this way, it was established that there was activation in particular parts of the brain approximately a second and a half before the subjects made the conscious decision to press the button. Furthermore, these researchers could predict the timing of the decision with an 80% accuracy (Fried *et al.*, 2011).

---

[3] In some trials the subjects reported that they had preplanned a range of clock times in which to act. In these cases, the readiness potential appeared approximately 0.8–1.0 seconds before the flexion. These values were similar to those reported by Kornhuber and Deecke.

In the disorder Tourette's syndrome, the *caudate nucleus* (which, as mentioned in Chapter 3 of Fokas, 2024, is part of the brain stem) exhibits heightened sensitivity to dopamine. Patients suffering from this syndrome make spontaneous *involuntary* movements, which, as expected, do not generate a readiness potential. On the other hand, such a potential does appear before these patients make an intentional movement.

Libet had speculated that, during the time interval that exists between the readiness potential that characterises the unconscious decision to act and the conscious decision to do so, it is possible to veto or at least modulate the unconscious decision. However, to the author's knowledge, there is *no* experimental evidence that this is indeed possible. Such modulation, especially in the form of inhibitory control, presumably takes place with respect to *future* actions. Indeed, a strong conscious experience of a person's own volition is relevant for the *next* time this person is in a similar situation. For example, the conscious intention to hit someone is not the cause of delivering that blow, but it certainly provides a vivid marker of how it felt and what the consequences of this action were. This affects future decisions.

Perhaps the main reason that Libet's findings created such interest is that the *experience* of performing a specific action gives rise to the feeling of *ownership of this action*. It is certainly reassuring for us to think that our actions are the result of our own free will. This feeling of ownership clearly distinguishes acts that are *made by us* as opposed to other events happening in the environment. Undoubtedly, it gives people great satisfaction to feel that their own actions yield tangible achievements. The experience of 'conscious will' associated with such achievements is often referred to as *perceived control* (Skinner, 1995) and apparently contributes towards psychological health (Haidt and Rodin, 1999). The prevailing legal and religious free will notions assume that the experience of 'conscious will' reflects a causal relation between a person's thoughts and actions. The work of Libet shows that *this causal relation is indeed valid if and only if the unconscious process that precedes the conscious awareness of a thought is included within the notion of 'thought', or, in other words, if and only if unconscious and conscious processes are treated as a continuum.*

Incidentally, the deep understanding of the notion of free will gained as a result of the proper analysis of the underlying biological mechanisms, emphasises the danger of using mathematical notions beyond their realm of applicability. The Russian mathematician Pavel Nekrasov erroneously attempted to prove free will using probability theory (Gamwell, 2015).

References

Feinstein, B., Alberts, W. W., Wright Jr., E. W., and Levin, G. (1960). A stereotoxic technique in man allowing multiple spatial and temporal. *Journal of Neurosurgery, 117*, 708–720.

Fokas, A. S. (2024). *Ways of Comprehending: The Grand Illusion and the Essence of Being Human.* Singapore: World Scientific.

Fried, I., Mukamel, R., and Kreiman, G. (2011). Internally generated preactivation of single neurons in human medial frontal cortex predicts volition. *Neuron, 69*(3), 548–562.

Gamwell, L. (2015). *Mathematics and Art: A Cultural History.* Princeton, New Jersey: Princeton University Press.

Haidt, J. and Rodin, J. (1999). Control and efficacy as interdisciplinary bridges. *Review of General Psychology, 3*(4), 317–337.

Jensen, A. R. (1979). *Bias in Mental Testing.* New York: Free Press.

Kornhuber, H. H. and Deecke, L. (1965). Hirnpotential ändrungen bei Willkürbewegungen und passiven Bewegungen des Menschen:. *Pflügers Archiv, 284*, 1–17.

Mlodinow, L. (2014). *Subliminal.* New York: Penguin.

Popper, K. (1992). *Unended Quest.* Oxfordshire, England: Routledge.

Skinner, B. F. (1995). *Perceived Control, Motivation, and Coping.* Thousand Oaks, CA: Sage.

Soon, C. S., Brass, M., Heinze, H. J., and Haynes, J. D. (2008). Unconscious determinants of free decisions in the human brain. *Nature Neuroscience, 11*(5), 543–545.

Swaab, D. F. and Hedley-Prole, J. (2014). *We Are Our Brains: A Neurobiography of the Brain, from the Womb to Alzheimer's.* New York: Spiegel & Grau.

Chapter 19

The Impact on Health of Largely Unknown Unconscious Processes

19.1 Placebo and Nocebo Effects

Placebo has a Latin origin, meaning '*I shall please*', which reveals its initial use. Namely, in the past, physicians often gave to anxious individuals substances which they knew were ineffective, simply in order to 'please' them. The effect of nocebo is the opposite of that of placebo. Nocebo and placebo are generated as a result of *expectations* that are based on information regarding a specific treatment, previous encounter with a drug or medical procedure, and the overall therapeutic milieu.

Enhancing the placebo and minimising the nocebo effects could have far-reaching health repercussions. The potential impact of deciphering the mechanisms related to placebo becomes evident by noting that, in many double-blind clinical trials evaluating specific treatments for psychiatric disorders, the benefit of the placebo effect is as important as the response of the active treatment (Weimer *et al.*, 2015). This is also true for the treatment of certain types of pain (Colloca and Barsky, 2020). Remarkably, as many as one quarter of patients receiving a placebo in clinical trials discontinue it because of perceived side effects (Rief *et al.*, 2006)!

Expectations affecting the course of a disease can be acquired in a number of different ways, including the following.

- *Prior experience of the effects of medication.* For example, a liquid with a characteristic taste and no biological effect that is given with a

295

drug that has prominent side effects can elicit those side effects when this liquid is given with a placebo (Rheker *et al.*, 2017).

- *Verbal instructions.* For example, in a study involving patients taking finasteride for benign prostatic hypertrophy, the incidence of erectile dysfunction was 43% among those men who were specifically told about potential sexual side effects, as opposed to 15% among those who were not informed about such side effects (Mondaini *et al.*, 2007). The corresponding numbers for sexual side effects in a study involving the beta blocker atenolol for cardiac disease were 31% and 16%, respectively (Silvestri *et al.*, 2003). In another study, falsely labelling a placebo as the serotonin receptor agonist rizatriptan reduced significantly its efficacy against migraine attacks (Kam-Hansen *et al.*, 2014). Half of the patients suffering with asthma who inhaled nebulised saline but were told that it was an allergen had dyspnoea and objective findings of an asthma attack (Luparello *et al.*, 1968).
- *Learning about the experiences of others.* For example, coverage in the media of side effects of statins led to a dramatic increase in reported adverse effects (Colloca and Barsky, 2020).

The course of a disease can be affected by many factors, including spontaneous fluctuations and a variety of psychological effects. Prominent among these effects are *expectations* regarding particular treatments. This implies that the only rigorous way to assess the effectiveness of a specific treatment is to use a control group, which is indeed the basis of the double-blind studies used nowadays.

What are the biological mechanisms responsible for the placebo and nocebo effects? It has been shown by many investigators that placebo effects are associated with the release of a variety of substances, including the following: endogenous opioids (Eippert *et al.*, 2009), dopamine (Lidstone *et al.*, 2010), oxytocin (Kessner *et al.*, 2013), vasopressin (Colloca *et al.*, 2016), and endocannabinoids (Benedetti *et al.*, 2011). The effect of each of these substances is specific to the target system and illness. For example, as will be discussed in the following, dopamine release plays a role in the placebo effect in patients treated for Parkinson's disease, but its effect on pain is only indirect, depending on the efficacy of the reward system (Zunhammer *et al.*, 2018). The nocebo effect is facilitated by a specific neurobiological network called the 'neuropeptide cholecystokinin system', which is of crucial importance for the transmission of pain.

The first study to elucidate the biological mechanism of placebo in analgesia was carried out in 1978 (Levine *et al.*, 1978). In this pioneering study, strong evidence was presented that the placebo effect of analgesia is due to the release of endogenous opioids. Specifically, it was shown that the administration of naloxone, which blocks the opioid receptors, eliminated the placebo effect. The final proof of the crucial role of endogenous opioids in the placebo effect in analgesia was achieved via the use of imaging techniques. Specifically, it was shown that following the administration of a placebo, there was activation of μ-opioid neurotransmission in the dorsolateral prefrontal cortex, the insula cingulate cortex, and the nucleus accumbens (Zubieta *et al.*, 2005).

Based on the suggestion that cholecystokinin facilitates the nocebo effect, it was hypothesised by a leading researcher in the placebo effect, the neuroscientist Fabrizio Benedetti, that the cholecystokinin antagonist proglumide can potentiate the placebo effect. This is indeed the case (Benedetti, 1996). Related to this interesting mechanism, it was later established that the exacerbation of experimentally induced pain via verbal suggestion is mediated by cholecystokinin, and this effect can be blocked by proglumide (Benedetti *et al.*, 2006). Incidentally, nocebo-induced hyperalgesia is blocked by the benzodiazepine diazepam, which suggests that the nocebo effect is exacerbated by anxiety (Benedetti, 2014).

After the administration of a placebo to patients suffering from Parkinson's disease, half of the patients had objective motor improvement. Recent PET studies have provided an explanation for this dramatic effect. The placebo led to dopamine release in the *striatum*, resulting in an increase by 200% of the extracellular dopamine concentration. After looking at the relevant images, a question arises as to why only half of the patients show improvement and not everyone. The answer is that in only half of the patients, there was a large amount of dopamine in the dorsal *motor striatum*, which is an area of the brain involved in movement (De la Fuente-Fernández *et al.*, 2001).

Interestingly, the placebo effect is crucially affected by the *efficiency* of the 'reward system', as well as by activation of the prefrontal cortex. For example, following a monetary reward, the larger the activation of the nucleus accumbens, which is a critical part of the reward system, the larger the response to a placebo (Scott *et al.*, 2007). The role of the prefrontal cortex is eloquently stated in Benedetti's book *Placebo Effects* (Benedetti, 2014): 'no prefrontal cortex no placebo'. Actually, by employing diffusion tensor MRI, which, as stated in Chapter 15 of Fokas (2024), is an imaging technique capable for assessing connectivity, it has been shown that the more impaired the functional connectivity involving the prefrontal cortex, the smaller the placebo effect (Stein *et al.*, 2012).

19.2 The Need for a Global as Opposed to a Local Approach to Patients

The psychiatrist Alastair Santhouse discusses the case of a 33-year-old unmarried man who had come to the UK from Gabon three years earlier and had found life in the UK difficult. His symptoms began with a cough and sore throat. He claimed that he continued to have problems with his breathing, which were getting worse and were affecting his ability to speak. An ear, nose, and throat (ENT) specialist diagnosed a problem with his vocal cords, a neurologist considered a rare neurological disorder, and Santhouse thought that this was a case of a 'functional disorder'. This terminology expresses the rather vague statement that an illness is caused by the dysfunction of an organ or system as opposed to a specific anatomical malfunction. The patient's cousin suggested that he was cursed and urged him to see a priest. After the administration of holy water by a priest who agreed that the patient was indeed cursed, a miraculous cure took place: all symptoms disappeared (Santhouse, 2021).

The way that individuals deal with health issues depends on cultural, religious, and political beliefs, as well as specific features of their personalities. For example, a study in New York concerning the way that patients deal with pain revealed that Italian Americans were concerned primarily with the pain itself, whereas Jewish patients focused on the cause of the pain and its impact on their health and future (Zborowski, 1952). Religious persons may consider an illness as a form of punishment. People believing in conspiracy theories and the diabolical role of 'big pharma' do not trust medical advice and often seek alternative therapies. Optimists tend to minimise the severity of their symptoms and expect quick improvement. This implies that there are myriad individual factors that affect a person's *illness*, which can be defined as *the collection of the subjective symptoms experienced by the patient*. Hence, in order to identify the *disease corresponding to the patient's symptomatology*, the physician must carefully analyse the personality of the patient (Sharpe and Greco, 2019). Paraphrasing the legendary physician Sir William Osler (1849–1919), the important question is 'not what disease the person has, but rather what person has the given illness'.

Physicians often order an exhaustive series of procedures in order to make sure that they do not miss a treatable cause of the given illness. Perhaps by focusing on the patient as a whole as opposed to their symptoms, the diagnosis will become apparent, thereby eliminating the need for further studies. In other words, following a *global* as opposed to a *local* approach should be most beneficial. Santhouse discusses the case of a young man who had been examined by a cardiologist, a rheumatologist, a neurologist, a gastroenterologist, and an ENT specialist in an attempt to diagnose the cause of his episodes of dizziness. Over a period of 18 months, the patient endured extensive blood tests, X-rays, and brain CT. Finally, it was realised that he was suffering from anxiety, which often causes hyperventilation (overbreathing) and hence dizziness (Santhouse, 2021).

A study carried out in a primary care in the US showed that among 550 new complaints of common symptoms, such as fatigue, dizziness, headache, back pain, chest pain, and numbness, a physical cause was found in only 16% of the cases (Kroenke and Mangelsdorff, 1989). Even in specialised clinics, the cause of an illness often cannot be identified. For example, only 38% of the cases with neurological symptoms could be explained by a neurologist and only 45% of cardiac symptoms by a cardiologist (Nimnuan *et al.*, 2001).

In chronic illnesses, such as the symptomatology associated with 'myalgic encephalomyelitis' (earlier known as 'chronic fatigue syndrome'), patients, after suffering for many years, often attempt to find relief in enzyme and coenzyme supplements, herbs, fish oils, a variety of stringent diets, etc. Unfortunately, many of these 'treatments' are, at best, expensive placebos and, at worst, harmful for the patient. Perhaps a better approach is to attempt to restore as much as possible a healthy way of living. Chronic illnesses are exacerbated by a range of physical and psychological factors, which include the following: poor sleep, physical inactivity, the stress associated with the relentless preoccupation with the symptoms of the given illness, and low mood or depression. Adhering to a regular sleep schedule, devising appropriate energy management/pacing approaches, developing distraction techniques to reduce the symptom-focusing preoccupation, and introducing appropriate antidepressant interventions are undoubtedly more beneficial than the unproven use of supplements. In such interventions, it is imperative that, in addition to the physician, specialists, including physiotherapists, occupational therapists, and dietitians, be involved. Importantly, each disorder should be managed differently. For example, post-exercise muscle biopsies in myalgic encephalomyelitis /post-COVID-19 patients showed severe muscle fibre destruction after just 15 minutes of exercise. Hence, for such patients, the earlier recommendation of an 'exercise program involving an incremental increase in daily activities' is inappropriate. Finally, efforts should be made to have the patients participate in local community activities.

The philosopher Ian Hacking has introduced the concept of the 'looping effect', which is associated with labelling the illness of a patient. After assigning to a given patient a specific syndrome, the patient tends, unconsciously, to adopt as many features of this syndrome as possible. So, labelling a given illness 'changes (i.e. has a significant effect on) the patient'. Also, since the patient may express new symptoms, the patient 'changes (enriches) the features of the named syndrome'. In this regard, I am concerned with the current tendency of excessive use of labelling and the possible resulting harm of medicalisation. For example, children with mild manifestations of Attention Deficit Hyperactive Disorder (ADHD) may benefit from the extra help. Perhaps, in such cases, sufficient support should be provided without assigning an explicit medical diagnosis since the labelling may lead to a behaviour consistent with the strongest form of ADHD. Unfortunately, currently, it is not possible to allow affected children to be offered extra help, such as more time during examinations, without first assigning to them the diagnosis of ADHD.

In the UK, during the period 1998–2018, there was almost an 800% increase in autism diagnoses. The number of women diagnosed with ADHD has also increased dramatically. Several authors have expressed concern about the tendency for overdiagnosis and medicalisation, claiming that this, apparently, has led to increased unhappiness instead of happiness; see, for example, O' Sullivan (2025), Shrier (2024), and Timini (2025).

19.3 A Variety of Mysterious Medical Syndromes

The onslaught of information on our brain is increasing. This makes it harder for the brain to assess, reassess, draw inferences, comprehend, and choose a proper course of action. Often, consciousness, in its relentless attempts to come up with a logical and complete understanding of a given situation, makes errors. This may lead to conflicts, disappointments, pain, and, ultimately, pathological behaviour. Moreover, unconscious processes often fail to cope with a plethora of contradictory factors in a given social-cultural-political environment. This failure is occasionally expressed by the emergence of 'mass hysteria' and 'mysterious illnesses'. In many such situations, it is relatively simple to identify the trigger. However, for a deeper understanding of such illnesses, the 'global' approach discussed in an earlier section, particularly the analysis of the social-cultural-political environment, becomes indispensable.

Sigmund Freud and the physician Josef Breuer, in their 1895 publication *Studies in Hysteria*, claimed that a variety of neurological symptoms, such as seizures, paralysis, and various disabilities, had their origin in suppressed psychological trauma of sexual nature that occurred in childhood. Their conclusion was that the 'hysteric suffers mainly from reminiscence', and they advocated the treatment of hypnosis.[1] Following these studies, psychosomatic manifestations are often called 'hysteria'. Nowadays, the names 'conversion' or 'functional' disorders are more prevalent.

There are many well-documented cases of what is often called 'mass hysteria'. The laughter epidemic of Tanganyika began in Kashasha in present-day Tanzania at a mission-run boarding school for girls. It started when three pupils began laughing, and it quickly spread throughout the school, affecting 95 of the 159 girls, aged 12–18. The first phase of the outbreak lasted 48 days, forcing the school to close. After it reopened about two months later, a second outbreak occurred, involving an additional 57 girls. The epidemic spread to another village, where several of the girls lived. Additional villages up to a 100-mile radius were affected, involving more than a thousand people. The epidemic disappeared in about 18 months.

In 2015, a student at a Remembrance ceremony in Ripon, North Yorkshire, UK, collapsed in an overheated assembly hall; this was quickly followed by the collapse of an additional 40 children, all of whom recovered by the next day.

In 2019, in Malaysia, a schoolgirl started screaming, which was almost immediately followed by screams from other girls in the same class, which then 'contaminated' neighbouring classes.

---

[1] Later, Freud abandoned hypnosis and advocated 'free associations'. It is worth noting that Freud treated the famous patient Anna O for six years with morphine injections, which did not alleviate her symptoms.

Unfortunately, mass hysteria can have long-lasting and occasionally irreversible effects. In 1518, a dance hysteria appeared in Strasbourg. A woman began dancing in the street, apparently unable to stop until she collapsed from exhaustion. After resting, she continued this compulsive behaviour for days. Soon, 30 more people joined her. The authorities thought that the problem would subside if the dancers were moved into proper dancing halls accompanied by music, but this exacerbated the problem. Hundreds of residents were driven to dancing beyond their endurance, with an estimated mortality rate of 15% due to cardiac manifestations or sheer exhaustion. The epidemic disappeared after a few months as mysteriously as it had appeared.

In 1774, J. W. von Goethe published *The Sorrows of the Young Werther*. It has been assumed that this highly successful novel led to an epidemic of 'imitation suicides'. It is argued in Thorson and Öberg (2003) that, although there were indeed some suicides apparently associated with this book, no epidemic occurred. However, there is rigorous documentation that the 'Werther effect' is significant; namely, suicides can increase as a result of imitation. In particular, it is shown in Kim *et al.* (2023) that the suicide rates in South Korea increased by 1.22, 1.30, and 1.28 after the suicides of three Korean idol singers, which occurred in the period 2017–2018.

The negative effects of the media

Katie was a senior at a high school in the small town of Le Roy, New York State, which is located 70 miles east of Niagara Falls. Katie was a straight-A student and a cheerleader, with many friends. In October 2011, after she woke up from a nap, she realised that she had developed certain involuntary movements and verbal outbursts, similar to those seen in the well-known disease Tourette's syndrome. In particular, her jaw would go into spasms and her face twisted; she twitched and often produced involuntary shouts. After a couple of weeks, her close friend and fellow cheerleader, Thera, developed almost identical symptoms. Soon afterwards, the illness spread, first to the girl's close friends and then to other girls in the same school.

Apparently, the looping effect, discussed in the previous section, is also valid regarding the manifestation of mysterious illnesses: the new individuals 'contaminated' by the given epidemic present new symptomatology. Ten more girls were affected, presenting symptomatology, which varied from Tourette-type symptoms to convulsions and absence seizures; some could not even walk. Eleven of the twelve girls affected by this mysterious illness were investigated by the same neurological team in Buffalo. After excluding environmental toxins and infectious agents, and after a series of tests, all of which were normal, a diagnosis of a functional neurological disorder was made, under the name of 'conversion disorder'. The situation soon stabilised. However, in January 2012, the Department of Health held a meeting in Le Roy, where it announced to a wider audience their findings. Until then, only the affected girls and their families knew details of the illness. Soon thereafter, Katie, Thera, and their mothers appeared on *NBC's Today Show*. A media frenzy followed. Many 'experts' began challenging the official diagnosis, offering all sorts of absurd theories. For example, it was claimed that a Jell-O factory, in addition to bringing wealth to Le Roy, also contaminated the water. It was claimed that the creek that ran through the town used to change colour according to the flavour manufactured in the Jell-O factory on a given day. Others claimed that the girls were suffering from PANDAS, which, as stated in Chapter 11, is a rare autoimmune disease believed to be triggered by a streptococcal infection. The celebrity investigator Erin Brockovich, whose 1993 landmark case against Pacific Gas and Electric in California was popularised in a 2000 Hollywood movie, appeared in early 2012 on *ABC News*. In an interview with celebrity doctor Drew Pinsky, they both questioned the medical diagnosis. Brockovich sent a team to investigate and raised the possibility that a train crash that took place in 1970, 3.5 miles east of the school, spilled cyanide, which due to heavy rains in 1999, caused a plume of toxin to contaminate the creek running through Le Roy. Six months after the start of this investigation, representatives of the team of Brockovich organised a town meeting in Le Roy, where they announced that the train crash was not the cause of the illnesses. It was confirmed that the creek and the underground water system flowed directly away from the train-crash site, in the opposite direction to the school's location! Incidentally, Brockovich did not appear in the town meeting and never accepted that her speculation was wrong. Instead, she continued to doubt the medical conclusion, claiming that she would continue to investigate (which apparently, she did not). Finally, the media stopped reporting on these events, and this was followed by the steady improvements of the 12 girls initially affected by this illness and of the seven additional girls who got 'contaminated' following the extensive media exposure.

304 The Embodied Mind

The detrimental impact of fear and animosity

In December 2014, President Barack Obama and Raul Castro announced that relations between the US and Cuba were improving. A year later, they reopened embassies in the respective countries. US diplomatic staff moved to Havana in 2015. In November 2016, Donald Trump was elected president of the US, which raised serious questions regarding the process of defrosting of the relations between the US and Cuba.

An atmosphere of uncertainty, lack of trust, and hostility naturally generates anxiety. A month after President Trump's election, several American diplomats presented with an illness involving similar symptomatology, which included the following: earache, tinnitus, hearing impairment, headache, visual disturbances, memory problems, difficulty concentrating, and fatigue. Interestingly, nearly all patients reported hearing an unusual noise before the appearance of their symptoms. This noise was described in different forms, such as a loud ringing or a high-pitched noise. One patient described the noise as following them around their home. Another said that they woke in the middle of the night with a loud ringing in their ears. In September 2017, BBC reported that 'US reveals details of recent sonic attack on Cuba diplomats', and the following January, ABC told its viewers that 'US […] open the door of a viral or an ultrasound cause'. Senator Marco Rubio asserted that 'an attack was a given', and President Trump added that 'some very bad things are happening in Cuba'.

It is remarkable that such claims were made, ignoring the following facts: no sonic weapon of the type required for such an attack is known to exist; the sound described by the patients is not associated with a brain injury that would justify the associated symptoms; in general, sound in the audible range is not known to cause persistent injury in the central nervous system; and sound loud enough to damage the ears would be heard within a large perimeter, but no one else heard such sounds except for the patients. Despite the above facts, the US government, convinced that these symptoms were the result of an attack, removed half of its staff from the Havana Embassy, leaving a highly reduced staff.

According to media reports, the first patient who developed symptoms after hearing a strange noise was trained and sensitised to covert operations. It is not important if this first patient actually heard a specific noise or if the noise was an illusion; what is important is that either the first patient or someone among the individuals that the first patient discussed their illness with, suggested a possible attack. Then, unconscious mechanisms, crucially affected by the existing political environment, gave rise to the so-called 'Havana syndrome'. In this sense, it is not surprising that diplomatic staff in China were also 'contaminated'.

In my opinion, the relevant medical publications in the respected medical journal *JAMA* were not very helpful. The first publication (Swanson *et al.*, 2018), although it reported that 'all brain scans were normal', it concluded that 'these individuals appear to have sustained injury to widespread brain networks without an associated history of head trauma'. But what was the evidence of the 'injury'? It is well known that, in contrast to disorders which are due to a specific anatomical malfunction, in 'functional disorders', even if a dysfunction of a specific subnetwork can be established, no 'injury to widespread brain networks' can be identified. In a subsequent article (Veram *et al.*, 2019), more detailed neuroimaging studies were performed. It was claimed that, in comparison to normal controls, these patients exhibited, among others, significant differences in whole-brain white matter, as well as abnormalities regarding functional connectivity in the auditory and visuospatial subnetworks. Although it was noted that the clinical importance of these findings was not clear, further studies were advocated. I consider it unfortunate that the title of the second article alludes to the possibility that this syndrome was indeed due to an actual attack. Also, in addition to the non-specificity of the neuroimaging findings, it should have been noted that diplomats have specific characteristics, which include engaging in long-haul travel and perhaps consuming more alcohol. Thus, the affected patients should have been compared with non-affected diplomats instead of a generic control group.

The remarkable ability to resist

Between 2003 and 2005, 424 children in several Swedish towns went to bed and did not get up the next morning. In a span of 20 years, hundreds of children of ages 9–16 were affected. Some were ill for a few months, but many did not get up for years. Typically, the illness had an insidious onset: a child would stop playing with other children, would withdraw to themselves, and finally would stop going to school; they would speak less and less and finally would stop speaking; eventually, they would refuse to get out of the bed, would not open their eyes, would be immobile, and would not respond to pain. All blood tests and the results of the analysis of the spinal fluid were normal; the EEG showed the cycles of wake and sleep expected in normal children; there was a slightly elevated pulse; upon physical examination, there were normal reflexes, but the children would resist the attempt of a physician to open their eyes, and also they clenched their teeth. Originally, this illness was referred to as the 'sleeping syndrome' but was later given the better name 'resignation syndrome'. This illness affected exclusively the children of asylum seekers. It is important to note that the parents of the affected children do not read Swedish, so the immigration papers are read by the children, who translate them to their parents. Apparently, after their parent's application for asylum is rejected, the resignation syndrome expresses the children's powerful response to resist deportation.

In 2018, several cases consistent with the resignation syndrome were reported in an island, Nauru, in the Pacific, which was used as a holding location for refugees seeking asylum in Australia.

Perhaps, the powerful mechanism of resisting leaving one's home is also of crucial importance for attempting to explaining another mystery illness, which, between 2010 and 2015, affected 130 individuals, almost all of whom were residents of the small town of Krasnogorsk in Kazakhstan. The symptoms of this mystery illness and of the resignation syndrome have many similarities. During Soviet times, Krasnogorsk was a thriving town due to a uranium-mining facility; it had several thousand residents. When the Soviet Union collapsed, the mine stayed open for a while, but eventually it closed down. This was a devastating blow to the town, whose population, by the time that the epidemic of the 'sleeping syndrome' appeared, was reduced to 300. Although the World Health Organization has reported that high levels of radon are often detectable in uranium-mining areas, all tests for the patients were within normal range. After the closing of the mines, the Kazakhstan government attempted to relocate the residents to other towns; perhaps this decision led to the 'sleeping syndrome'.

Additional cases and the need for an explanation

Several other cases of mystery illnesses are described in the book *The Sleeping Beauties* (O' Sullivan, 2021). They include several epidemics of the illness, *grisi siknis*, first described by the anthropologist Philip Dennis, which appeared in an indigenous tribe in Nicaragua in 2003, 2009, and 2019. Symptoms included tremors, difficulty breathing, trance-like states, and convulsions. A prevalent visual hallucination is the appearance of a frightening dark stranger wearing a hat, who carries the patient away, intending to have sex with them. School children are especially vulnerable. Typically, an epidemic starts with one child, and then it can 'contaminate' the entire school. Traditional ways of healing can be very effective.

Another interesting 'epidemic' occurred in El Carmen, Columbia. In 2014, on a hot day, a girl in a school complained that she was unable to breathe, becoming increasingly distressed until, apparently, she fainted. Almost immediately 15 more girls collapsed, some convulsing. The resulting pandemonium alerted girls in other classes. By the end of the first day, approximately 50 girls had collapsed, most of them convulsing; by the end of the first week, at least 100 girls were sick. Consistent with Hacking's looping effect, the illness evolved, presenting with a wide range of symptomatology, from visual blurring, headache, and breathing difficulties, to chest pain and hair loss. Until 2019, approximately 1,000 girls, among a population of 120,000, had been ill. Most recovered, but not all. Many of the girls involved with the beginning of this 'epidemic' believed that this was caused by the Human Papillomavirus Vaccine (HPV), despite the fact that the second dose was given a month earlier.

There is no doubt that acute as well as chronic stress can have a detrimental impact on our health. In some cases, this impact can be recognised immediately. Such a case is 'takotsubo cardiomyopathy', first reported in Japan in 1990. It is named after the shape of a pot used to trap octopuses. Apparently, following a sudden surge of stress hormones, especially adrenalin, the myocardium (heart muscle) weakens, and the heart takes the shape of a *takotsubo*. Among reported triggers are the following: the anxiety of a public lecture, violent arguments, intense fear, sudden emotional or physical shock including a surprise party, and the chronic stress associated with bereavement or financial losses. It is more prevalent in women, especially after menopause. Noting that a given physiological quantity, such as the excreted adrenalin, varies within a wide range of values, and taking into consideration the fact that there exists a plethora of hormones affecting our organism, it is expected that the overall impact of hormonal variations can be expressed in many different forms, from takotsubo cardiomyopathy to very subtle manifestations. The variety of symptoms associated with the 'mysterious illnesses' described earlier not only illustrates the highly complex interactions between a given cultural-social-political environment and important physiological processes, but they also reveal our current ignorance regarding the mechanisms that regulate these interactions.

References

Benedetti, F. (1996). The opposite effects of the opioid antagonist naloxone and the cholecystokinin receptor antagonist proglumide on placebo analgesia. *Pain, 64*, 535–543.

Benedetti, F. (2014). *Placebo Effects.* Oxford: Oxford University Press.

Benedetti, F., Amanzio, M., Rosato, R., and Blanchard, C. (2011). Nonopioid placebo analgesia is mediated by CB1 cannabinoid receptors. *Nature Medicine, 17*(10), 1228.

Benedetti, F., Amanzio, M., Vighetti, S., and Asteggiano, G. (2006). The biochemical and neuroendocrine bases of the hyperalgesic nocebo effect. *Journal of Neuroscience, 26*(46), 12014–12022.

Colloca, L. and Barsky, A. (2020). Placebo and nocebo effects. *New England Journal of Medicine, 382*(6), 554–561.

Colloca, L., Pine, D., Ernst, M., Miller, F., and Grillon, C. (2016). Vasopressin boosts placebo analgesic effects in women: A randomized trial. *Biological Psychiatry, 79*(10), 794–802.

De la Fuente-Fernández, R., Ruth, T., Sossi, V., Schulzer, M., Calne, D., and Stoessl, A. (2001). Expectation and dopamine release: Mechanism of the placebo effect in Parkinson's disease. *Science, 293*(5532), 1164–1166.

Eippert, F., Bingel, U., Schoell, E., Yacubian, J., Klinger, R., Lorenz, J., and Büchel, C. (2009). Activation of the opioidergic descending pain control system underlies placebo analgesia. *Neuron, 63*(4), 533–543.

Fokas, A. (2024). *Ways of Comprehending: The Grand Illusion and the Essence of Being Human.* Singapore: World Scientific.

Kam-Hansen, S., Jakubowski, M., Kelley, J., Kirsch, I., Hoaglin, D., Kaptchuk, T., and Burstein, R. (2014). Altered placebo and drug labeling changes the outcome of episodic migraine attacks. *Science Translational Medicine, 6*(218), 218ra5–218ra5.

Kessner, S., Sprenger, C., Wrobel, N., Wiech, K., and Bingel, U. (2013). Effect of oxytocin on placebo analgesia: A randomized study. *JAMA, 310*(16), 1733–1735.

Kim, L. H., Lee, G. M., Lee, W. R., and Yoo, K. B. (2023). The Werther effect following the suicides of three Korean celebrities (2017–2018): An ecological time-series study. *BMC Public Health, 23*(1), 1173.

Kroenke, K. and Mangelsdorff, A. D. (1989). Common symptoms in ambulatory care: Incidence, evaluation, therapy, and outcome. *The American Journal of Medicine, 86*(3), 262–266.

Levine, J., Gordon, N., and Fields, H. (1978). The mechanism of placebo analgesia. *Lancet, 312*(8091), 654–657.

Lidstone, S., Schulzer, M., Dinelle, K., Mak, E., Sossi, V., Ruth, T., . . ., and Stoessl, A. (2010). Effects of expectation on placebo-induced dopamine release in Parkinson disease. *Archives of General Psychiatry, 67*(8), 857–865.

Luparello, T., Lyons, H., Bleecker, E., and McFadden Jr., E. (1968). Influences of suggestion on airway reactivity in asthmatic subjects. *Biopsychosocial Science and Medicine, 30*(6), 819–825.

Mondaini, N., Gontero, P., Giubilei, G., Lombardi, G., Cai, T., Gavazzi, A., and Bartoletti, R. (2007). Finasteride 5 mg and sexual side effects: How many of these are related to a nocebo phenomenon? *The Journal of Sexual Medicine, 4*(6), 1708–1712.

Nimnuan, C., Hotopf, M., and Wessely, S. (2001). Medically unexplained symptoms: An epidemiological study in seven specialities. *Journal of Psychosomatic Research, 51*(1), 361–367.

O' Sullivan, S. (2021). *The Sleeping Beauties: And Other Stories of Mystery Illness.* New York: Picador.

O' Sullivan, S. (2025). *The Age of Diagnosis.* UK: Hodder Press.

Rheker, J., Winkler, A., Doering, B., and Rief, W. (2017). Learning to experience side effects after antidepressant intake–Results from a randomized, controlled, double-blind study. *Psychopharmacology, 234*(3), 329–338.

Rief, W., Avorn, J., and Barsky, A. (2006). Medication-attributed adverse effects in placebo groups: Implications for assessment of adverse effects. *Archives of Internal Medicine, 166*(2), 155–160.

Santhouse, A. (2021). *Head First: How the Mind Heals the Body.* New York: Avery.

Scott, D., Stohler, C., Egnatuk, C., Wang, H., Koeppe, R., and Zubieta, J. (2007). Individual differences in reward responding explain placebo-induced expectations and effects. *Neuron, 55*(2), 325–336.

Sharpe, M. and Greco, M. (2019). Chronic fatigue syndrome and an illness-focused approach to care: Controversy, morality and paradox. *Medical Humanities, 45*(2), 183–187.

Shrier, A. (2024). *Bad Therapy: Why the Kids Aren't Growing Up.* New York: Sentinel.

Silvestri, A., Galetta, P., Cerquetani, E., Marazzi, G., Patrizi, R., Fini, M., and Rosano, G. (2003). Report of erectile dysfunction after therapy with beta-blockers is related to patient knowledge of side effects and is reversed by placebo. *European Heart Journal, 24*(21), 1928–1932.

Stein, N., Sprenger, C., Scholz, J., Wiech, K., and Bingel, U. (2012). White matter integrity of the descending pain modulatory system is associated with inter-individual differences in placebo analgesia. *Pain, 153*(11), 2210–2217.

Swanson, R. L., Hampton, S., Green-McKenzie, J., Diaz-Arrastia, R., Grady, M. S., Verma, R., and Smith, D. H. (2018). Neurological manifestations among US government personnel reporting directional audible and sensory phenomena in Havana, Cuba. *JAMA, 319*(11), 1125–1133.

Thorson, J. and Öberg, P.-A. (2003). Was there a suicide epidemic after Goethe's Werther? *Archives of Suicide Research, 7*(1), 69–72.

Timini, S. (2025). *Searching for Normal: A New Approach to Understanding Mental Health, Distress and Neurodiversity.* HB: Fern Press.

Veram, R., Swanson, R. L., Parker, D., Ismail, A. A., Shinohara, R. T., Alappatt, J. A., and Smith, D. H. (2019). Neuroimaging findings in US government personnel with possible exposure to directional phenomena in Havana, Cuba. *JAMA, 322*(4), 337–347.

Weimer, K., Colloca, L., and Enck, P. (2015). Placebo effects in psychiatry: Mediators and moderators. *The Lancet Psychiatry, 2*(3), 246–257.

Zborowski, M. (1952). Cultural components in responses to pain. *Journal of Social Issues, 8*(4), 16–30.

Zubieta, J., Bueller, J., Jackson, L., Scott, D., Xu, Y., Koeppe, R., . . ., and Stohler, C. (2005). Placebo effects mediated by endogenous opioid activity on μ-opioid receptors. *Journal of Neuroscience, 25*(34), 7754–7762.

Zunhammer, M., Gerardi, M., and Bingel, U. (2018). The effect of dopamine on conditioned placebo analgesia in healthy individuals: A double-blind randomized trial. *Psychopharmacology, 235*(9), 2587–2595.

Chapter 20

The Crucial Importance
of Sleep

In 1952, William Dement, Nathaniel Kleitman, and Eugene Aserinsky of the University of Chicago observed that there exist two basic phases of sleep. These two phases give rise to different types of EEG recordings, which is why EEG is indispensable for the study of sleep. In the phase called rapid eye movement (REM), EEG recordings indicate that the associated brain activity is remarkably similar to that of the awake state. Furthermore, as suggested by its name, the eyes rapidly dart from side to side underneath the lids. Dreams occur mostly during this phase. For a long time, it was assumed that dreams can occur only in the REM stage. However, it is now known that dreams can also occur outside the REM phase (Siclari *et al.*, 2017). In REM, there occurs an almost complete shutdown of dopamine, serotonin, and histamine, but there is an increase in the production of acetylcholine (Greenfield, 2016). In the other basic sleep phase, called non-rapid eye movement (NREM), the recorded brain waves show global synchronisation, and the eyes are motionless. NREM sleep is divided into light NREM, which is the very first stage of sleep, and deep NREM. The former is further divided into stages 1 and 2, whereas the latter is divided into stages 3 and 4. The deep NREM sleep is also referred to as 'slow-wave' sleep since this brain rhythm is of the delta type, generating an EEG pattern of two to four waves per second.

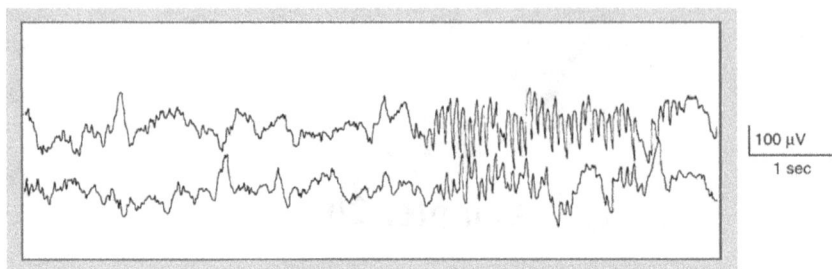

Figure 20.1.   Sleep spindles.

EEG recordings of light NREM sleep occasionally exhibit the so-called 'sleep spindles' (Figure 20.1), namely, a pattern of sudden bursts of very fast oscillations superimposed on the basic slower waves.[1]

Taking into consideration that EEG recordings of awake individuals are similar to those of individuals in REM sleep, how can these two states be distinguished? REM sleep (in contrast to NREM) is characterised by *atonia*, which refers to complete paralysis of all voluntary muscles. Involuntary muscles, which control autonomic operations such as heartbeat and breathing, continue to function properly. This paralysis is necessary in order to prevent individuals from 'acting out' their dreams. During the movement-rich experience of dreams, the brain continuously generates motor commands. However, these commands are not executed because disabling signals have already been sent from the pons to the spinal cord, and these signals prevent the 'fictional movements' from becoming reality. In 1965, the French pioneer of sleep studies, Michel Jouvet, destroyed some of the cells responsible for causing paralysis during sleep in a cat, which then acted out its dream by attacking imaginary prey. The pathological condition in which an individual is able to act out vivid dreams is called 'REM sleep behaviour disorder'. A famous case involving such an act occurred in 2008, when BT and his wife spent a night in their van in a public car park. At night, BT dreamed that a man had broken into his van and was on top of his wife. He dreamed he was grappling with the intruder, but when BT woke up, he discovered that he had strangled his wife. BT was acquitted.

---

[1] This shows that the underlying phenomenon is mathematically characterised by two different 'scales': a slow scale describing the slow waves and a fast scale describing the sleep spindles.

During NREM sleep, the thalamus imposes a sensory blackout in the cortex. On the other hand, during REM sleep, while the thalamus continues to shut out all external sensory input, it does allow various endogenous signals to reach the cortex, including those of emotions, motivations, and goals, as well as past and recent memories.

NREM and REM phases form a 90-minute cycle, which, during a typical night's sleep, repeats itself five times. The lengths of the two sleep phases change as the night progresses. Only less than 20% of the first cycle is spent in the REM phase, but this proportion increases to about 60% by the fifth cycle.

Taking into consideration that REM sleep is accompanied by atonia, it follows that 'somnambulism', i.e. 'sleepwalking', can occur only during NREM sleep. In normal sleep, the brain transitions from the sleep state to the awake state either from REM or from stages 1 and 2 of NREM. However, in somnambulism, the brain attempts to make this transition from stages 3 or 4 of NREM. This, apparently, forces the brain into an unfamiliar stage, resulting in rare cases where somnambulism is associated with committing violent acts. The most famous case of homicidal somnambulism happened in 1987 involving KP, a married 23-year-old Canadian man with a five-month-old daughter, who was suffering from anxiety and sleep deprivation due to gambling problems. This led to embezzlement from his work, for which he was fired. While asleep, he drove 24 km to the house of his in-laws. There, he murdered his mother-in-law and then attempted to strangle his father-in-law, who survived the attack. Subsequent EEG studies showed that his brain was trying to wake from deep NREM sleep 10–20 times a night. On the basis of these studies, KP was acquitted.

After the REM phase is completed, normally the brain releases the body from complete paralysis. However, sometimes this paralysis lingers on after consciousness has returned, causing 'sleep paralysis'. This is the frightening loss of the ability to talk or move upon waking from sleep. Patients suffering from narcolepsy often exhibit episodes of sleep paralysis. Narcolepsy is a sleep disorder which causes a person to fall asleep frequently and unexpectedly. It is associated with excessive daytime sleepiness and disrupted nigh-time sleep, which reduces quality of life. It is characterised by reduced production of the hypothalamic messenger *orexin*, also called *hypocretin*. Patients receiving oveporexton, an oral orexin2-receptor agonist, had significantly improved measures of wakefulness (Dauvillers *et al.*, 2025). Narcolepsy is further discussed in the following chapter in connection with the imaging technique of optogenetics.

The sleep–wake cycle is determined by two basic factors: the 'circadian rhythm' and the neurotransmitter adenosine. The circadian rhythm (*circa* means 'about', and *dia* means 'day') is imposed endogenously. This important fact was established in 1938 by Nathaniel Kleitman and Bruce Richardson, who spent 32 days in Mammoth Cave in Kentucky, one of the deepest caverns on Earth. In this completely dark environment, these researchers established that humans generate their own circadian rhythm in the absence of sunlight. Indeed, while living in the cave, their biological rhythms of sleep and wakefulness did not become chaotic. Instead, these two stages exhibited a predictable pattern of approximately 15 hours of wakefulness paired with approximately 9 hours of sleep. Interestingly, the average duration of adults' endogenous circadian 'clock' is not 24 hours, but 24 hours plus 15 minutes. Remarkably, more than two hundred years earlier, the French geophysicist Jean-Jacques d'Ortous de Mairan established that the plant *Mimosa pudica* generates its own endogenous circadian rhythm. Mairan placed the plant in a sealed box for 24 hours and occasionally observed the plant in its completely dark environment. During the day, the plant behaved as if it were bathed in the sun, and its leaves expanded; as night approached, its leaves began to retract, and finally they closed, remaining so for the duration of the night.

In humans, the circadian rhythm is determined by the 'master biological clock', which consists of the so-called *suprachiasmatic nucleus*. This is a collection of approximately 50,000 neurons located above the crossing point of the two optic nerves. In 2017, three US scientists were awarded the Nobel Prize in Medicine for deciphering how this clock works in the fruit fly *Drosophila*. The relevant mechanism is very complicated; in humans, it involves the activation and de-activation of 20 genes, which results in a 24-hour cycle of protein production and breakdown. In turn, this cycle acts as a signal to switch on and off countless genes, resulting in the circadian production of a plethora of proteins, which regulates rhythmic physiology.

Consistent with the notion of local *versus global processes* emphasised in Fokas (2024) and in this book, it turns out that, in addition to the 'global' clock, there also exist 'local' clocks in every organ and tissue of the body. Apparently, the master clock acts as a pacemaker to coordinate the circadian activity of billions of local clocks, each of which drives the activity of the associated tissue (Foster, 2022). In the same way that the combination of different rhythms used in a musical composition achieves syncopation in music, the local and global clocks characterise the 'syncopation' of life. This is achieved through the combinations of heartbeat, the frequency of breathing, the rhythms of the electrical impulse which determine the actions of the nervous system, the specific rhythms associated with the secretion of each particular hormone, etc.

Our circadian rhythm is also affected by a small number of cells at the back of the eye. As shown by Russell Foster's group in studies performed at Imperial College, the latter cells are not involved in processing visual images, but instead respond to the brightness of the environment (Freedman *et al.*, 1999). The above two sets of cells analyse the light signals sent from each eye and communicate this information to the *pineal* gland, which, during the night, excretes melatonin. In this way, a specific sleep–wake cycle is determined by the associated dark–light cycle, which is the reason for the sleep disturbances caused by jet lag. In other words, for us, like most plants and animals, light is the most important 'entrainment' signal that aligns the 'internal day' with the external world. However, the time duration of wakefulness–sleep is determined endogenously. Simple unicellular organisms, such as bacteria, also have active and passive phases associated with the light–dark cycle of our planet. Perhaps, this provided the evolutionary precursor of the wakefulness-sleep cycle. For most humans, the maximum performance during the circadian rhythm, at least regarding physical activities, is achieved in the mid-afternoon. This coincides with the maximum likelihood of breaking Olympic records (Walker, 2017, p. 12).

The local clocks govern every aspect of our physiological regulation, including daily variations of blood pressure, heart rate, the levels of various hormones, and exercise capacity. For example, the 'dawn phenomenon' occurring in patients with diabetes, where glucose levels are high in the morning, is due to the increased production of glucose in the liver during sleep. The knowledge of the precise effect of the circadian rhythm on specific processes can have useful clinical implications. For example, the activity of a key enzyme (HMG-CoA) in cholesterol synthesis peaks at night; this led to the recommendation that short-term-acting statins should be administered in the evening.

Adenosine is the core ingredient of adenosine triphosphate (ATP), which is the main source of energy for most cellular processes. After being released, ATP can quickly lose its three phosphates, becoming the inhibitory neurotransmitter adenosine. The longer an individual is awake, the more ATP is released, the higher the accumulation of adenosine, and the stronger the desire to sleep. The molecular structure of caffeine is similar to that of adenosine, which allows caffeine to bind to adenosine receptors. Thus, much fewer receptors are available for adenosine, causing inhibition of its action. This is the reason that, in moderate amounts, namely 40–300 mg, caffeine reduces fatigue and enhances wakefulness.[2]

Caffeine blood levels peak after about 0.5–2 hours following oral administration, and its effect lasts for many hours. In addition to caffeinated drinks, decaffeinated drinks should also be avoided in the evening because one decaf cup may contain up to 20% of the dose of a regular cup of coffee. Caffeine is removed via the process of degradation achieved via a specific liver enzyme, called cytochrome P450 1A2. Some individuals have a more efficient version of this enzyme, and hence caffeine can be rapidly cleared from their bloodstream. This explains why some individuals can have coffee at dinner and still fall asleep at night without any difficulty.

Incidentally, drinking three of four standard cups of coffee daily is associated with a reduced risk of several chronic diseases and reduced mortality (Freedman *et al.*, 2012). In particular, there is a strong inverse association between caffeine intake and the risks of Parkinson's disease, endometrial cancer, and hepatocellular carcinoma. There is also a slightly reduced risk for prostate cancer, breast cancer, melanoma, liver fibrosis, and cirrhosis (van Dam *et al.*, 2020). In addition, coffee improves energy balance by reducing appetite and increasing the basal metabolic rate. Coffee contains hundreds of active phytochemicals, which reduce oxidative stress and may also have a positive impact on the gut microbiome. Interestingly, despite the fact that caffeine intake raises blood pressure in the short term, no substantial negative effect was found in trials with caffeine, even among hypertensive individuals (Mesas *et al.*, 2011). Importantly, although caffeine raises epinephrine levels, several clinical studies found that the consumption of up to five cups of coffee per day was not associated with any negative cardiovascular effects, including atrial fibrillation (Ding *et al.*, 2014). Perhaps this can be explained by the fact that, although caffeine increases the pulse rate through the increase of epinephrine, it compensates by dilating both the coronary arteries and the bronchi (the bronchodilator theophylline contains caffeine). On the

---

[2]Caffeine blocks the so-called A_{2A} receptors. Caffeine 'wakes up' neurons via a highly complicated molecular signalling pathway that includes a specific positive feedback loop, which further enhances the effect of caffeine inside the cell (Lindskog *et al.*, 2002).

negative side, coffee can cause anxiety, but this is rarely found below a daily consumption of 400 mg of caffeine (a standard cup of coffee contains 135 mg of caffeine).

What is the required duration of sleep? After a large number of rigorous studies, it has become clear that the brain needs between seven and nine hours of sleep each night (Walker, 2017). For example, it was shown in a relevant study that after 10 days with only seven hours of sleep per night, individuals performed as they would have performed after sleep deprivation of 24 hours. Furthermore, it took more than three good nights' sleep to recover from the sleep deficit. Of course, there exist individual variations. About 1% of the population can function normally with five or six hours of sleep. Most of these individuals possess a sub-variant of the gene BHLHE4I, also known as DEC2.

In order to delineate functional differences between the REM and NREM phases, it is useful to examine their duration across the lifespan. The construction of the complex neural connectivity characterising the human brain is achieved via a highly dynamic procedure that involves both creating new connections and pruning existing ones. The former process is called synaptogenesis. Although these two opposite processes take place throughout the lifespan, their occurrence varies according to age. Synaptogenesis is dominant during foetal development, when a very large number of connections are established. During this period, most of the sleep is of the REM type. In the last week before birth, REM sleep reaches a lifetime maximum of 12 hours daily. By the time the infant becomes six months old, the total of 14 hours of sleep needed is equally divided between the two phases of sleep. A five-year-old needs 11 hours of sleep, of which 70% is NREM type. As the need for pruning increases, so does the amount spent in the NREM phase. By the time children are 10 years old, the ratio of the duration of their NREM and REM sleep is the same as in adults; namely, only 25% is spent in REM.

Not only the duration but also the intensity of NREM sleep changes dramatically during the lifespan. Irwin Feinberg documented that, as children make the transition through adolescence to adulthood, there is a remarkable change in the EEG recordings of their deep NREM sleep. In the middle and late childhood, the associated slow waves are moderately deep, but in early adulthood, when shedding connections becomes dominant, the slow waves become very deep (Campbell *et al.*, 2016). Feinberg also observed that cognitive and developmental milestones are preceded by several weeks or months by notable changes in the recordings of the slow waves. This suggests that deep NREM sleep is the driving force of brain maturation. This has been confirmed by several other studies (Frank *et al.*, 2001).

Feinberg also observed that during adolescence, maturation progresses from the back of the brain, which is involved in visual and spatial perception, towards the front. The last step in this maturation procedure occurs at the tip of the frontal lobe, which characterises the 'adult' functions of critical thinking and decision-making. Several studies have followed neural development in young teenagers using CT. Such studies have associated abnormal pruning, particularly in the frontal part of the brain, with the later occurrence of schizophrenia (Sarkar *et al.*, 2010). This suggests that schizophrenia is associated with abnormalities in neural maturation, and hence it is not surprising that in this mental disorder, there is a two- to three-fold reduction in deep NREM sleep.

In summary, it appears that REM sleep is crucial for establishing new connections, whereas deep NREM sleep is important for the maturation of the brain occurring via the remodelling of connections, which is a procedure aided by neural pruning.

The brain waves of NREM sleep signify highly synchronous brain activity, which expresses a 'global co-operation' among different parts of the brain. A subset of such waves, apparently, reflects the transformation of fragile short-term memories to stable long-term memories. As noted in Fokas (2024), short-term memories are 'labile'. After several hours from the time of their formation, short-term memories become stabilised via complicated processes requiring neurons to synthetise new RNA and proteins. Matthew Walker has shown that the mechanism of stabilising memories is accompanied by specific characteristics of the electrical activity recorded in EEG, expressed in a particular pattern of sleep spindles. Apparently, this reflects the transfer of memories from the hippocampus (short-term deposition) to the cerebral cortex. It is interesting that this process is prevalent in stage 2 of light-NREM sleep of late-morning hours; these periods are sandwiched between long periods of REM sleep (Walker, 2017, p. 110).

20.1 The Detrimental Effects of Sleep Deprivation

Memory

The relation between sleep and memory formation was investigated using fMRI, performed on two groups of subjects during the process of retrieving memories. One group was allowed to sleep after learning verbally a list of facts, whereas the other group was kept awake. In the group that had slept, retrieving memories was associated with activation of the cortex, whereas in the other group, there was increased activation in the hippocampus (Walker, 2017, p. 114). These studies suggest that sleep had transferred the memories to the cortex; hence, retrieving these memories activates this part of the brain instead of the hippocampus. Experiments performed much earlier (Jenkins and Dallenbach, 1924) had already established that the overall performance in memory tasks of the group that had slept was better by 20–40% than that of the group which was kept awake. Several additional studies support a role of sleep in consolidating an array of memory and learning tasks (Stickgold *et al.*, 2001).[3] In addition, sleep can also 'rescue' memories that were lost during the day (Fenn *et al.*, 2003).

The statement that older adults need less sleep is not accurate. However, whereas the quality of REM sleep remains stable as humans age, the quality of deep NREM sleep deteriorates. Indeed, by the fourth decade of life, the deep NREM brainwaves are smaller in amplitude, and their frequency is reduced. This is consistent with the known fact that slow brainwaves originate from the middle of the frontal lobe. Unfortunately, this is precisely the area in the brain that suffers the most dramatic deterioration with ageing. Another important fact affecting sleep patterns in the elderly is that their melatonin is released earlier in the evening and also follows a flat pattern instead of the usual sharp rise in its concentration at night followed by a rapid fall during the day. These changes are associated with the reduced ability of older individuals to learn and to consolidate new memories. The earlier-than-normal release of melatonin can be partially counteracted by bright-light exposure in the late-afternoon hours. Some studies suggest that prescription melatonin can revitalise the blunted melatonin pattern, which improves both sleep quality and morning alertness (Wade *et al.*, 2007).

Incidentally, certain medications, including beta-blockers, reduce melatonin levels; in a rigorous study, melatonin supplementation in hypertensive patients treated with beta-blockers increased sleep time by 36 minutes and reduced the time to fall asleep by 14 minutes (Scheer *et al.*, 2012).

---

[3] Evidence reviewed by Siegel (2001) suggests that "sleep is clearly important for optimum acquisition and performance of learned tasks". On the other hand, in the same article, the puzzling statement is made that "a major role for memory consolidation is unproven".

In patients with Alzheimer's disease, the deterioration of deep sleep is more pronounced. Sleep disturbances precede the onset of this devastating condition by several years. Remarkably, just one night of sleep deprivation increases the levels of the beta-amyloid protein in the brain, whose accumulation is one of the hallmarks of Alzheimer's disease (Shokri-Kojori *et al.*, 2018). Also, obstructive sleep apnoea, which affects the quality and quantity of sleep, is associated with increased beta-amyloid deposition (Cordone *et al.*, 2019).

These findings suggest that there exists a relation between sleep and Alzheimer's disease. This is further supported by the remarkable discovery of Maiken Nedergaard of the so-called 'glymphatic system' (Jessen *et al.*, 2015). Namely, studies in mice show that during NREM sleep, their glial cells shrink in size by up to 60%, creating conduits for the elimination of toxic debris, including the beta-amyloid protein. Assuming that the human brain possesses a similar system, it follows that the reduction in NREM sleep contributes to the accumulation of beta protein. Since the slow waves start from the middle of the frontal cortex, it follows that the greater the accumulation of beta protein in the frontal cortex, the stronger the reduction of NREM sleep, and hence the greater the further accumulation of this protein.

A recent study, after reviewing relevant investigations into the role of sleep deprivation in the pathogenesis of Alzheimer's disease, concluded that sleep deficiency promotes the development of this devastating disease. In particular, insufficient sleep was associated with beta-amyloid protein deposition, the creation of tau protein tangles, neuroinflammation, and oxidative stress, all of which are well-established risk factors for Alzheimer's disease. In addition, sleep deficiency increases glucocorticoid levels, reduces the number of synapses in the central nervous system, and decreases brain-derived neurotrophic factor levels (Lv *et al.*, 2022).

Sleep is very important for procedural memory. In a relevant study, a group of right-handed individuals were trained to learn to type on a keyboard a specific sequence of numbers as fast and as accurately as possible using their left hand. These individuals were tested 12 hours later; half of them learned this task in the morning and were tested in the evening, whereas the other half learned the task in the evening and were tested the following morning after a normal night's sleep. The latter group had a better performance in both speed and accuracy by 20% and 35%, respectively (Walker, 2017, p. 127). Imaging studies showed that this improvement was due to the fact that sleep shifted this memory from the cortex to sub-cortical structures, and in this way procedural memory became unconscious. Sleep is also important for the recovery of motor function following a stroke (Siengsukon and Boyd, 2009).

Alcohol is a powerful suppressor of REM sleep and, certainly, has a detrimental effect on memory integration and learning. In one study, participants learned an artificial grammar in the morning of day 1. This type of memory task makes vital use of associations, and for this reason it is greatly facilitated by REM sleep. Then, the participants were divided into three groups. The first group was the control; the second group was given two drinks on the first evening after the daytime learning; and the third group was given two drinks on the evening of day 3. All participants were tested on day 7. Remarkably, not only the second group but also the third had lost 40% of the knowledge learned on day 1 (Zarcone, 1978). Taking into consideration that the subjects of the third group had two full nights of sleep after the day of learning, this study suggests that memories remain vulnerable to disruption of sleep (in this case caused by alcohol) for a long period. The effect of alcohol on REM sleep has also been documented in pregnancy. In one study, mothers were asked to drink non-alcoholic fluids on day 1 and two glasses of wine on day 2. Alcohol significantly reduced the duration of REM sleep in the near-term foetus. Incidentally, there were additional negative effects, including a marked depression in breathing during the REM phase (Mulder *et al.*, 1998).

Emotional and mental health

REM sleep is of crucial importance for emotional and mental health (van der Helm and Walker, 2009). Both positive and negative emotions, including anxiety and fear, are common during dreams. Taking into consideration the crucial importance of the amygdala in emotions, the recent discovery that REM sleep in mice is regulated by a dopaminergic circuit involving the amygdala emphasises further the importance of the relationship between emotions and sleep (Burgess and Scammell, 2022).

During REM sleep, the release of the molecule noradrenaline in the brain is completely shut down. This implies that painful emotional experiences revisited during dreaming appear less traumatic without the stress-causing noradrenaline. This suggestion is supported by an experiment that studied the response of a group of healthy individuals to a set of emotional images. The subgroup of these subjects that was allowed to sleep before viewing these images again exhibited a milder emotional response. Also,

MRI images of the amygdala of these subjects showed decreased activation in comparison to the subgroup that remained awake.

In patients suffering from 'post-traumatic distress syndrome', the release of noradrenalin is not shut down during REM sleep, which presumably explains the severity of their nightmares. Remarkably, some of these patients were cured after they were treated with the medication prazosin, which blocks noradrenalin (Raskind *et al.*, 2018).

The crucial importance of REM sleep for mental health was established in the 1960s. Young adults had EEG electrodes placed on their heads and were awakened as soon as they entered the REM phase. The goal was to study the effect of REM sleep deprivation. Although the study was supposed to last for a week, it was interrupted on the third day because, by then, the participants showed early signs of psychosis; namely, they were anxious and moody, began to hallucinate, and became paranoid!

Insufficient sleep disrupts the proper communication between the frontal cortex and the amygdala, affecting the proper emotional response. It is well documented that lack of sleep has a detrimental effect on the relapse of addiction disorders (Brower and Perron, 2010). Also, sleep deprivation is a causal trigger of manic or depressive episodes.

A relationship between dreams and emotions was established by Robert Stickgold and collaborators. (Wamsley and Stickgold, 2011). Over a two-week period, these investigators correlated 299 dream reports of healthy participants with detailed notes of their daily activities. They concluded that in almost half of these dreams, emotional themes and concerns of the participants present in their daily lives resurfaced in their dreams. These reports are consistent with MRI studies performed during dreaming, which show activation in the limbic system, especially the amygdala, which controls emotions. In addition, there was activation in the visuospatial region at the back of the brain, associated with complex visual perception; in the motor cortex, responsible for initiating movement; and in the hippocampus and surrounding areas, involved in autobiographical memory. Interestingly, there was a dramatic de-activation of the frontal cortex responsible for rational thinking.

A relationship has been established between autism and sleep. The autistic spectrum disorder usually appears in the second or third year of life. Autism is characterised by abnormal synaptogenesis, namely the presence of excess synapses in some areas and deficiencies in others. In addition, the rate of growth of the brain in autistic children is abnormal.

The size of their brains is slightly smaller than average at birth, larger than average from ages two to five, and again average by adulthood. An important feature of autism is the occurrence of a deficit in emotional intelligence. In particular, autistic individuals have difficulty deciphering the emotional state of others. Taking into consideration the role of REM in both synaptogenesis and emotional maturity, it is not surprising that in autistic individuals there is a reduction of 30–50% in REM sleep. Several additional studies have reported that infants and young children who show signs of autism do not have normal sleep patterns or duration. Among several observed abnormalities is the presence of flat patterns of melatonin (Cohen *et al.*, 2011).

General health

Sleep deprivation has various detrimental effects on health. For example, it affects the proper function of the sympathetic system, which in turn raises the pulse and causes an increase in blood pressure (Tochikubo *et al.*, 1996). It also causes a chronic increase in the stress hormone cortisol. In addition, sleep deprivation causes a marked decline in testosterone levels, reducing sperm count, libido, and the ability to concentrate.

Sleep deprivation, via increased stress hormone levels and other effects, increases the risk for cardiovascular diseases. Remarkably, there is a spike in heart attacks the week after the switch to daylight savings time in March, when most people lose an hour of sleep; the opposite effect is observed in the autumn when the clocks move forward (Manfredini *et al.*, 2018). A similar effect is reported for the incidence of strokes. Also, remarkably, the week after the switch to daylight savings time, there is a 6% increase in fatal car accidents (Fritz *et al.*, 2020).

Reduced sleep affects the concentration of the satiety-signalling hormone leptin, which causes weight gain. This effect is further enhanced by the increase of endogenous endocannabinoids, which stimulate appetite. The weight gain may cause further reduction of sleep since overweight individuals are at high risk for developing sleep apnoea, which causes chronic, severe sleep deprivation.[4] In turn, chronic sleep deprivation reduces the effectiveness of insulin, which increases the risk for type 2 dia-

---

[4]In children, an under-diagnosed cause of sleep deprivation is obstructive sleep apnoea, usually caused by enlarged tonsils and adenoids.

betes. Healthy subjects who were forced to sleep only four hours a night for six consecutive nights were 40% less effective in absorbing a standard dose of glucose compared to the period when they were sleeping normally. Sleep deprivation is now recognised as one of the major factors contributing to the escalation of diabetes.

Replenishing components of the immune system requires the consumption of a large amount of energy, and hence sleep provides an ideal state for this process. Several studies have shown that the less the duration of sleep, the higher the probability that an individual will be infected by the common virus causing a cold. An indirect proof of the correlation between sleep and the immune system was presented in 2002 when it was shown that subjects who obtained seven to nine hours of sleep in the week before receiving a flu shot generated a powerful antibody response, indicating a healthy immune system (Walker, 2017, p. 183). As discussed in Chapter 8, there is now renewed excitement about the effectiveness of immunotherapy against various forms of cancer (June and Sadelain, 2018; Morschhauser *et al.*, 2018). This further underlines the importance of the immune system for preventing and fighting a variety of diseases. In particular, natural killer cells and macrophages play an important role in this process. Remarkably, Michael Irwin *et al.* (1996) has shown that a single night of four hours of sleep eliminated 70% of natural killer cells in normal participants! Furthermore, David Gozal (Hakim *et al.*, 2014) has shown that sleep-deprived mice exhibited a 200% increase in the speed and size of their cancer growth. In these mice, as a result of sleep deprivation, a specific type of macrophage, called M_1, that fights cancer was reduced, whereas another type, called M_2, that promotes cancer was increased.

Derk-Jan Dijk has examined gene expression in a group of healthy young men and women after restricting them to six hours of sleep for one week. He identified 711 genes whose activity was distorted (Möller-Levet *et al.*, 2013). In about half of them, sleep deprivation had increased their expression, while in the remaining, it had the opposite effect. The former genes included those promoting chronic inflammation, cellular stress, and cardiovascular diseases, whereas the under-expressed genes are important for maintaining stable metabolism and an optimal immune system.

Taking into consideration the detrimental effects of sleep deprivation, it is not surprising that studies on aeroplane cabin crews who frequently fly long-haul routes have shown that their short-term memory is impaired. In addition, there is atrophy in the parts of their brains involved with learning and memory. Furthermore, some studies have reported higher rates of both cancer and type 2 diabetes (Walker, 2017, p. 26).

References

Brower, V. K. and Perron, B. E. (2010). Sleep disturbance as a universal risk factor for relapse in addictions to psychoactive substances. *Medical Hypotheses*, *74*(5), 928–933.

Burgess, C. and Scammell, T. (2022). Catching up on REMs. *New England Journal of Medicine*, *386*(20), 1950–1952.

Campbell, I., Kraus, A., Burright, C., and Feinberg, I. (2016). Restricting time in bed in early adolescence reduces both NREM and REM sleep but does not increase slow wave EEG. *Sleep*, *39*(9), 1663–1670.

Cohen, S., Conduit, R., Lockley, S. W., Rajaratnam, S. M., and Cornish, K. M. (2011). The relationship between sleep and behavior in autism spectrum disorder (ASD): A review. *Journal of Neurodevelopmental Disorders*, *1*, 44.

Cordone, S., Annarumma, L., Rossini, P., and De Gennaro, L. (2019). Sleep and β-amyloid deposition in Alzheimer disease: Insights on mechanisms and possible innovative treatments. *Frontiers in Pharmacology*, *10*, 695.

Dauvilliers, Y., Plazzi, G., Mignot, E., *et al.* (2025) Oveporexton, an oral receptor 2-selective agonist, in narcolepsy Type 1. *New England Journal of Medicine*, *392*, 1905–1916.

Ding, M., Bhupathiraju, S., Satija, A., van Dam, R., and Hu, F. (2014). Long-term coffee consumption and risk of cardiovascular disease: a systematic review and a dose–response meta-analysis of prospective cohort studies. *Circulation*, *129*(6), 643–659.

Fenn, K., Nusbaum, H., and Margoliash, D. (2003). Consolidation during sleep of perceptual learning of spoken language. *Nature*, *425*(6958), 614–616.

Fokas, A. (2024). *Ways of Comprehending: The Grand Illusion and the Essence of being Human.* Singapore: World Scientific.

Foster, R. (2022). *Life-Time.* London: Penguin Books.

Frank, M. G., Issa, N. P., and Stryker, M. P. (2001). Sleep enhances plasticity in the developing visual cortex. *Neuron*, *30*(1), 275–287.

Freedman, M. S., L. R., Soni, B., von Schantz, M., Muñoz, M., David-Gray, Z., and Foster, R. (1999). Regulation of mammalian circadian behavior by non-rod, non-cone, ocular photoreceptors. *Science*, *284*(5413), 502–504.

Freedman, N., Park, Y., Abnet, C., Hollenbeck, A., and Sinha, R. (2012). Association of coffee drinking with total and cause-specific mortality. *New England Journal of Medicine*, *366*(20), 1891–1904.

Fritz, J., VoPham, T., Wright, K., and Vetter, C. (2020). A chronobiological evaluation of the acute effects of daylight saving time on traffic accident risk. *Current biology*, *30*(4), 729–735.

Greenfield, S. (2016). *A Day in the Life of the Brain*. London: Allen Lane.

Hakim, F., Wang, Y., Zhang, S., Zheng, J., Yolcu, E., Carreras, A., ..., and Gozal, D. (2014). Fragmented sleep accelerates tumor growth and progression through recruitment of tumor-associated macrophages and TLR4 signaling. *Cancer Research*, *74*(5), 1329–1337.

Irwin, M., McClintick, J., Costlow, C., Fortner, M., White, J., and Gillin, J. (1996). Partial night sleep deprivation reduces natural killer and cellular immune responses in humans. *The FASEB Journal*, *10*(5), 643–653.

Jenkins, J. G. and Dallenbach, K. M. (1924). Obliviscence during sleep and waking. *American Journal of Psychology*, *35*, 605–612.

Jessen, N. A., Munk, A. S., Lundgaard, I., and Nedergaard, M. (2015). The glymphatic system: A beginner's guide. *Neurochemical Research*, *40*(12), 2583–2599.

June, C. and Sadelain, M. (2018). Chimeric antigen receptor therapy. *New England Journal of Medicine*, *379*(1), 64–73.

Lindskog, M., Svenningsson, P., Pozzi, L., Kim, Y., Fienberg, A., Bibb, J., ..., and Fisone, G. (2002). Involvement of DARPP-32 phosphorylation in the stimulant action of caffeine. *Nature*, *418*(6899), 774–778.

Lv, Y., Cui, Y., Zhang, B., and Huang, S. (2022). Sleep deficiency promotes Alzheimer's disease development and progression. *Frontiers in Neurology*, *13*, 1053942.

Manfredini, R., Fabbian, F., De Giorgi, A., Zucchi, B., Cappadona, R., Signani, F., ..., and Mikhailidis, D. (2018). Daylight saving time and myocardial infarction: Should we be worried? A review of the evidence. *European Review for Medical and Pharmacological Sciences*, *22*(3), 750–755.

Mesas, A., Leon-Muñoz, L., Rodriguez-Artalejo, F., and Lopez-Garcia, E. (2011). The effect of coffee on blood pressure and cardiovascular disease in hypertensive individuals: A systematic review and meta-analysis. *The American Journal of Clinical Nutrition*, *94*(4), 1113–1126.

Möller-Levet, C., Archer, S., Bucca, G., Laing, E., Slak, A., Kabiljo, R., ..., and Dijk, D. (2013). Effects of insufficient sleep on circadian rhythmicity and expression amplitude of the human blood transcriptome. *Proceedings of the National Academy of Sciences*, *110*(12), E1132–E1141.

Morschhauser, F., Fowler, N., Feugier, P., Bouabdallah, R., Tilly, H., Palomba, M., ..., and Haioun, C. (2018). Rituximab plus lenalidomide in advanced untreated follicular lymphoma. *New England Journal of Medicine*, *379*(10), 934–947.

Mulder, E. J., Morssink, L. P., van der Schee, T., and Visser, G. H. (1998). Acute maternal alcohol consumption disrupts behavioral state organization in the near-term fetus. *Pediatric Research*, *44*(5), 774–779.

Raskind, M., *et al.* (2018). Trial of prazosin for post-traumatic stress disorder in military veterans. *New England Journal of Medicine, 378*, 507–517.

Sarkar, S., Katshu, M. Z., Nizamie, S. H., and Praharaj, S. K. (2010). Slow wave sleep deficits as a trait marker in patients with schizophrenia. *Schizophrenia Research, 124*(1), 127–133.

Scheer, F., Morris, C., Garcia, J., Smales, C., Kelly, E., Marks, J., …, and Shea, S. (2012). Repeated melatonin supplementation improves sleep in hypertensive patients treated with beta-blockers. *Sleep, 35*, 1395–1402.

Shokri-Kojori, E., Wang, G., Wiers, C., Demiral, S., Guo, M., Kim, S., …, and Miller, G. (2018). β-Amyloid accumulation in the human brain after one night of sleep deprivation. *Proceedings of the National Academy of Sciences, 115*(17), 4483–4488.

Siclari, F., Baird, B., Perogamvros, L., Bernardi, G., LaRocque, J., Riedner, B., …, and Tononi, G. (2017). The neural correlates of dreaming. *Nature Neuroscience, 20*(6), 872.

Siegel, J. M. (2001). The REM sleep-memory consolidation hypothesis. *Science, 294*, 1058–1063.

Siengsukon, C. and Boyd, L. A. (2009). Sleep enhances off-line spatial and temporal motor learning after stroke. *Neurorehabilitation & Neural Repair, 4*(23), 327–335.

Stickgold, R., Hobson, J., Fosse, R., and Fosse, M. (2001). Sleep, learning, and dreams: Off-line memory reprocessing. *Science, 294*(5544), 1052–1057.

Tochikubo, O., Ikeda, A., Miyajima, E., and Ishii, M. (1996). Effects of insufficient sleep on blood pressure monitored by a new multibiomedical recorder. *Hypertension, 27*(6), 1318–1324.

van Dam, R., Hu, F., and Willett, W. (2020). Coffee, caffeine, and health. *New England Journal of Medicine, 383*(4), 369–378.

van der Helm, E. and Walker, M. P. (2009). Overnight therapy? The role of sleep in emotional brain processing. *Psychological Bulletin, 135*, 731–748.

Wade, A. G., Ford, I., Crawford, G., *et al.* (2007). Efficacy of prolonged release melatonin in insomnia patients aged 55–80 years: Quality of sleep and next-day alertness outcomes. *Current Medical Research and Opinion, 23*(10), 2597–2605.

Walker, M. (2017). *Why We Sleep.* New York: Scribner.

Wamsley, E. and Stickgold, R. (2011). Memory, sleep, and dreaming: Experiencing consolidation. *Sleep Medicine Clinics, 6*(1), 97–108.

Zarcone, V. (1978). Alcoholism and sleep. *Advances in Bioscience and Biotechnology, 21*, 29–38.

Chapter 21

Mapping and Acting on the Brain

21.1 Electroencephalography and Magnetoencephalography

The important functional imaging techniques of PET, SPECT, and fMRI were discussed in detail in Fokas (2024). A serious limitation of these techniques is their relatively poor temporal resolution. Due to the slow nature of the haemodynamic response, the temporal resolution of the above techniques is of the order of a few seconds. This is to be contrasted with the high temporal resolution of electroencephalography (EEG) and magnetoencephalography (MEG), which is less than 1 millisecond. In other words, the latter techniques provide, essentially, real-time information about mental processes.

EEG and MEG are based on the fact that a specific mental process is associated with brain activation of a particular form. In turn, this expresses itself via the generation of a specific *neuronal electric current*. As discussed in Part III, Maxwell showed that the precise relationship between electricity and magnetism is captured by the equations of electromagnetism. According to these fundamental mathematical equations, an electric current gives rise to an electric potential as well as to a magnetic field. EEG is based on the measurement of the electric potential on the scalp, whereas MEG relies on the measurement of the magnetic field outside the head.

Both the electrical potential and the magnetic field are generated by cortical neuronal currents, which are mainly due to the summation of the ionic currents flowing across the sodium and potassium ionic gates located at the neuron's dendrites. The currents affecting EEG and MEG measurements are mostly due to the activation of the pyramidal neurons.

EEG

EEG recordings depend on the location of the electrodes and the specific mental function performed by the subject during the recordings. The brains of normal subjects exhibit several characteristic rhythms, which are distinguished by their frequencies. Specifically, the 'delta rhythm', as noted in the previous chapter, involves high-amplitude and slow waves, namely waves of frequency less than 4 Hz, i.e. less than 4 cycles per second. This is the characteristic rhythm occurring during deep sleep; these waves are best obtained from recordings on the frontal part of the head. Waves of the 'theta rhythm' have a frequency between 4 and 7 Hz and occur when subjects try to repress a specific response or action; these waves are found in regions not related to the task at hand. The frequency of 'alpha waves' is between 8 and 15 Hz, and such waves occur during relaxation. The frequency of the 'beta waves' is in the range of 16–32 Hz, and these waves are associated with active thinking. Finally, the 'gamma rhythm' involves waves of high frequency, namely waves of frequency higher than 32 Hz. They typically occur during the processing of at least two different types of sensory information.

In most clinical EEG applications, the electric potential on the scalp is measured via 19 electrodes. For research purposes, there exist EEG systems which use 256 electrodes. The measured potential is in the micro-volt range (a microvolt is a millionth of a volt). Scalp voltage recordings were first measured in humans by Hans Berger (1873–1941). In 1924, this German psychiatrist, extending earlier works by Richard Caton (1842–1926) and others, invented what he called the 'electroencephalogram'. EEG is an inexpensive, minimally invasive technique, which can provide real-time information about the brain's functions in both normal and pathological conditions. In particular, it is an indispensable tool in sleep studies, as well as for the diagnosis of epilepsy.

The neurological disorder of epilepsy includes 'focal seizures' and 'generalised seizures'. The latter are further subdivided into 'tonic-clonic seizures' and 'absence seizures.' Tonic-clonic seizures, previously called 'grand mal', are the seizures that most people associate with epilepsy. Absence seizures, previously called 'petit mal', are more common in children than in adults. In absence seizures, a child may suddenly stop talking or cease walking; the child stares emptily and does not respond to stimuli. After one to several seconds following these short 'absences', the child regains consciousness and has no memory of the episode.

There is a rather rare type of functional neurological disorder called 'functional seizure' (also known as 'dissociative seizure' or 'psychogenic nonepileptic seizure'). In a typical episode, the eyelids are forcefully closed and there occur intermittent movements of the upper body, as well as non-rhythmic tremulous movements of the hands. Remarkably, these movements become less intense when the physician places their hand on the patient's shoulder; the movement resumes at the previous intensity after the hand is removed. The patient does not speak but can follow commands, such as an instruction to squeeze the physician's hand. The episodes often last more than five minutes.

Electroencephalograms obtained during different types of seizures exhibit characteristic patterns, which can aid the diagnosis of the particular type of disorder. For example, in generalised seizures, there is a pattern of unison hyperactivity, which can be contrasted with the highly complex pattern of a normal electroencephalogram. The regularity of the EEG patterns observed in generalised seizures suggests that the loss of consciousness accompanying these seizures is the result of a dramatic reduction in the complexity of the brain's neural states.

An emerging use of EEG involves identifying patients who are thought to be in minimally conscious state, vegetative state, or coma, but surprisingly, they can perform cognitive tasks. It was discussed in Chapter 15 of Fokas (2024) that using fMRI, the group led by Owen in Cambridge established the remarkable fact that one in five patients thought to be in the vegetative state, after being given a verbal command, could imagine playing tennis or navigating in their house. It is now established that EEG can also be used for such a task: among 353 patients, data were available from fMRI only or from EEG only for 65% of the participants, and for the remaining participants, data were available from both fMRI and EEG. Remarkably, approximately one in four participants who, based on their behaviour, were given the diagnosis of a minimally conscious state, vegetative state, or coma, performed a cognitive task (Bodien *et al.*, 2024).

MEG

The magnetic field generated by neuronal currents is extremely weak. It is of the order of femto-tesla, i.e. approximately 10^9 times smaller than the magnetic field at the surface of the Earth. Thus, MEG requires the use of an extremely sensitive apparatus to measure this field. Such a tool, called magnetometer, was invented in the early 1970s by David Cohen using the exquisitely sensitive device called superconducting quantum interference device (SQUID). This Canadian physicist built a magnetically shielded room at MIT and measured MEG signals at a single point outside the head. The SQUID had just been developed at the Ford Motor Company by James Zimmerman (Zimmerman *et al.*, 1970). Remarkably, MEG signals are almost as clear as those obtained via EEG. Present-day technology allows the measurement of the magnetic flux at about 300 points in a helmet-shaped apparatus. Furthermore, in a recent technological breakthrough, a new type of magnetometer was developed, which takes the form of a portable helmet. This device, called a *spin exchange relaxation-free magnetometer*, does not require a bulky cooling system for its operation. Hence, it is small and portable, and the sensors are placed on the scalp instead of being located a few millimetres away from the head. The proximity of the sensors to the brain gives rise to less noisy signals and, importantly, allows the subject to move.

In addition to their important role in the investigation and treatment of epilepsy, EEG and MEG have numerous additional applications in the study of physiology, psychiatry, and cognition. These applications range from the investigation of the somatosensory system and the relation between brain rhythms and cognitive processes, to the study of Alzheimer's disease and the identification of neural abnormalities in various psychiatric and neurological diseases.

The inverse problems associated with EEG and MEG

Recordings obtained from EEG and MEG were originally used to aid the diagnosis of several neurological conditions. Later, it was realised that these recordings could have an additional use. Namely, they could be employed to *determine the electric current that gave rise to these EEG and MEG signals*. For EEG, this new application involves solving the following *mathematical inverse problem*: given the electric potential on the scalp, determine the electric current that generated this potential. Similarly, for MEG, the relevant inverse mathematical problem is the determination of the electric current that generates the magnetic field measured outside the head.

For the solution of these inverse problems, the head-brain system is usually modelled by four different compartments, corresponding to the cortex, cerebrospinal fluid, skull, and scalp.[1] Unfortunately, the inverse problems of EEG and MEG cannot be solved in a unique way. Namely, given the EEG or the MEG recordings, it is impossible to determine uniquely the neuronal current that gave rise to these recordings. In other words, there are *many* different currents that could generate the same EEG and MEG signals. This is the Achilles' heel of EEG and MEG. Although this problem was identified by Helmholtz in 1853, the question of determining which part of the current *can* be determined uniquely from given EEG or MEG data remained open until 2009. That year, first, I presented a precise characterisation of this non-uniqueness problem; and second, I derived an algorithm for computing the part of the current that *can* be determined uniquely either from EEG or MEG data (Dassios and Fokas, 2020). For EEG, the relevant algorithm has been jointly implemented with Georgina Paraskevopoulou. Importantly, this new approach, which is based on the assumption that the physiological neuronal current is such that energy consumption is minimised, yields more accurate results than the commercially available algorithm, called LORETA.

It should be noted that there exist particular idealised situations where the non-uniqueness difficulty can be bypassed, and the current *can* be computed uniquely. Such a situation arises when the current is so well localised that it can be adequately approximated by the assumption that it is localised at a *single* point. In this highly idealised case, the current can be represented in terms of a mathematical entity known as 'Dirac's delta function', introduced by the theoretical physicist Paul Dirac. The associated current is called a *dipole*. There exists a commercial software that can determine a suitable dipole from the given data (Mosher *et al.*, 1992). The dipole approximation is very useful for the study of focal epilepsy, which remains the most important clinical application of EEG and MEG.

In patients with drug-resistant epilepsy, a viable treatment option is surgery to remove the brain tissue that generates seizures. Drug-resistant

---

[1] The analysis of the EEG inverse problem requires the computation of certain complicated auxiliary functions. These functions depend on the geometry of the four compartments and on their conductivities but are independent of the current. A novel approach for the fast determination of these auxiliary functions based on the new technique of 'deep learning' is presented in Hashemzadeh *et al.* (2020).

epilepsy is defined as the failure of an adequate control of epilepsy after trials of two appropriately chosen drugs with acceptable side effects (Kwan *et al.*, 2010). In a rigorous study, 77% of 57 patients who underwent surgery and continued to receive pharmacological treatment became seizure-free after one year. In comparison, only 7% of those receiving pharmacological treatment but without surgery were seizure-free (Dwivedi *et al.*, 2017).

In order to identify the 'epileptogenic focus', namely the brain tissue that gives rise to the epilepsy, the usual pre-operative evaluation involves a combination of long-term video-EEG (based on the one-dipole approximation) and MRI. If the epileptogenic focus identified by EEG is not in agreement with the results of MRI, MEG is used to identify the focus (under the assumption of a single dipole). Due to recent technological developments, it is possible to combine MEG with a newer MRI, called ultralow-field MRI, and this significantly improves the localisation capability of MEG (Vesanen *et al.*, 2013). Comparisons for the determination of epileptic foci via the dipole approximation using either MEG or EEG are presented in Pataraia *et al.* (2004).

Another physiologically important situation where the electric current can be computed uniquely, via either EEG or MEG recordings, is when the data are mainly produced by pyramidal neurons situated *perpendicularly* to the cortical surface. It turns out that, under the assumption that the current is *localised on the cortical surface and is normal to this surface*, the current can be computed uniquely (Fokas *et al.*, 2020).[2]

---

[2]In this case, the direction of the current vector is known, and thus the inverse problem involves determining a single unknown scalar function (the magnitude of the current), defined on a surface. For EEG, the electric potential is measured on the scalp, which is a surface. Hence, both the unknown current and the EEG data are represented by single functions defined on a surface. Therefore, it is not surprising that in this case, there exists a unique relationship between them. In the case of MEG, it turns out that the magnetic field outside the head satisfies the so-called Laplace equation. Since the solution of this equation can be uniquely determined from the knowledge of the function on the boundary of the domain, it follows that MEG data can also be considered as given on a surface.

21.2   Resonance and the Possibility of Affecting Unconscious Processes

It is widely known that any physical system that can exhibit an oscillatory motion possesses its own *intrinsic* frequency. If such a system is excited with an *external* frequency that is identical to its intrinsic frequency, then the system oscillates with maximum amplitude, a phenomenon called 'resonance'. Although this phenomenon has numerous implications in physics, it has not been sufficiently explored in neuronal oscillations. It is suggested in the following that by combining existing technology with transcranial electrical stimulations (TES), reviewed in Fokas (2024), it may be possible to exploit the phenomenon of resonance in order to affect unconscious processes.

Specific forms of thinking give rise to particular 'signatures' in the associated EEG recordings. Taking into consideration that TES provides a non-invasive approach for exciting a chosen part of the brain with a specific frequency, it follows that the combination of EEG with TES can be used to produce *resonant neuronal oscillations*. This could provide a powerful approach for both investigating and, more importantly, interfering with various mental functions.

One possible application involves learning. For example, by analysing recordings made while rats were learning how to navigate a given maze, investigators identified particular electrical oscillations associated with this activity. Remarkably, the same pattern reappeared during their sleep, suggesting that the rats were 'replaying' these memories (Walker, 2017). Such oscillations occur both during REM sleep and NREM sleep. Interestingly, during the former phase, the pattern involved half or a quarter of the speed of the NREM-sleep. This observation perhaps provides an explanation for why we feel that our dreams last longer than they actually do. A possible way to explore this concrete unconscious mechanism is to stimulate the rats' brains with the same electrical oscillations and check whether this interference improves their performance.

Another possible application of resonance mechanisms concerns memory. As discussed in the previous chapter, sleep seems to enhance procedural memory by transferring brain activity from cortical to sub-cortical areas. Actually, it has been observed in sleep studies that during this transfer, there is an increase in fast spindles during stage 2 NREM sleep. These changes are detected in the EEG electrodes placed in the region of the scalp above the motor cortex. Spindles are generated by neurons in the reticular thalamic area and are due to synchronous brain waves of 12–16 HZ, which propagate in the thalamocortical loop. The greater the increase in sleep spindles in this area, the better the performance following sleep. Spindles are thought to reflect appropriate synaptic changes needed to elicit long-term potentiation, which, as noted in Chapter 10 of Fokas (2024), is crucial for learning. Perhaps, using external brain stimulation to excite this part of the brain with a similar oscillatory pattern as that observed in the associated EEG recordings, may further facilitate the underlying process of learning involved in procedural memory (unfortunately, at the moment, it is not known how to influence spindle generation using TES).

It appears that the brain's electrical stimulation can also be used for *forgetting* certain experiences. This is motivated by a study performed in 2009, in which young graduate students were presented a list of words on a large screen. After each word, a large 'R' or 'F' was displayed, indicating to the participants that they should remember or forget the prior word. The subjects were told that regardless of the tag 'R' or 'F', they should try to remember as many words as possible. Half of the participants were allowed to have a 90-minute afternoon nap after the test, whereas the other half remained awake. Later that evening, all participants were tested. Remarkably, the students who were allowed to sleep had made an important *unconscious* choice. Namely, their brain preferentially boosted the retention of only those words previously tagged with 'R'. Furthermore, the larger the number of the fast sleep spindles in their NREM sleep, the greater the efficiency with which these participants strengthened words tagged for 'remembering' and actively eliminated words designated for 'forgetting' (Walker, 2017). Perhaps, by stimulating the brains of such participants with the specific electrical pattern of the associated fast spindles will further enhance their ability to forget the words tagged with 'F'. Related studies have shown that the more enhanced the activation in the dorsolateral prefrontal cortex during the process of directing subjects to forget, the more suppressed the ongoing activity in the hippocampus, where memories are typically stored.

The hypothesis that there exist specific unconscious mechanisms responsible for 'forgetting' is strengthened by the existence of the so-called 'transient global amnesia' (TGA). In this benign condition, subjects, after waking up, are somewhat confused, repeating questions. The only neurological deficit is anterograde amnesia of several hours. Affected individuals realise that they have a complete or partial loss of memory of several hours prior to falling asleep. Conditions that may precipitate TGA include physical exertion, emotional stress, sexual intercourse, swimming in cold water, and marijuana intoxication. Individuals suffering from migraines and temporal lobe epilepsy have a slightly increased risk of TGA. This benign condition should be differentiated from temporal epilepsy—and especially from a 'transient ischaemic attack', which is a dangerous vascular episode that increases the risk of a stroke. TGA attracted media attention in 1987 when it was reported that it occurred to three neuroscientists, travelling from New York to Europe, who attempted to minimise the effects of jet lag by taking 0.5 mg of the sleeping pill triazolam and also consumed a small quantity of alcohol.

TGA confirms that memories remain vulnerable for several hours. Also, since memory consolidation occurs during NREM sleep, it is natural to speculate that during this phase, it should be possible to wipe out painful memories. Perhaps such mechanisms could be explored for eliminating the memories associated with those causing the 'post-traumatic distress syndrome'.

Incidentally, measuring brain activity via fMRI during the sleep-onset period and using machine learning models, it is possible to predict the contents of visual imagery (Horikawa *et al.*, 2013). I expect that the correspondence between fMRI data and visual contents of dreams will be further delineated using deep learning techniques.

Insomnia

Brain stimulation may also be useful in combating insomnia, which is a disorder defined as a sleep disturbance that occurs three or more nights a week and persists for more than three months. Chronic insomnia is a serious health problem worldwide, which is associated with increased risks of major depression (Hertenstein *et al.*, 2019), cardiovascular diseases (Bertisch *et al.*, 2018), and Alzheimer's disease (Shi *et al.*, 2018). Contributing factors to the inability to initiate and maintain sleep include an overactive sympathetic system, which, in addition to enhancing the action of the alertness-promoting cortisol and the neurotransmitters adrenaline and noradrenaline, raises the metabolic rate. The latter effect gives rise to a higher core body temperature; this contributes to insomnia since the core temperature must be lowered by a few degrees for sleep to be initiated. Another important factor contributing to insomnia is a lack of reduction in activity in the brain areas involved with emotions and memory, especially the amygdala and the hippocampus. Reducing the excitation of these areas is a necessary condition for initiating sleep.

There exist several well-known steps for combating insomnia. These steps include avoiding late afternoon naps and coffee, maintaining a regular sleep schedule, and going to bed only when sleepy, thus avoiding remaining in bed awake for a prolonged period. In addition, a hot bath is advisable since it causes surface veins to dilate, which helps radiate more heat after getting out of the bath, which lowers the core temperature. Also, it should be taken into consideration the fact that different foods have different effects on sleep. For example, a high-carbohydrate, low-fat diet enhances dreaming but reduces deep NREM sleep. Low-fibre foods result in more awakening during the night. Several studies suggest that exercise (not close to sleep time) increases total sleep time, particularly deep sleep, in younger healthy adults and, to a lesser extent, in older adults.

The medical community has endorsed cognitive behavioural therapy as a possible treatment of insomnia. In this connection it is noted that: *If an individual succeeds in concentrating on only one item, despite the continuous bombardment of many other thoughts, sleep will come quickly even if the brain appears to be fully alert.*

Many individuals use sleep medications, which include antihistamines (such as diphenhydramine and hydroxyzine), GABA A receptor agonists, orexin receptor antagonists (such as suvorexant and daridorexant), and sedating antidepressant drugs. There are two types of GABA A agonists: benzodiazepines (such as astriazolam, temazepam, and clonazepam) and the Z-drugs (such as zolpidem, zaleplon, and eszopiclone). Among the sedating antidepressants, although only the tricyclic drug doxepin has received FDA approval for insomnia, the heterocyclic drug trazodone is extensively prescribed. EEG recordings show that brain activity during the night affected by medications is different from the brain activity of normal sleep, particularly with respect to deep sleep (Arbon *et al.*, 2015). Also, even with the newer, shorter-acting sedative medications, there still exist side effects, including next-day sleepiness and daytime forgetfulness. Melatonin or GABA A receptor agonists are usually the medication of choice for insomnia with predominately sleep-onset symptoms and when short-term use is likely (for example, in response to acute or periodic stressors). Low-dose heterocyclic drugs or orexin antagonists are preferable in treating patients with disturbances predominantly related to sleep maintenance or early awakening. This is also the preferred treatment for patients with substance use disorders (Morin and Buysse, 2024).

By using TMS to dissociate the amygdala and the hippocampus, perhaps one of the main obstacles to initiating sleep could be reduced. There is also an obvious need for developing pharmacological or brain-stimulating approaches to selectively counteract the overactivity of the sympathetic system.

As discussed in the previous chapter, older individuals have a deficit in the quality of NREM sleep. Preliminary studies suggest that TES may restore some of the sleep difficulties of older individuals (Walker, 2017, p. 104). Perhaps, by stimulating the brain with the typical deep-sleep pattern of young adults, there will be a significant improvement.

Future developments

There exists a variety of simple devices used for bio-brain feedback. For example, an individual can observe the activation of their frontal lobe in the process of resisting the temptation to eat a chocolate cake (Frazzeto, 2013). I expect that soon, by using suitable prostheses, it will be possible to begin exploiting unconscious mechanisms before fully comprehending them. Several laboratories are already exploring such possibilities. For example, specific wearable devices are being designed in the MIT Media Laboratory within the context of 'designing the future'. One of them is a glove-like device, named 'dormion', designed to enhance creativity by affecting the state of semi-consciousness occurring between wakefulness and unconsciousness, which is known as *hypnagogia*.

21.3 Optogenetics

PET, SPECT, fMRI, EEG, and MEG provide information about the integrated behaviour of a very large number of neurons. For example, fMRI tracks the energy consumption of hundreds of thousands of neurons (Logothetis, 2010). Thus, these techniques, although very useful for delineating the involvement of particular areas of the brain during specific mental processes, cannot monitor the behaviour of individual neurons. Until recently, the latter could be achieved only via single-neuron measurements. However, there now exists a new technique which, in addition to the ability to follow the activation of single neurons, has the additional advantage of monitoring a *predetermined* number of neurons. This breakthrough is based on the synergy of molecular biology and optical fibre technology. In 2002, the biophysicists Peter Hegemann, Ernst Bamberg, and Georg Nagel, along with their collaborators, isolated a gene called ChR2, which is found in the single-cell green algae (Nagel *et al.*, 2002).[3] This gene produces a protein which is a specific photoreceptor that converts blue light into an *excitatory* electric signal. By using a virus as a carrier of ChR2, this gene can be inserted into a cell *in vivo*; in this way, many such photoreceptors can be incorporated into a neuron. By using an optic fibre, it is possible to flash a blue light on such a neuron. This light, via the collective action of the above photoreceptors, generates an action potential.

This technique provides investigators with the unprecedented capability of controlling the electrical activity of a predetermined number of individual neurons. For example, using a virus, it is also possible to insert into neurons another type of gene, NpHR, which produces a photoreceptor that converts yellow light into an *inhibitory* signal. A yellow light flashed simultaneously with a blue light blocks the generation of an action potential. By employing blue and yellow colours of light, *this provides a two-knob remote control for increasing or decreasing the activity of specific neurons*. It is shown in Zhang *et al.* (2007) that in neurons carrying both genes activation of the Ch R2 gene with blue light caused a large increase in calcium, but when both blue and yellow light were used at the same time, no calcium increase was observed.

In a particular study, ChR2 was inserted in those neurons of the hypothalamus of mice that produce orexin, which, as discussed in the earlier chapter, is a hormone promoting wakefulness. Its deficit gives rise to narcolepsy, which, as noted in the earlier chapter, is a sleep disorder characterised by excessive sleepiness. Normal mice are awakened by blue light after about one minute, but in mice carrying ChR2, the same light woke them up in about half this duration. The action potentials in the modified neurons caused by the blue light potentiated the release of orexin, which caused a faster awakening.

Optogenetics provides unlimited possibilities for the understanding of how specific neural circuits affect behaviour, as well as for the precise manipulation of specific circuits. For example, by using optogenetics, the

---

[3] This invention was chosen in 2010 by the journal *Nature Methods* as the 'method of the year' across all fields of science and engineering.

role of cholinergic interneurons in the nucleus accumbens has been clarified. This nucleus is a collection of neurons producing dopamine, the neurotransmitter crucial for anticipating pleasure. Experiments in mice have shown that, although the cholinergic interneurons represent less than 1% of the neurons in the nucleus accumbens, they are critical for cocaine conditioning (Tecuapetla *et al.*, 2010). This suggests that these interneurons should be targeted for the pharmacological treatment of this type of addiction. Another example involves the use of optogenetics for the study of fear. The importance of the interaction between the amygdala and the frontal cortex in fear conditioning has been known for a long time. Using optogenetics, this relationship has been further clarified. Namely, experiments in mice show that the critical step in this process is the occurrence of oscillations with a frequency of 4 Hz in specific interneurons connecting the baso-lateral part of the amygdala with the dorsal-medial part of the prefrontal cortex (Karalis *et al.*, 2016).

In an important study, optogenetics was used to confirm the results of Logothetis and his collaborators regarding the origin of fMRI signals mentioned in Chapter 16 of Fokas (2024). By employing a viral vector, two specific genes were introduced in rat brain cells (Lee *et al.*, 2010). One of them (originating from glowing jellyfish) encodes a fluorescent protein, and one (from a species of green alga) produces a light-sensitive membrane-associated protein. Using optical fibre illumination, the investigators could excite a particular network in the motor cortex of the rats. By simultaneously measuring electrical and haemodynamic responses in this optically excited network, they were able to confirm the robust correlation between neural activity and fMRI signals first observed by Logothetis' group.

21.4 Further Developments in Mapping the Brain

A key aspect of the analysis of any organism is the elucidation of the relationship between structure and function. In neuroscience, this task is particularly complicated for a variety of reasons, including those described in the following (Lichtman and Denk, 2011). First, the immense anatomical diversity of cell types and the astonishing variety of their functions. Second, the difficulty of simultaneously imaging electrical and chemical activities. Third, the incongruity of the length scales characterising various neurons. For example, although the cell body of a neuron is about 0.02 millimetres (similar to the width of a human hair), the length of the branches of some neurons can exceed a centimetre in a mouse and a meter in the human brain. In particular, a pyramidal neuron in the cerebral cortex can have an axon that crosses into the other hemisphere or one that goes down into the brain stem, or even to the spinal cord; the dendrites of such a cell may cover a region of about one cubic millimetre. Fourth, the inability of 'light microscopy' to visualise specific components of neurons. For example, axons connecting to dendritic spines thin down to less than 100 nanometres, well beyond the spatial resolution of light microscopy.

Despite these difficulties, it is now possible to monitor the activity of a large number of neurons in awake animals. For this purpose, the technique of 'two-photon calcium imaging' is of vital importance. This technique is based on the understanding that neural activity is accompanied by an increase in intracellular calcium levels. Hence, the first step is to load the neurons under consideration with a calcium-sensitive dye or make these neurons express a genetically encoded calcium indicator. Then, it becomes possible to monitor neural activity by following the magnitude of the fluorescence signal emitted from the excited neurons. Using the resulting light, two-photon microscopes scan across the tissue under scrutiny, and in this way neural activity is monitored. After the network of neurons participating in a given task is identified, the corresponding volume of the brain tissue is prepared for 'electron microscopy', which can accurately delineate synaptic connections. In this way, definitive data can be generated, which can be used to elucidate the relationship between structure and function in different areas of the brain. In particular, electron microscopy is an ideal tool for characterising highly interconnected networks, such as those existing in the cortex.

It is worth noting that, in contrast to traditional 'light microscopy', 'serial-section electron microscopy' can be used to delineate the three-dimensional structure of neuronal membranes. This makes it possible to trace particular components of a given neuron, such as a single dendrite. The relevant procedure is quite cumbersome, but it can yield very useful information. For example, it has been used to obtain a complete 'wiring diagram' for the common worm *C. elegans* (White *et al.*, 1986).

Electron microscopy has the disadvantage that only a maximum of 100 neurons can be analysed simultaneously. To circumvent this and other limitations, several new techniques have been developed, including techniques based on DNA sequencing (Zador, 2016). Using these new developments, it is now possible to delineate a variety of neuronal diagrams. In 2005, Olaf Sporns and Patrick Hagmann, appreciating the importance of such diagrams, independently proposed the term 'connectome' to refer to the complete map of the brain's neural connections. This term was popularised by Sebastian Seung' book *Connectome*, which summarises ongoing efforts to build a three-dimensional map of the brain tissue at the micro and macro scales, as well as global efforts to map the human connectome. Further developments led to the initiation of the Human Connectome Project (HCP; http://www.humanconnectome.org), which has the ambitious goal of mapping the large-scale circuitry of the brains of 1,200 healthy adults. It is important to emphasise that the HCP adopted a macro, as opposed to a micro, approach. Among the techniques used by this ambitious project are MRI, EEG, and fMRI. The concentration on a macro, as opposed to a micro, approach was necessitated by the fact that the latter approach requires the unambiguous determination of synaptic connections, which can be achieved only via electron microscopy, and this is, of course, impossible for live human brains.

The use of preliminary results of the HCP has begun to have an impact on the diagnosis of certain neurological conditions. Traditionally, the basis of clinical neurology has been the analysis of single lesions. This analysis associates a particular symptom or sign with a specific focal lesion in the brain. However, sometimes this approach fails because lesions in *different* parts of the brain can yield *similar* symptoms. For example, most lesions that disrupt memory occur in areas *outside* the hippocampus, the traditional 'memory area'. In cases when there does not exist a one-to-one relationship between the area of the lesion and the resulting symptomatology, it is useful to identify the *neural network* responsible for the given deficit. The human connectome can assist in identifying such a network and hence can provide an explanation for the lack of the one-to-one correspondence between symptoms and the affected brain area. This new approach begins with the determination of the locations of possible lesions, by using CT or MRI. Next, these results are placed onto a standard brain atlas. Then, connectome data that are co-registered in the same atlas are used to identify a *functional network* connected to each lesion location (Fox, 2018).

In summary, before the existence of connectome information, neural networks were identified by employing only *anatomical* connections, but now the human connectome allows the use of *functional* connectivity, which incorporates wider networks. Indeed, anatomical networks are based on point-to-point connectivity, as opposed to *polysynaptic* connectivity captured by the functional connectivity of the connectome.

The higher the complexity of a subject, the more apparent the importance of interdisciplinarity. A concrete illustration of this fact is provided by the creation of the Allen Institute, founded by Paul Allen, the cofounder of Microsoft. In 2006, the interdisciplinary scientific team of this institute created the first atlas of *gene expression* of a complete mammalian brain (which is available online at www.brain-map.org). In 2010, the same institute created the *Allen Human Brain Atlas*, which contains a combination of anatomical and genetic information (Koch and Reid, 2012). Another example of an interdisciplinary approach is the Janelia Research Campus set up by the Howard Hughes Medical Institute, which develops tools and protocols for research in mechanistic cognitive neuroscience. It focuses on attempts to answer the question of 'how does the brain enable cognition'? (https://www.janelia.org/about-us).

These exciting developments, together with remarkable advances in brain imaging, motivated the birth of the new field of *neuro-informatics*. This field combines computer technologies and mathematical algorithms for organising and analysing data involving *brain architecture*, *gene expression*, and *imaging information*.[4] The Human Brain Project (HBP, www.humanbrainproject.eu) funded by the European Community, aims to facilitate further progress in this area.

References

Arbon, E., Knurowska, M., and Dijk, D. (2015). Randomised clinical trial of the effects of prolonged-release melatonin, temazepam and zolpidem on slow-wave activity during sleep in healthy people. *Journal of Psychopharmacology*, *29*(7), 764–776.

Bertisch, S., Pollock, B., Mittleman, M., Buysse, D., Bazzano, L., Gottlieb, D., and Redline, S. (2018). Insomnia with objective short sleep duration and risk of incident cardiovascular disease and all-cause mortality: Sleep Heart Health Study. *Sleep*, *41*(6), zsy047.

---

[4]These developments, as correctly emphasised by the neuroscientist Alan Evans of the Montreal Neurological Institute, have made it imperative that a standardised three-dimensional coordinate framework be adopted, both for data reporting and analysis.

Bodien, Y., Allanson, J., Cardone, P., Bonhomme, A., Carmona, J., Chatelle, C., . . ., and Heinonen, G. (2024). Cognitive motor dissociation in disorders of consciousness. *New England Journal of Medicine, 391*(7), 598–608.

Dassios, G. and Fokas, A. (2020). *Electroencephalography and Magnetoencephalography: An Analytical-Numerical Approach.* De Gruyter.

Dwivedi, R., Ramanujam, B., Chandra, P., Sapra, S., Gulati, S., Kalaivani, M., . . ., and Sagar, R. (2017). Surgery for drug-resistant epilepsy in children. *New England Journal of Medicine, 377*(17), 1639–1647.

Fokas, A. (2024). *Ways of Comprehending: The Grand Illusion and the Essence of Being Human.* Singapore: World Scientific.

Fokas, A. S., Hashemzadeh, P., and Leahy, R. M. (2020). Which Part of the Neuronal Current can be Determined by Electroencephalography? In Papanicolaou, A. C., Roberts, T. P., and Wheless, J. W. (eds.), *Fifty Years of Magnetoencephalography: Beginnings, Technical Advances, and Applications.* Oxford Scholarship Online.

Fox, M. (2018). Mapping symptoms to brain networks with the human connectome. *New England Journal of Medicine, 379*(23), 2237–2245.

Frazzeto, G. (2013). *How We Feel.* London: Black Swan.

Hashemzadeh, P., Fokas, A., and Schonlieb, C.-B. (2020). A Hybrid Analytical-Numerical Algorithm for Determining the Neuronal Current via EEG. *Journal of the Royal Society Interface, 17*(163), 20190831.

Hertenstein, E., Feige, B., Gmeiner, T., Kienzler, C., Spiegelhalder, K., Johann, A., . . ., and Baglioni, C. (2019). Insomnia as a predictor of mental disorders: a systematic review and meta-analysis. *Sleep Medicine Reviews, 43*, 96–105.

Horikawa, T., Tamaki, M., Miyawaki, Y., and Kamitani, Y. (2013). Neural decoding of visual imagery during sleep. *Science, 340*(6132), 639–642.

Karalis, N., Dejean, C., Chaudun, F., Khoder, S., Rozeske, R., Wurtz, H., . . ., and Herry, C. (2016). 4-Hz oscillations synchronize prefrontal–amygdala circuits during fear behavior. *Nature Neuroscience, 19*(4), 605.

Koch, C. and Reid, R. (2012). Observatories of the mind. *Nature, 483*, 397–398.

Kwan, P., Arzimanoglou, A., Berg, A., Brodie, M., Allen Hauser, W., Mathern, G., . . ., and French, J. (2010). Definition of drug resistant epilepsy: Consensus proposal by the ad hoc Task Force of the ILAE Commission on Therapeutic Strategies. *Epilepsia, 51*(6), 1069–1077.

Lee, J., Durand, R., Gradinaru, V., Zhang, F., Goshen, I., Kim, D., . . ., and Deisseroth, K. (2010). Global and local fMRI signals driven by neurons defined optogenetically by type and wiring. *Nature, 465*(7299), 788–792.

Lichtman, J. and Denk, W. (2011). The big and the small: Challenges of imaging the brain's circuits. *Science, 334*, 618–623.

Logothetis, N. (2010). Bold claims for optogenetics. *Nature, 468*(7323), 788–792.

Morin, C. and Buysse, D. (2024). Management of insomnia. *New England Journal of Medicine, 391*(3), 247–258.

Mosher, J., Lewis, P., and Leahy, R. (1992). Multiple dipole modeling and localization from spatio-temporal MEG data. *IEEE Transactions on Biomedical Engineering, 39*(6), 541–557.

Nagel, G., Ollig, D., Fuhrmann, M., Kateriya, S., Musti, A., Bamberg, E., and Hegemann, P. (2002). Channelrhodopsin-1: A light-gated proton channel in green algae. *Science, 296*(5577), 2395–2398.

Pataraia, E., Simos, P., Castillo, E., Billingsley, R., Sarkari, S., Wheless, J., . . ., and Breier, J. (2004). Does magnetoencephalography add to scalp video-EEG as a diagnostic tool in epilepsy surgery? *Neurology, 62*(6), 943–948.

Shi, L., Chen, S. M., Bao, Y., Han, Y., Wang, Y., Shi, J., . . ., and Lu, L. (2018). Sleep disturbances increase the risk of dementia: A systematic review and meta-analysis. *Sleep Medicine Reviews, 40*, 4–16.

Tecuapetla, F., Patel, J., Xenias, H., English, D., Tadros, I., Shah, F., . . ., and Koos, T. (2010). Glutamatergic signaling by mesolimbic dopamine neurons in the nucleus accumbens. *Journal of Neuroscience, 30*(20), 7105–7110.

Vesanen, P., Nieminen, J., Zevenhoven, K., Dabek, J., Parkkonen, L., Zhdanov, A., . . ., and Ahonen, A. (2013). Hybrid ultra-low-field MRI and magnetoencephalography system based on a commercial whole-head neuromagnetometer. *Resonance in Medicine, 69*(6), 1795–1804.

Walker, M. (2017). *Why We Sleep.* London: Penguin Books.

White, J., Southgate, E., Thomson, J., and Brenner, S. (1986). The structure of the nervous system of the nematode Caenorhabditis elegans. *Philosophical Transactions of the Royal Society London B: Biological Sciences, 314*(1165), 1–340.

Zador, A. (2016). The connectome as a DNA sequencing problem. In Marcus, G. and Freeman, J. (eds.), *The Future of the Brain.* Princeton University Press.

Zhang, F., Wang, L., Brauner, M., Liewald, J., Kay, K., Watzke, N., . . ., and Deisseroth, K. (2007). Multimodal fast optical interrogation of neural circuitry. *Nature, 446*(7136), 633–639.

Zimmerman, J. E., Thiene, P., and Harding, J. T. (1970). Design and operation of stable rf-biased superconducting point-contact quantum devices, and a note on the properties of perfectly clean metal contacts. *Journal of Applied Physics, 41*, 1572.

Epilogue

In this book, emphasis has been placed on science and technology. However, as stated in the Introduction, the role of arts and letters for the preservation of humanity remains paramount. In what follows, brief remarks are made about philosophy, language, and literature.

Philosophy and Sciences

The relationship between philosophy and the sciences continues to be debated. For example, Francis Crick (Nobel Prize in Medicine, 1962) has stated that 'the philosophers often ask good questions but do not have techniques for getting the answers' (Blackmore, 2005, p. 74). In his conversation with Susan Blackmore, Crick claimed that the only tool at the philosophers' disposal is 'thought experiments', which, according to him, are of limited use.[1] The same claim is made by Christof Koch, who was the closest collaborator of Crick in the field of neuroscience (Blackmore, 2005, p. 129). Also, Wittgenstein, as well as several distinguished physicists, including Richard Feynman (Nobel Prize in Physics, 1965), Steven Weinberg (Nobel Prize in Physics, 1979), and Edward Witten, have claimed that philosophy does not affect scientists or mathematicians. Echoing Crick, many scientists and mathematicians believe that philosophy is good

---

[1] Crick notes that the best use of thought experiments was made not by a philosopher, but by a physicist, namely, Einstein. As it was mentioned in Chapter 18 of Fokas (2024), Alan Turing also had the remarkable ability of making crucial use of thought experiments.

348   *The Embodied Mind*

for clarifying concepts and for posing questions of 'why' but not for answering questions of 'how', which is the prerogative of science.

In my opinion, the deepest results elucidating the relationship between science and philosophy were achieved by the Vienna Circle and its contemporaries (which is discussed in detail in my forthcoming book, *From Ancient Greece to Viennese Modernism: Lessons in Creativity*). The philosophical approach of the Vienna Circle illustrates that there exists *continuity* between science and philosophy, in the sense that the former often leads to the latter. Indeed,

> The search for truth via scientific inquiry often raises more and more complex questions. In turn, these questions create the need for a unified approach, inevitably leading to philosophy.

In this process, mathematicians and physicists construct, consciously or unconsciously, their own philosophical framework, which of course crucially affects their subsequent work. In other words:

> There is a continuous, dialectic interaction between the process of searching for mathematical and scientific truths and the creation of individual philosophical frameworks.

A clear illustration of this interaction is provided by the writings of Albert Einstein. In 1905, the great man placed emphasis on the significance of physical observations, writing that 'only those entities that are observable are important in physics'. But in 1927, he wrote, 'What is observable? Theory decides what is observable'. Clearly, by 1927, theory, in comparison to observations, had reached a prime status in Einstein's brain. This was the result of his immersion in mathematical studies during his development of the theory of general relativity and of the philosophical implications of this cognitive activity.

The Importance of Language in Today's World

The continuous preoccupation with digitalised forms of interactions, in addition to threatening the development of normal personal relations and eroding the time available for the vital process of reflection, may also affect the richness of language. The agony of the danger of the potential down-grading of language was beautifully expressed by the polymath George Steiner with the quote, 'O Word, O Word that I lack', from Schoenberg's masterpiece, the unfinished opera *Moses and Aron* (1932).

Despite the enormous intellectual wealth documented in Steiner's books and the central role of language in generating this wealth, surprisingly, Steiner claimed that, as a result of huge scientific and technological developments, language has become *anachronistic*. He writes:

"Our 'time' and 'space' are archaic, almost allegoric banalities out of touch with relativistic algorithms. From the perspective of the theoretical and exact sciences we speak a kind of Neanderthal babble".

He further adds:

"What is certain is that our ordinary vocabulary, our common grammars have ceased to speak the world as the scientist, or the engineer conceive and manipulate it" (Steiner, 2014, p. 196).

I disagree with this assessment. In my opinion, language will continue to play a vital role as an essential ingredient of humanity. Actually, I believe that the enormous advances in mathematics, the sciences, engineering, and technology, provide opportunities for enriching language, literature, and philosophy in an unprecedented way. In this respect, it is noted that, first, the high complexity and computational power of mathematics have created trepidation for the layperson and the illusion of omnipotence. This is also true for various sciences and for technology. However, many of these developments can be understood in simple linguistic terms. For example, the most important laws of physics, from those of Newton, Maxwell, and Einstein to quantum mechanics and particle physics, are reviewed in Part III without the use of any mathematical formulas. Second, despite the unambiguous precision and economy of the mathematical language, many difficult physical concepts are elucidated only after they are expressed in linguistic terms. For example, the understanding of the deeper meaning of quantum mechanics was significantly enhanced by the famous debates between Niels Bohr and Albert Einstein. Furthermore, the writings of the philosopher Moritz Schlick had a significant impact in elucidating the fundamental aspects of the theory of general relativity; both the Nobel Laureates, Albert Einstein and Max Born, underlined this fact. Third, there is now a unique opportunity to employ

350 The Embodied Mind

newly discovered techniques and the deeper insight gained regarding the function of the brain (including the introduction of the concept of meta-representations) to revisit literature and philosophy. Such techniques, including quantitative approaches based on computational algorithms, are already used in linguistics.

The Crucial Role of Humanities for Avoiding a Catastrophe

It is noted in the introduction that research in AI may lead to conflicts between individual researchers and the society at large. Is there a more profound exposition of such conflicts than Sophocles' *Antigone*, surpassing even the plays of William Shakespeare? According to Hegel, this tragedy was 'in every respect the most consummate work of art human effort has ever brought forth'. It is stated by the Hellenophile philosopher that this play centred around 'the noblest of figures that ever appeared on earth'. The great admiration of Hegel for Sophocles becomes evident by his belief that the death of Antigone defines heroism beyond Golgotha: Jesus could anticipate resurrection, whereas Antigone chose freely the abyss, which was made even more terrifying by her doubt of whether her stance was consistent with the wills of Gods.[2] The work elucidates the dialectic conflict of the individual and the state, as well as the agonising antithesis between civil jurisdiction and the primordial dictates of familiar love. Sophocles advocates the need for a balanced approach, where conflicts can be resolved only within the norms provided by the society. Incidentally, Antigone's conflict reminds me of the clash of the King with the Grand Inquisitor in *Don Carlos* (1867), my favourite among all of Verdi's operas. During their meeting, which I believe provides the peak of the opera both musically and dramatically, they discuss the need for the King to sacrifice his rebellious son (like God did) for the benefit of the state.

Does there exist a more iconic prediction of collective human suffering than *Antigone's* 'The Ode on Man' (sung by the chorus), particularly

---

[2]Of course, the scope of Jesus' sacrifice is much larger since, according to the Christian dogma, it absolved the original sin. For this to occur, Jesus had to suffer as a man, which implies that he could not use the knowledge of resurrection to ease his pain.

of the words 'immense the suffering' (*polla ta deina*)[3]? Is there a more emotionally charged exposition of the appalling implementation of this prediction to millions of Jewish people than the poetry of Paul Celan (Celan, 1955)? In Celan's poems, the economic power of poetry is leveraged in a superb way. Celan, who was born Paul Antschel to a Jewish family in the Kingdom of Romania, wrote poetry in German. Along with Goethe, Hölderlin, and Rilke, this tragic figure is considered one of the preeminent poets in the German language. While he was under the Nazi captivity, Celan learned that his father had perished and that his mother had been killed a year earlier in a different concentration camp. Although aware of Heidegger's adherence to Nazism, he studied his works, and even met the philosopher at Heidegger's hut in 1967. Celan committed suicide in 1970.

It is only through the immersion in such tragedies and poetry that one can hope that *polla ta deina* will never be repeated.

References

Blackmore, S. (2005). *Conversations on Consciousness: What the Best Minds Think about the Brain, Free Will, and What It Means to Be Human.* Oxford: Oxford University Press.

Celan, P. (1955). *Selected Poems.* London: Penguin Books.

Fokas, A. (2024). *Ways of Comprehending: The Grand Illusion and the Essence of Being Human.* Singapore: World Scientific.

Fokas, A. S. (in preparation). *The Golden Ages of Athens and Vienna.*

Heidegger, M. (1959). The Ode on Man in Sophocles' Antigone. In *An Introduction to Metaphysics.* Connecticut: Yale University Press.

Steiner, G. (2014). *The Poetry of Thought: From Hellenism to Celan.* New York: New Directions.

---

[3] The word '*deina*' has a double interpretation: 'terrible' but also 'beautiful'. Heidegger has extensively analysed *Antigone* in his essay *The Ode on Man in Sophocles' Antigone* (Heidegger, 1959) and in his 1942 series of lectures, *Hölderlin's Hymn, The Ister*. The translation of *Antigone* by Hölderlin influenced Heidegger significantly.

Index

* 9 7 8 1 8 0 0 6 1 8 4 8 0 *